Diving Birds of North America

Paul A. Johnsgard

Diving Birds

of North America

University of Nebraska Press: Lincoln & London

Copyright 1987 by the
University of Nebraska Press
All rights reserved
Manufactured in the United
States of America

The paper in this book meets
the minimum requirements of
American National Standard
for Information Sciences –
Permanence of Paper for
Printed Library Materials,
ANSI Z39.48–1948.

Library of Congress
Cataloging-in-Publication Data
Johnsgard, Paul A.
Diving birds of North America.
Includes index.
1. Divers (Birds) 2. Alcidae.
3. Birds – North America.
I. Title.
QL681.J63 1987   598.29'24'097
86-6896
ISBN 0-8032-2566-0

Contents

Illustrations and Tables

Preface

Considering the great nostalgic attraction of such birds as the common loon for people who have lived at least part of their lives around the lakes of Canada and the northern United States, and given the endearing visual appeal of species like puffins and auklets, it is rather surprising that there are so few books on these groups of aquatic birds. During my childhood summers at our Minnesota lake cottage I used to spend hours watching loons and red-necked grebes, and I marveled at their wonderful diving ability and powerful voices. Much later, the wild puffins and massed breeding colonies of murres along the rugged coasts of Scotland and Alaska were some of the most memorable sights of my entire life. In trying to learn more about these wonderful species I found myself forced repeatedly to turn to A. C. Bent's classic *Life Histories of Northern American Diving Birds.* This book, published in 1919, was the first in Bent's long and distinguished series of authoritative references. Yet it provided a frustratingly small amount of biological information, since at the time very little was known of the breeding biology and ecology of most of these elusive species.

At about the same time Bent's book was published, ornithologists taxonomically isolated the auks from the loons and grebes. Researchers thus ceased to deal collectively with these three groups, and the birds' generally remote nesting sites, as well as the minimal economic significance of most, caused them to be relatively neglected by ornithologists. Each summer as I returned to Minnesota and was excited by the sights and sounds of breeding grebes and loons, I wondered if a book dealing with them might not be worthwhile. But it was not until the late 1970s, after I had seen several of the auks on their nesting grounds, that I began to think in terms of dealing collectively with all the North American "diving birds," in spite of the artificial "lumping" this approach would require. As I considered it further, it

seemed that such coverage would emphasize the impact of convergent and parallel evolution better than would dealing with the patterns of adaptive radiation within a single phyletic group as has been the typical approach of my earlier books.

This vision became stronger after a trip to the seabird colonies on the Pribilof Islands in 1982, and soon thereafter I began organizing references for my book. In attempting to learn if any works on the same subject area were in the offing, I eventually discovered that Dr. Asa C. Thoreson had nearly finished a book on the auks of the world. When I wrote to him it became apparent that our planned books were oriented in rather different ways. In any case, none of the comparative aspects of diving bird biology would be dealt with by his book, which consists almost exclusively of species accounts. Thus I went ahead with my project and was further aided by Dr. Thoreson, who very kindly sent me a photocopy of his text, which greatly helped me in my own work and allowed me to avoid unnecessary overlap with his.

The literature on the North American loons, grebes, and auks is diverse and also is very unequally distributed among the thirty-one species concerned, largely as a reflection on the varied distribution patterns and relative abundance of each of the species, as well as differing amounts of public interest in them. Thus, of 237 literature references in my loon card file, 41 percent relate to the common loon, 31 percent to the arctic loon, 16 percent to the red-throated loon, and 12 percent to the yellow-billed loon. Of 292 available references to North American grebes, 30 percent refer to the eared grebe, while the horned, red-necked, and western grebes each account for 20–25 percent, 16 percent refer to the pied-billed, and 5 percent refer to the least grebe, including duplicate coverage. Of 755 references in my auk file, 25 percent relate to the common murre, 14 percent to the Atlantic puffin, 13 percent to the razorbill, 8 percent to the black guillemot, 6 percent each to the dovekie and thick-billed murre, from 2 to 3 percent each to the Craveri and Cassin auklets, ancient and marbled auklets, and tufted puffin, with less than 2 percent each pertaining to the remaining ten species. There is thus roughly five times as much information available on the eared grebe as on the least grebe and more than ten times as much literature on the common murre as on nearly half of the other North American auks. I have not attempted to provide complete bibliographies for all these groups, but Thoreson (in press) listed 570 references relating to the Alcidae. Likewise Clapp et al. (1982) have provided an extensive literature listing for all the species of loons and grebes that I consider North American, with the sole exception of the yellow-billed loon. I have included in this book all the references that I consider of general North American significance to this species. Cramp and Simmons (1977) and Bauer and Glutz (1966) have provided additional references relative to this species in the Palearctic.

I began writing this book in 1983 and continued through the 1984–85 academic year. A summer fellowship from the University of Nebraska's Research Council enabled me to travel in 1983 to the Peabody Museum of Yale University to do preliminary library work as well as to begin related work in that museum and the American Museum of Natural History. I obtained specimens on loan from the Field Museum of Natural History in Chicago and the Nebraska State Museum in Lincoln as well as from the two museums previously mentioned. I observed alcid diving behavior at Sea World in San Diego, whose staff allowed me access to special viewing tanks. I also appreciate the loan of library materials from the Van Tyne Memorial Library and access to the library and museum collections at the University of Kansas Museum, Lawrence, the British Museum (Natural History), Tring, and the California Academy of Science, San Francisco. I offer my sincere thanks to all these institutions as well as to the persons who helped me, in particular Luis Baptista, John Fitzpatrick, Ian Galbraith, Robert Mengel, Charles G. Sibley, Lester Short, Jr., Frank S. Todd, and David Willard.

A very large number of people assisted me with various kinds of information or provided help on illustrations. Chief among the latter were Jon Fjeldså, who provided the splendid painting of downy grebes, Mark E. Marcuson, who painted the least grebe and auklet scenes, and John Felsing, Jr., who did the superb horned puffin painting. Photographs were offered or provided by Robert Armstrong, Thomas Cardamone, Robert Day, Kenneth W. Fink, Greg Hiemenz, Stuart Johnson, Alan Nelson, Gary Nuechterlein, David G. Roseneau, Asa Thoreson, Frank S. Todd, and C. Fred Zeillemaker. Other kinds of assistance were provided by James Bellingham, Scott Drieschmann, Bruce Elliott, Kenneth W. Fink, Otto Höhn, Scott Johnsgard, Ned Johnson, Brina Kessel, Judith McIntyre, David Rimlinger, John Rogers, George Watson, Duff Wehle, and the Bird Banding Laboratory at Patuxent, Maryland. The entire manuscript was critically read by C. Fred Zeillemaker and Kenneth W. Fink, and parts were also read by David Rimlinger.

Comparative Biology

1. General Attributes and Evolutionary Relationships

Loons, grebes, and auks comprise a rather large number of species of aquatic diving birds that are fairly unfamiliar to most people, inasmuch as they tend to spend much of their time well away from shore and, when approached on the water, usually dive inconspicuously and reappear a considerable distance away. Thus in many areas grebes, simply called "helldivers," are often confused with coots or even diving ducks. Many people know loons only by their wild, penetrating cries and romantically associate them with northern woods and lakes, while auks are symbolic of arctic coastal cliffs. But ornithologists can find fascinating examples of convergent or parallel evolution in avian locomotory and foraging behaviors among these bird groups. They also offer a host of problems of more general ecological and behavioral interest to biologists. Nevertheless, all these groups have been largely neglected in terms of their comparative biology, and not a single inclusive book has dealt with them since Bent's 1919 monograph on their "life histories."

Before dealing with the many specific attributes of loons, grebes, and auks that are of special interest to ornithologists, let me define each of them in a formal but rather broad manner and point out their overall similarities as well as some of their differences.

Loons are large diving birds having long, straight, and acutely pointed bills, with feathers covering the lores (region between eye and bill) and extending to the linear nostrils. There are 10 functional primaries and 22–23 secondaries, the inner secondaries shorter than the primaries. Molting of the remiges (flight feathers) is simultaneous, and the wings are not used for underwater propulsion. The 16–20 rectrices (tail feathers) are short and stiff. The body feathers are shiny and water-resistant; aftershafts and adult down feathers are present. The legs are set extremely far back on the body, making standing very difficult and walking or takeoff

from land nearly impossible. The tarsus is laterally compressed, with reticulated (networklike) scales, and is unserrated behind. The lobed hallux (hind toe) is long and slightly elevated, the front toes are fully webbed, and all the toes have sharp claws. The wings are long and pointed, and the body is somewhat elongated, with a fairly long neck. The adults have white underparts, and in most species both sexes are spotted or striped with black and white on the back and neck during the breeding season. All species are monogamous, with distinct breeding and wintering plumages that lack sexual dimorphism. The nests are built of vegetational debris and placed at the water's edge. The eggs (usually 2) are spotted and are elliptical to ovate, and the young are unpatterned and nidifugous, with two successive coats of uniformly brownish down. The family has a Holarctic distribution, and all the species are migratory, wintering primarily on salt water and breeding solitarily on freshwater lakes or large tundra ponds. Their foods are primarily fish and aquatic invertebrates, but they also eat small amounts of plant materials. Four extant species are usually recognized, all having breeding ranges that include North America, primarily in boreal and arctic regions (table 1).

Grebes are small to medium-sized diving birds having bills that vary from short and rather blunt tipped to long and acutely pointed, with bare lores and head feathers not extending to the linear or oval nostrils. There are 11 functional primaries and 17–22 secondaries, the inner secondaries longer than the primaries. Molting of the primaries is simultaneous; the wings are not used for underwater propulsion, which is provided by the feet. The rectrices are soft, rudimentary, and hidden. The body feathers are small, water-resistant, and shiny. Aftershafts and adult down feathers are present. The legs are set extremely far back on the body, making walking difficult and takeoff from land impossible. The

Table 1: Taxonomy and Geographic Distributions of the Loons and Grebes of the World

Scientific and Vernacular Names	Distribution	References
Gavia		
stellata (red-throated loon)	Circumpolar Holarctic tundra	This work
arctica (arctic loon)[a]	Circumpolar Holarctic tundra and taiga	This work
immer (common loon)	Boreal Nearctic and western Palearctic	This work
adamsii (yellow-billed loon)	Circumpolar Holarctic tundra	This work
Rollandia		
rolland (white-tufted grebe)	South America	Fjeldså 1985
microptera (Titicaca grebe)	South America (Titicaca area)	Fjeldså 1985
Tachybaptus		
novaehollandiae (Australian dabchick)	East Indies, Australia	Frith 1976
ruficollis (little grebe)	Eurasia, Africa, Madagascar, East Indies	Cramp and Simmons 1977
rufolavatus (Alaotra grebe)	Madagascar (Lake Alaotra)	Voous and Payne 1965
pelzelni (Madagascan grebe)	Madagascar	Voous and Payne 1965
dominicus (least grebe)	North America, South America	This work
Podilymbus		
podiceps (pied-billed grebe)	North America, South America	This work
gigas (giant pied-billed grebe)	Guatemala (Lake Atitlan)	Bowes 1969
Poliocephalus		
poliocephalus (hoary-headed grebe)	Australia	Fjeldså 1983b
rufopectus (New Zealand dabchick)	New Zealand	Storer 1971
Podiceps		
major (great grebe)	South America	Storer 1963
auritus (horned grebe)	North America, Eurasia	This work
grisegena (red-necked grebe)	North America, Eurasia	This work
cristatus (great crested grebe)	Eurasia, Africa, Australia, New Zealand	Cramp and Simmons 1977
nigricollis (eared grebe)[b]	North America, South America, Eurasia, Africa	This work
occipitalis (silvery grebe)	South America	Fjeldså 1982a
taczanowskii (puna grebe)	Peru (Lake Junin)	Fjeldså 1982a
gallardoi (hooded grebe)	Argentina	Storer 1982
Aechmophorus		
occidentalis (western grebe)[c]	North America	This work

NOTE: Taxonomy of Storer 1979; brackets connect probable superspecies groups.

[a] The form *pacifica* has recently been recognized by the AOU as a distinct species (*Auk*, 102:680).

[b] The now apparently extinct South American form (*andinus*) is sometimes considered a distinct species.

[c] The form *clarkii* has recently been recognized by the AOU as a distinct species (*Auk*, 102:680).

tarsus is laterally compressed, with a scutellated scale pattern, and is serrated behind. The hallux is long and elevated; it and the front toes are separately lobed, and the claws are flat and naillike. The wings are somewhat rounded to rather elongated, and most species have a white patch or "speculum" on the secondaries. The body is short, with a variably long neck and silvery white plumage on the underparts. Facial tufts or crests are usually present in both sexes during the breeding season. All species are monogamous and lack apparent sexual dimorphism; most have distinct breeding and wintering plumages. Their nests are built of floating and emergent vegetation anchored in shallow water. The eggs (3 to 9) are unspotted whitish and are elliptical to nearly fusiform. The young are covered with dense down and typically have complex color patterning. As with loons, the young have extended fledging periods and often are carried about on their parents' backs. The family has a nearly cosmopolitan distribution, and the species are mostly migratory, usually wintering on salt water and breeding colonially or solitarily in shallow and reedy freshwater habitats. Their foods include fish, aquatic invertebrates, and some plant materials; for uncertain reasons feathers are also often swallowed. There are at least twenty species, six (or seven) of which breed in part or entirely in North America, mainly in temperate-latitude marshes (table 1).

Auks are small to medium-sized diving birds having bills that are variably pointed and compressed but never acuminate and are sometimes covered with colorful horny sheaths in breeding adults. Feathering densely covers the lores and often extends to the nostrils, which vary from linear to oval. There are 10 functional primaries and 15–19 secondaries; the greater secondary and primary coverts are usually lengthened. Molting of the primaries is usually simultaneous but is gradual in some species; the wings are used for underwater propulsion, and the feet are then used mainly for steering. The 12–18 rectrices are short and normal in shape. The feathers are dense, water-resistant, and shiny; aftershafts and adult down feathers are present. The legs are set fairly far back on the body; walking is done in an erect posture with the weight on the toes (digitigrade). Takeoff from level ground is difficult and infrequent; the birds usually take flight from cliffs or into the wind from water. The tarsus is compressed laterally, is reticulated or scutellated, and is usually shorter than the middle toe. The hallux is absent or vestigial, and the front toes are fully webbed. The wings are relatively short (one recently extinct species and several fossil species were flightless), bowed, and pointed, and the body is robust, with a short neck and large head. Adults have white to dark grayish underparts and usually are black-ish dorsally. Adults of most species exhibit crests, facial tufts, or other distinctive plumage or horny bill adornments in both sexes during the breeding season. All species are monogamous and monomorphic, often having distinct breeding and wintering plumages. The eggs are laid on rock ledges, in crevices or burrows, or rarely among the branches of trees. The eggs (1 or 2) are often spotted and are pyriform to ovate. The young are down-covered, unpatterned to bicolored, and are nidifugous to seminidicolous. The family is Holarctic in distribution, and the species are entirely marine in winter but usually breed colonially (sometimes solitarily) along coastlines. Their foods mainly consist of fish, plankton, and other marine fauna. There are twenty-two extant species, twenty of which breed in North America, particularly along the northern Pacific coast (table 2).

The relative evolutionary relationships among these three groups of diving birds have been a source of continuing controversy, which has not slackened but indeed has intensified as more recent information has become available. There are, to be sure, many similarities that unite each of the three groups with one or both of the other two groups (table 3), but there are also substantial numbers of unique or nearly unique characteristics exhibited by each of the three (table 4). During the late 1800s and until the early decades of the current century, all three groups were usually placed in a single order "Pygopodes." This group was formed in 1880 by W. Sclater to accommodate the earlier family "Colymbidae" (a family erected by T. Huxley that included the loons and grebes) and also the Alcidae. Coues (1882) characterized the Pygopodes in some detail, assigning each of the three groups considered in this book to separate suborders but excluding the penguins, which some had considered part of the same general assemblage. In 1919 Ridgway separated the auks from the loons and grebes, assigning them to a suborder (Alcae) of the Charadriiformes. He said that the anatomical evidence indicated a close relationship of the auks with the gulls rather than with the loons, though he admitted that the loons and auks evidently are also fairly closely related.

Later studies by authors such as Shufeldt (1904) still supported the general position that loons and grebes are closely related; indeed, Shufeldt regarded the two groups as part of a single family (Podicipidae) and the sole component of his "Pygopodes," an order he considered most probably related to the gull-like birds. However, Stolpe (1935) undertook an anatomical comparison of the hind-limb structure of loons, grebes, and the fossil diving bird *Hesperornis* and concluded that inasmuch as loons and grebes differ by so many of their hind-limb structures and their leg movements during swimming, they are not closely related to one another.

Table 2: Taxonomy and Geographic Distributions of the Auks of the World

Scientific and Vernacular Names	Distribution	References
Tribe Allini		
Alle		
alle (dovekie)	North Atlantic and adjacent areas	Glutz and Bauer 1982
Tribe Alcini		
Uria		
aalge (common murre)	Circumpolar Holarctic	Glutz and Bauer 1982
lomvia (thick-billed murre)	Circumpolar Holarctic	Glutz and Bauer 1982
Alca		
torda (razorbill)	North Atlantic and adjacent areas	Glutz and Bauer 1982
Pinguinus		
impennis (great auk)	Extinct since 1844	Bengtson 1984
Tribe Cepphini		
Cepphus		
grylle (black guillemot)	North Atlantic and adjacent areas	Glutz and Bauer 1982
columba (pigeon guillemot)	North Pacific and Bering Sea	This work
carbo (spectacled guillemot)	Coasts of northeastern Asia	Thoreson (in press)
Tribe Brachyramphini		
Brachyramphus		
marmoratum (marbled murrelet)	Coastlines of North Pacific	This work
brevirostris (Kittlitz murrelet)	Coastlines of North Pacific	This work
Tribe Synthliboramphini		
Synthliboramphus		
hypoleucus (Xantus murrelet)	Pacific Coasts of California and Baja California	This work
craveri (Craveri murrelet)	Gulf of California	This work
antiquus (ancient murrelet)	North Pacific and Bering Sea	This work
wumizusume (Japanese murrelet)	Coasts of Japan	Thoreson (in press)
Tribe Aethini		
Ptycoramphus		
aleuticus (Cassin auklet)	Aleutian Islands to Baja California	This work
Cyclorrhynchus		
psittacula (parakeet auklet)	North Pacific and Bering Sea	This work
Aethia		
pusilla (least auklet)	North Pacific and Bering Sea	This work
pygmaea (whiskered auklet)	Commanders, Kuriles, Aleutians	This work
cristatella (crested auklet)	North Pacific and Bering Sea	This work
Tribe Fraterculini		
Cerorhinca		
monocerata (rhinoceros auklet)	North Pacific	This work
Fratercula		
cirrhata (tufted puffin)	North Pacific and Bering Sea	This work
arctica (Atlantic puffin)	North Atlantic and adjacent areas	Glutz and Bauer 1982
corniculata (horned puffin)	North Pacific and Bering Sea	This work

NOTE: Taxonomy of AOU *Check-list* 1983; brackets connect superspecies.

Table 3: Characteristics That Variably Serve to Associate Loons, Grebes, and Auks

Traits That Associate All Three Groups

1. Consume aquatic animal life
2. Prey captured by extended dives
3. Bill sharply pointed in most species; skull with schizognathous palate
4. Legs variably placed toward rear, affecting standing and walking efficiency, pelvis variably narrowed
5. Predominantly temperate to arctic in North American breeding distribution
6. Mostly migratory; primarily marine in winter
7. Sexes monogamous and monomorphic
8. Distinctive nuptial and winter plumages in most species

Traits That Associate Two of the Three Groups

A. *Traits shared by loons and grebes*
   1. Hallux present and paddlelike
   2. Pelvis strongly narrowed; underwater propulsion by feet alone
   3. Nest situated very close to water
   4. Young tended for extended periods and carried on backs of both parents

B. *Traits shared by loons and auks*
   1. Rectrices normally developed
   2. Toes palmate and sharply clawed
   3. Lores fully featured

C. *Traits shared by grebes and auks*
   1. Nuptial head tufts present in many species
   2. Variably colonial nesting frequent

Table 4: Characteristics That Variably Serve to Separate Loons, Grebes, and Auks

Traits of Loons	Traits of Grebes	Traits of Auks
1. Foot-propelled divers	1. Foot-propelled divers	1. Wing-propelled divers
2. Carpometacarpus elongated and narrow	2. Caropometacarpus short and narrow	2. Carpometacarpus long and robust
3. Long cnemial process on tibiotarsus; no separate patella	3. Long cnemial process on tibiotarsus; separate patella	3. Tibiotarsus lacking cnemial process
4. Large foot area relative to body weight	4. Large foot area relative to body weight	4. Small foot area relative to body weight
5 Synsacrum longer than sternum; very narrow at acetabulum	5. Synsacrum as long as sternum; narrow at acetabulum	5. Synsacrum shorter than sternum; broad at acetabulum
6. Two generations of natal down	6. One generation of natal down	6. One generation of natal down
7. 16–20 normally developed rectrices	7. Rectrices vestigial	7. 12–18 normally developed rectrices
8. 10 functional primaries; 22–23 secondaries	8. 11 functional primaries; 15–21 secondaries	8. 10 functional primaries; 15–19 secondaries
9. Large supraorbital glands	9. Small supraorbital glands	9. Large supraorbital glands
10. Skeleton nonpneumatic	10. Skeleton virtually nonpneumatic	10. Skeleton slightly pneumatic
11. 14–15 cervical vertebrae	11. 17–21 cervical vertebrae	11. 15 cervical vertebrae

This conclusion has been rather widely adopted and has been fairly recently supported by Storer (1960), who went so far as to question the possibility that the loons, grebes, and hesperornithiform birds even had a common swimming ancestor. Storer's study and other historically important studies on the loon-grebe question have been admirably summarized by Sibley and Ahlquist (1972).

These authors furthermore reported on their own egg-white protein studies, indicating that the starch-gel electrophoretic patterns of loons and grebes do not provide unequivocal evidence of their evolutionary relationships. However, more recent and unpublished work on DNA hybridization conducted in their laboratory supports the position that the loons share a common if distant ancestry with the penguins and tube-nosed swimmers, while the grebes appear to be extremely isolated from all other extant bird groups. As expected, the alcids exhibit close DNA affinities with gulls and terns (C. Sibley, pers. comm.). Korzun (1981) has also recently concluded from anatomical evidence that any evolutionary connection between the loons and grebes is probably a very ancient one.

Recent but still unpublished osteological studies by Boertmann (1980), whose conclusions have been summarized by Glutz and Bauer (1982), point to an even closer relation between loons and shorebirds than has been judged by Sibley and Ahlquist or by earlier authors. Boertmann has recommended merging the loons into the order Charadriiformes and reducing them to family status within the superfamily Laroidea. Within this superfamily he recognizes five distinct families: Glareolidae, Dromaidae, Laridae, Alcidae, and Gaviidae.

Thus several recent morphological studies strongly support the existence of an evolutionary affinity between the loons and the gull-like birds and a separate but essentially unknown evolutionary origin of the grebes. However, Cracraft (1982) has recently generated a new cycle of controversy by taking the position not only that are loons and grebes monophyletic but that this evolutionary assemblage also includes the flightless Cretaceous divers *Hesperornis* and *Baptornis*. He has proposed that loons and grebes be regarded only as families in the order Gaviiformes and that this order and the extinct Hesperornithiformes be encompassed within a superorder Gaviomorphae. He did not deal with the anatomical evidence suggesting a charadriiform relationship of the loons but rather offered the suggestion that the Gaviomorphae are instead most closely related to the Sphenisciformes (penguins) and less closely related to the Procellariiformes (tube-nosed swimmers) and Pelecaniformes (totipalmate swimmers).

Unfortunately the fossil record does little to help re-solve this general controversy. The fossil record of grebes is especially weak, with the genus *Podiceps* extending back only to the Lower Miocene, while in Lower Pliocene strata the related *Pliodytes* also has been found (Brodkorb 1963).

The fossil record of loons is appreciably better, with *Gavia* remains extending back to the Miocene and with the related *Gaviella* of Oligocene age. Even older is the Eocene *Colymbiodes*, which some have believed represents an intermediate link between the loons and grebes (Howard 1950). However, Storer (1956) judged *Colymboides* to be an ancestral type of loon, with no affinities to grebes but instead with probable charadriiform affinities. Some even older loonlike fossils of Upper Cretaceous age, now assigned to the family Lonchodytidae (Brodkorb 1963), have also fairly recently been found, suggesting that separation of the loons from the typical charadriiform assemblage must have occurred in pre-Cenozoic times.

The fossil record of the alcidlike group is also moderately extensive (Brodkorb 1967) and provides at least some clues to the early radiation of the auks. The earliest known alcid fossils are the Eocene *Nautilornis* and *Hydrothericornis*, known from Utah and Oregon respectively. Both had well-developed wings and were thus probably excellent fliers. However, reduction of wing surface for more efficient diving probably began fairly early in auk evolution and led to an early separation of a group of flightless "Lucas auks" that are generally recognized as the Mancallidae but sometimes are considered a subfamily of the Alcidae (Brodkorb 1967). This fossil lineage extends forward only to the Pliocene, when the group became extinct, possibly through competition with more typical alcid diving specialists. These perhaps included the ancestors of *Penguinus*, which had already become differentiated by this time.

An attempt has been made to present some of this information in graphic form (fig. 1), as a hypothetical evolutionary dendrogram that represents my interpretation of the information available on the probable relation of the loons and grebes to auks, as well as the more definite affinities of the genera of living auks. Much of the basis for the evolutionary organization of the auks is dependent on information to be discussed in the species accounts, and at this point it is only important to note that the surviving auks seem to consist of three fairly recognizable morphological types. First, there are the relatively generalized forms that have not greatly modified their wings for diving or their legs for walking. These include the auklets, murrelets, and guillemots and the similar but rather distantly related dovekie. Second, there are the forms that have become better adapted for walking at the expense of diving profi-

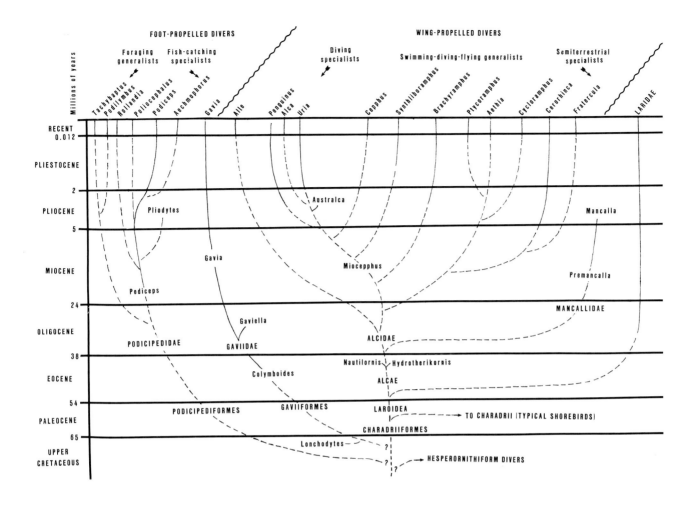

1. Hypothetical evolutionary dendrogram of the loons, grebes, and auks, including known fossil representatives of these groups.

ciency; these consist of the puffins and the rhinoceros auklet. Last, there are the auks that have modified their body and wing configuration to facilitate diving, at the expense of increased wing loading and decreased ability for walking. These include the murres, the razorbill, and especially the great auk, which became completely flightless and eventually extinct.

On the basis of anatomical evidence, Stettenheim (1959) concluded that the alcids evolved from an undifferentiated shorebird ancestral stock that was as distantly related to the gulls and terns as to the more typical shorebirds. He judged that these may have been wading rather than plunge-diving birds, which initially used diving to escape from danger and only later began to forage by this means. Kozlova (1961), however, judged that gulls and alcids shared a common ancestry and that

the earliest alcids specialized on marine fish and only later began to eat various marine invertebrates. She stated that, at least on the basis of skull conformation, the guillemots, murrelets, murres, razorbill, and dovekie are the least modified (most gull-like), whereas in the auklets and puffins the skull has become greatly modified, apparently in part at least because of its role (in typical puffins) in digging nesting sites. Such a role for the bill might help explain the original source of selection for increased bill size, which might later have been supplemented by the influence of sexual selection. Kozlova outlined a highly plausible evolutionary history of alcids and their geographic dispersion during Cenozoic times.

Relationships within the grebe family are less clear-cut, but recent studies by Fjeldså (1982a, 1983a) support

9

the general view that the genus *Rollandia* approaches the ancestral stem of grebe evolution, from which the dabchicks (*Poliocephalus*) and pied-billed grebes (*Podilymbus*) have diverged in one direction, while the remaining genera have diverged in another direction, toward the relatively specialized *Podiceps* and *Aechmophorus* types that largely constitute the North American grebe fauna. Fjeldså (1982a) has published a provisional phylogeny of all the grebe species that has an independent origin from figure 1 and differs from it in minor respects.

2. Comparative Distributions and Structural Adaptations

Distributions

The geographic distributions of the loons, auks, and grebes are primarily reflections of the evolutionary histories of each of the groups, past climatic and geologic phenomena, and present-day climatic and ecological conditions. Thus all the loons and auks are Northern Hemisphere birds, which presumably have never been able to bridge the tropical barrier into the Southern Hemisphere, where seemingly suitable breeding habitat might exist in, for example, southern South America and Tierra del Fuego. Indeed, the breeding distribution of the Northern American loons is distinctly arctic oriented (fig. 2), with the greatest species density north of the boundaries of arctic tundra in Canada and Alaska and no breeding occurring south of the limits of continental glaciation (see fig. 4).

On the other hand, the breeding distributions of the grebes are distinctly more southerly; indeed, more than half the species of grebes are equatorial or Southern Hemisphere in occurrence. In North America the breeding distributions of grebes (fig. 3) are more closely related to topography and ecology than to climate, with the greatest species density occurring in the grasslands of southern Canada and the adjacent northern United States, in a general east-west band more or less approximating the distribution of the "prairie pothole region" (fig. 4) of Pleistocene glacial till that until recently was primarily grassland covered, but with abundant potholes and marshes. In such areas all but one of the North American grebes can sometimes be found breeding in a single marsh, which suggests that there may be substantial selective pressures for ecological segregation of foraging niches and other aspects of niche adaptation as well. Some of these questions of niche segregation will be dealt with in chapter 4.

Finally, the breeding distributions of the North American auks are, like those of the loons, distinctly arctic in orientation (fig. 3). Species densities reach a maximum in the Aleutian Islands, where as many as twelve to fourteen species might be found breeding. This is substantially greater than on the eastern coast of North America, where there is a maximum density of five sympatric breeding species. Udvardy (1979) analyzed the distribution of the Pacific alcids in considerable detail; he correlated the present-day distribution of these species with dispersal waves extending back to mid-Eocene times and reflecting a Pacific basin origin of the family centering on the Bering Sea. Udvardy imagined a series of five dispersal waves between the Pacific and Atlantic areas, reflecting the five interglacial periods when the Bering Strait was open. During at least the latest glacial period the Sea of Okhotsk was perhaps the most important and also the northernmost refuge, and many of the Pacific Ocean alcids still are essentially limited to glacial refuge areas. The present-day subtropical ranges of various southern endemics such as Craveri and Xantus murrelets were considered to be a result of their being Tertiary relicts, whose ranges and populations farther north were eliminated by climatic changes.

Beyond these historical effects, the current distribution of the alcids is also strongly affected by the present-day availability of marine food resources, ranging in size from planktonic invertebrates to small fish. To illustrate this relationship, the densities of zooplankton surrounding North America are illustrated in figure 4 (based on the *Atlas of the Living Resources of the Sea*, 1981, published by the Food and Agriculture Organization of the United Nations). Here it may be seen that the largest of the North American alcid colonies are all associated with adjacent ocean areas that are relatively rich in planktonic life, and doubtless also with other associated food resources of higher trophic levels.

2. Species density map of breeding loon distributions in North America.

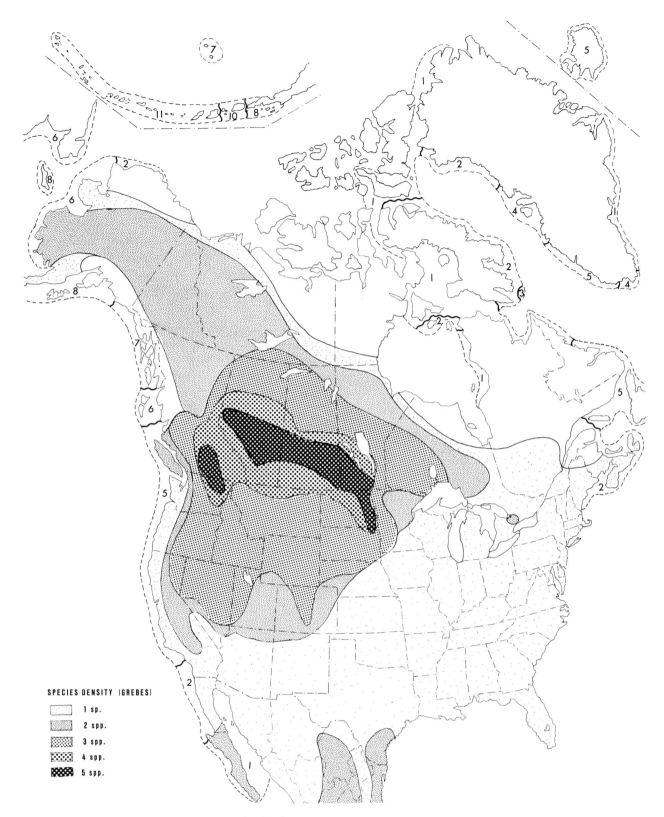

SPECIES DENSITY (GREBES)

- 1 sp.
- 2 spp.
- 3 spp.
- 4 spp.
- 5 spp.

3. Species density map of breeding grebe (*shading*) and
auk (*numerals*) distributions in North America.

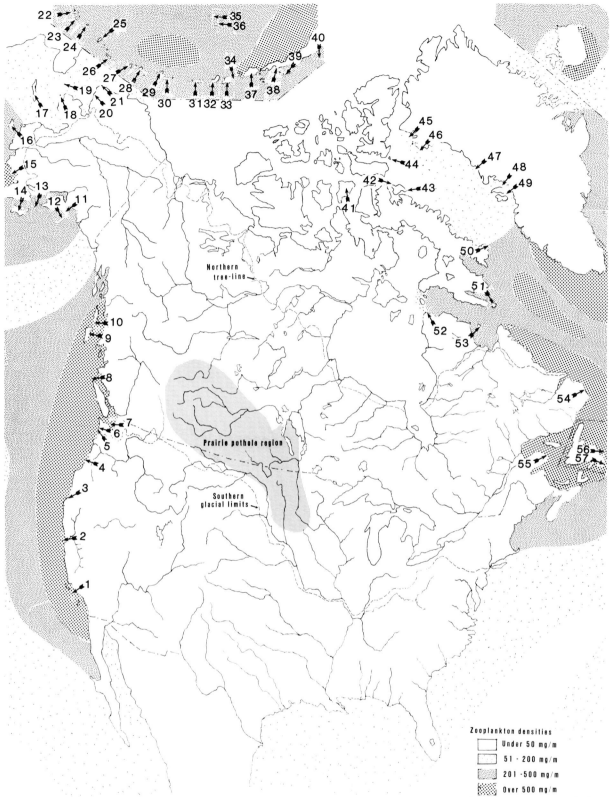

4. Distribution of major auk colonies in North America in relation to zooplankton densities of adjoining oceans. See appendix 3 for identification of colonies. Limits of arctic tree line, southern continental glaciation boundaries, and "prairie pothole" region are also indicated.

Although not shown on this map, the nonbreeding distributions of high-latitude auks and loons are strongly affected by patterns of sea-ice development during winter, which impose migrations of varied lengths and magnitudes on these forms. Winter distributions of these species are still very poorly known, since few efforts have been made to establish the distribution and abundance patterns of marine birds in North America. However, efforts to that end have begun in Alaska (Gould, Forsell, and Lensink 1982; Forsell and Gould 1981), and similar studies have been under way in eastern Canada for some time (Brown et al. 1975; Gaston 1980). Important information on the winter distributions of loons and grebes in the southeastern states and Gulf of Mexico has been provided by Clapp et al. (1982). Unfortunately there is as yet no corresponding

Table 5: Seasonal and Ecological Distribution of Loons in the Gulf of Alaska and Eastern Bering Sea, Based on Shipboard Surveys

| Species | Gulf of Alaska | | | | Eastern Bering Sea | | |
	Bay	Shelf	Shelfbreak	Oceanic	Bay and Shelf	Shelfbreak	Oceanic
Common							
Spring	5.8	+	+	+	0	+	0
Summer	0	+	0	0	0	0	0
Fall	+	0	+	0	+	0	0
Winter	1.8	0	0	0	−	−	−
Yellow-billed							
Spring	+	+	0	0	0	0	0
Summer	0	0	0	0	+	0	0
Fall	+	+	0	0	0	0	0
Winter	0	0	0	0	−	−	−
Arctic							
Spring	2.9	+	+	+	+	0	0
Summer	0	+	0	0	+	0	0
Fall	+	+	+	+	8.14	0	0
Winter	0	0	0	0	−	−	−
Red-throated							
Spring	0	0	0	0	0	0	0
Summer	+	0	0	0	+	0	0
Fall	+	0	0	0	+	0	0
Winter	0	0	0	0	−	−	−
Loon (sp.?)							
Spring	17.4	15.8	5.7	0	+	0	0
Summer	+	+	+	0	+	0	0
Fall	3.56	+	+	+	8.14	0	0
Winter	3.6	1.37	0	0	−	−	−

NOTE: Calculated number of birds/100 km^2, based on data of Gould, Forsell, and Lensink 1982.
+ = small number present.   0 = not observed.   − = not surveyed.

Table 6: Seasonal and Ecological Distribution of Auks in the Gulf of Alaska and Eastern Bering Sea, Based on Shipboard Surveys

Species	Gulf of Alaska				Eastern Bering Sea		
	Bay	Shelf	Shelfbreak	Oceanic	Bay and Shelf	Shelfbreak	Oceanic
Common murre							
Spring	93.4	79.1	34.3	+	94.2	10.8	4.1
Summer	85.0	93.9	50.2	+	380.9	6.8	4.2
Fall	466.4	143.8	8.9	2.0	16.3	+	0
Winter	373.1	30.1	19.8	0.3	–	–	–
Thick-billed murre							
Spring	8.7	15.8	+	9.8	134.6	43.4	2.7
Summer	+	+	+	0	105.8	20.3	12.5
Fall	+	+	2.2	0.7	132.6	+	1.5
Winter	7.3	4.1	2.2	0	–	–	–
Murre (sp.?)							
Spring	21.2	743.5	127.3	21.9	1,460	130.1	29.7
Summer	124.7	415.7	228.8	1.5	1,206	81.2	22.2
Fall	192.2	239.6	98.5	3.4	407	120.5	5.9
Winter	414.9	412.4	695.2	4.9	–	–	–
Pigeon guillemot							
Spring	69.6	15.8	+	0	13.5	0	0
Summer	102.1	13.4	0	0	+	+	0
Fall	18.8	+	+	0	+	0	0
Winter	0	0	0	0	–	–	–
Marbled murrelet							
Spring	46.4	15.8	5.7	0	0	0	0
Summer	73.7	13.4	5.6	0	0	0	0
Fall	14.2	+	0	0	+	0	0
Winter	54.6	0	0	0	–	–	–
Kittlitz murrelet							
Spring	2.9	+	+	0	+	0	0
Summer	5.7	+	0	0	+	0	0
Fall	+	+	+	0	0	0	0
Winter	0	0	0	0	–	–	–
Murrelet (sp.?)							
Spring	2.9	15.8	5.7	0	0	0	0
Summer	232.5	40.2	0	0	0	0	0
Fall	39.2	+	0	0	+	0	0
Winter	0	0	0	0	–	–	–

(continued)

Table 6: (*Continued*)

Species	Gulf of Alaska				Eastern Bering Sea		
	Bay	Shelf	Shelfbreak	Oceanic	Bay and Shelf	Shelfbreak	Oceanic
Ancient murrelet							
Spring	2.9	47.5	+	0	356.7	173.4	68.8
Summer	17.0	53.6	39.1	1.5	126.9	74.5	0
Fall	0	+	+	−	16.3	+	0
Winter	5.5	0	0	0.3	−	−	−
Cassin auklet							
Spring	+	15.8	+	+	6.7	0	0
Summer	17.0	80.5	5.6	+	+	+	0
Fall	53.4	30.0	29.1	4.0	+	0	0
Winter	9.1	0	0	0.3	−	−	−
Parakeet auklet							
Spring	0	+	+	0	40.4	243.9	0
Summer	+	13.4	5.6	+	+	6.8	1.4
Fall	0	35.9	38.1	4.7	56.9	+	0
Winter	0	0	0	0	−	−	−
Crested auklet							
Spring	0	+	0	0	114.4	422.8	6.8
Summer	0	+	+	0	84.6	60.9	83.4
Fall	14.2	137.8	0	1.3	154.7	24.1	5.9
Winter	0	0	0	0	−	−	−
Least auklet							
Spring	0	+	0	0	666.3	1,067	159.3
Summer	0	+	0	0	63.5	94.8	23.6
Fall	0	0	0	0	58.8	0	1.5
Winter	0	0	0	0	−	−	−
Whiskered auklet							
Spring	0	284.7	17.2	0	652.8	5.4	0
Summer	0	+	0	0	21.2	13.5	0
Fall	0	0	0	0	+	0	0
Winter	0	0	0	0	−	−	−
Rhinoceros auklet							
Spring	0	+	+	0	0	0	0
Summer	+	+	+	5.9	0	0	+
Fall	0	5.6	0	4.7	0	0	0
Winter	0	0	0	0	−	−	−

(*continued*)

Table 6: (Continued)

Species	Gulf of Alaska				Eastern Bering Sea		
	Bay	Shelf	Shelfbreak	Oceanic	Bay and Shelf	Shelfbreak	Oceanic
Horned puffin							
Spring	11.7	15.8	5.7	0	6.7	5.4	0
Summer	51.0	80.5	111.6	1.5	42.3	40.6	9.7
Fall	64.1	89.8	29.1	4.0	16.3	+	0
Winter	0	16.4	6.6	15.0	–	–	–
Tufted puffin							
Spring	559.7	332.3	40	6.5	242.3	65.0	83.7
Summer	1,967	898.5	546.8	61.7	275.1	236.9	83.4
Fall	99.7	179.7	91.8	84.4	374.4	192.8	127.3
Winter	3.6	35.6	44.0	22.1	–	–	–
Alcid (sp.?)							
Spring	139.2	126.6	34.3	8.7	329.8	222.2	255.1
Summer	79.4	227.9	155.7	11.8	84.6	1,029	13.9
Fall	39.2	83.9	26.9	11.4	130.2	24.1	41.4
Winter	127.4	63.0	22.0	2.9	–	–	–

NOTE: Symbols and density estimates as in table 5.

information on the Pacific coastal wintering range from British Columbia south to southern California, which appears to provide major wintering sites for arctic and red-throated loons, western, eared, and red-necked grebes, and numerous species of auks.

To illustrate the importance of the Gulf of Alaska and the eastern Bering Sea as wintering areas for loons and auks, a tabular summary of the information provided by Gould, Forsell, and Lensink (1982) has been condensed and summarized in tables 5 and 6. This information suggests that the Bering Sea is used during spring and fall, with only very limited wintering occurring there. On the other hand, the data for auks suggest very high usage of the eastern Bering Sea by murres, auklets, and puffins for much of the year and similar year-round usage of the Gulf of Alaska for these same groups. In the vicinity of Kodiak Island alone it is probable that more than a million murres overwinter, a significant portion of the entire common murre nesting population of the Bering Sea and Gulf of Alaska (Forsell and Gould 1981). Similarly, perhaps the entire Gulf of Alaska population of crested auklets winters in a limited area of the Kodiak archipelago (not included in table 7). The high concentration of whiskered auklets

shown in table 7 was the result of finding large numbers in a few transects taken in the eastern Aleutian Islands and is thus an artifact of data gathering.

Gould, Forsell, and Lensink (1982) concluded that the major migratory point of entry and departure to and from the eastern Bering Sea is Unimak Pass and that some alcid species (thick-billed murres, parakeet auklet, least auklet, crested auklet) are usually many times more numerous in the Bering Sea than in the Gulf of Alaska, while the reverse is true for several others (rhinoceros auklet, Cassin auklet, ancient murrelet). Waters over the continental shelf typically show the greatest number of species, and in such areas the common murre reaches its greatest numbers, while tufted puffins are among the species dominating the avifaunas of bays and pigeon guillemots are especially characteristic of nearshore areas. Generally, seabirds reach their maximum numbers over continental shelf areas south of Kodiak Island and the Alaska Peninsula, both in summer and in winter, though during mild winters many birds may remain in the southern Bering Sea or the far western Gulf of Alaska. Grebes are generally limited to inland freshwater habitats and inshore marine areas of coastal Alaska, and the numbers recorded were too

Table 7: Estimated Breeding Auk Populations and Approximate Biomass Equivalents

Species	World Population[a]	North American Population[b]	World Biomass (tons)[c]
Dovekie	80,000,000	30,000,000	12,800
Thick-billed murre	20,000,000	11,000,000	20,000
Atlantic puffin	16,000,000	625,000	16,000
Common murre	10,000,000	6,500,000	10,400
Least auklet	10,000,000	6,000,000	9,200
Tufted puffin	8,000,000	4,035,000	6,400
Horned puffin	30,000,000	1,500,000	1,800
Crested auklet	2,500,000	2,000,000	715
Cassin auklet	1,800,000	1,800,000	306
Rhinoceros auklet	1,000,000	250,000	520
Parakeet auklet	1,000,000	800,000	285
Ancient murrelet	800,000	580,000	160
Pigeon guillemot	700,000	220,000	315
Marbled murrelet	500,000	300,000	110
Razorbill	416,000	40,000	287
Black guillemot	100,000	56,000	43
Kittlitz murrelet	100,000	50,000	22.5
Whiskered auklet	50,000	20,000	4.9
Xantus murrelet	20,000	20,000	3.2
Craveri murrelet	6,000–10,000	6,000–10,000	0.8–1.35

[a]After Thoreson, in press, with some revised estimates. Nettleship & Birkhead (1985) offer alternative estimates.
[b]Including Greenland.
[c]Metric tons (equals 2,204 pounds, or 1,000 kilograms).

small to warrant inclusion in tables 5 and 6. Common loons made up 56 percent of the total loons observed in the Gulf of Alaska and 24 percent of those seen in the eastern Bering Sea, while the arctic loon composed 39 percent and 59 percent respectively. Migrant loons were often observed as far as 300 kilometers from land and in water more than 100 meters deep, and during summer periods they were usually within 50 kilometers of land and in water no more than 50 meters deep.

It is still impossible to judge the numbers of North American loons and grebes, owing in part to their great breeding and wintering dispersion and to difficulties of censusing them on a continental basis. However, some efforts have been made to estimate continental populations of the alcids. Thus Nettleship (1977) provided estimates of the total numbers of alcids in eastern Canada, and Thoreson (in press) has done the same for all of North America and the world. With minor changes I have accepted Thoreson's estimated totals for North America and the world, and they are summarized in table 7, together with a calculated world biomass based on average adult weights shown elsewhere in this text. The enormous avian biomass represented by the auks becomes evident with such calculations, and thus the role of the alcids in coastal and marine ecosystems can be more readily visualized.

Structural Adaptations

The specializations of the loons, grebes, and auks for swimming, diving, and flight are among the most interesting for any birds. Storer (1960) has traced the general evolutionary patterns associated with the development of both the wing-propelled diving birds (auks in the Northern Hemisphere and penguins and diving petrels in the Southern Hemisphere), and the foot-propelled divers (loons, grebes, and the extinct hesperornithiform birds). These two groups have followed distinctly different pathways, though in some respects there are certain similarities imposed by the diving syndrome. According to Storer, the wing-propelled divers are all marine and largely pelagic forms, whereas foot-propelled divers are more associated with freshwater and littoral habitats. He judged that one possible reason for this is that the subsurface aquatic vegetation so typical of freshwater habitats might impede wing-propelled divers more than foot-propelled ones. A second disadvantage of wing-propelled diving is that there is an upper size limit on the birds, above which selection for the small wings associated with efficient diving results in flightlessness, a condition typical of all penguins, the great auk, and the extinct Lucas auks. Storer also notes that simultaneous molting of the flight feathers, as is common in aquatic birds, influences the diving efficiency of wing-propelled divers, with smaller wing areas needed for diving than for flying. Thus the smallest alcids (such as the least and whiskered auklets) molt their wing feathers gradually, since in these alcids the relatively small wing needed for flying more nearly approaches the optimal size for use underwater.

Wing shapes and wing areas of the loons, grebes, and auks thus differ substantially (fig. 5), with loons tending to have relatively long and narrow wings and in particular primaries that are relatively long and supported by a long and relatively weak carpometacarpus-phalanges component. They also have a very large number of secondaries, supported by a long and relatively weak ulna. In spite of their long wings, loons have a relatively high wing loading (table 8). This wing loading is substantially greater than that reported for any grebe, though total weight differences between loons and grebes make such comparisons questionable.

The wing shape of grebes varies from moderately long to distinctly short and elliptical (with an aspect ratio of only 2.51 in the least grebe, according to Hartman 1961). Like the loons, their primaries are supported by a relatively thin and (compared with loons) relatively short carpometacarpus-phalanges component, and their numerous secondaries are supported by a long and fairly weak ulna. In both loons and grebes the upper medial wing coverts are variably enlarged and partially cover the innermost secondaries, presumably providing additional structural support. The wing loading of grebes appears in general to be the lowest of the three groups under consideration here, though grebes are not considered either strong or rapid fliers. There is virtually no information available on the flight speeds of grebes, and also very little on wingbeat rate, which seems to be

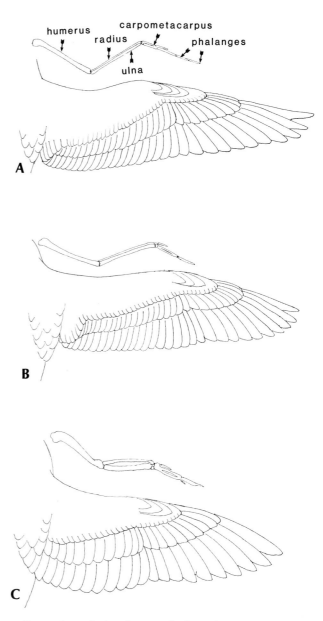

5. Comparison of wing shapes and relative lengths of arm and hand bones: A, loons (*Gavia*); B, grebes (*Podiceps*); C, auks (*Alca*).

somewhat higher than that of loons (table 9), as might be expected from their smaller size.

The wing shape of auks is relatively similar in all the flying species and may be characterized as relatively short, paddlelike, and with a short, stout forearm (Kozlova 1961). The bones supporting the primaries are broad and strong, making up approximately a third of the total wing skeleton. The ulna is short and robust; probably it and the radius are responsible for withstand-

1. Arctic loon, adult in breeding plumage. Photo by author.

2. Red-throated loon, nesting adult.
Photo by Kenneth W. Fink.

3. Yellow-billed loon, adult in breeding
plumage. Photo by Kenneth W. Fink.

4. Common loon, nesting adult. Photo
by Kenneth W. Fink.

5. Pied-billed grebe, adult in breeding
plumage. Photo by author.

6. Least grebe, adult and young. Painting by Mark E. Marcuson.

7. Red-necked grebe, nesting adult. Photo by author.

8. Eared grebe, adults with young. Photo
by Kenneth W. Fink.

9. Horned grebe, nesting adult. Photo by
Kenneth W. Fink.

10. Western grebe, adult with young.
Photo by Gary Nuechterlein.

Table 8: Estimates of Wing Loading and Foot Loading in Loons, Grebes, and Auks

Species	Weight (g)	Wing Area (cm²)	Foot Area (cm²)[a]	Foot/Wing Ratio (%)	Wing Loading (g/cm²)	Foot Loading (g/cm²)	References
Loons							
Common	2,425	1,358	130.4	9.6	1.78	18.6	Pool 1938
Grebes							
Least	—	—	—	—	0.67	—	Hartman 1961
Pied-billed	343.5	291	—	—	1.18	—	Pool 1938
Pied-billed (average of 4)	372	301	30.2	9.4	1.23	13.2	Various sources
Giant pied-billed	414	804.5	—	—	1.9	—	Bowes 1965
Horned	369.5	350	30.9	8.8	1.05	12.2	Pool 1938
Auks							
Dovekie	96	146	—	—	0.66	—	Pool 1938
Murres	982–1,069	ca. 400[b]	34.9	8.7	2.5–2.6	ca. 29	Kartashev 1960
Razorbill	774	ca. 340[b]	30.0	8.8	2.3	25.8	Kartashev 1960
Great auk	ca. 5,000	—	—	—	over 4.0	—	Kartashev 1960
Black guillemot	421	ca. 300[b]	—	—	1.4	—	Kartashev 1960
Pigeon guillemot	539	157	—	—	1.63	—	Lehnhausen 1980
Marbled murrelet	234	146	—	—	1.60	—	Stettenheim 1959
Parakeet auklet	247	149.6	—	—	0.83	—	Lehnhausen 1980
Least auklet	92	132.9	—	—	0.69	—	Sealy 1968
Tufted puffin	872	219.6	—	—	1.99	—	Lehnhausen 1980
Atlantic puffin	552	ca. 290[b]	25.5	8.8	1.9	21.6	Kartashev 1960

[a]Personal estimates.   [b]Estimated from graphic data of author cited.

ing much of the stress associated with wing-propelled diving. The total wing area of auks is relatively small, resulting in a high wing loading (1.9 g/cm² in the Atlantic puffin, compared with 1.05 in the comparably sized horned grebe), and the auks typically have a very rapid wingbeat in flight, especially in the smaller species.

A substantial difference between auks and the foot-propelled loons and grebes is to be expected in the relative development of their feet. In general, auks should exhibit relatively small foot surfaces in conjunction with their limited roles during diving (being used mainly for steering and to a limited extent for in-place paddling while bottom foraging). Major differences in foot shape and leg skeletal structure are indicated in figure 6, but comparisons are easier in table 8, where the estimated surface areas of the two feet (toes maximally expanded) are shown for six species. There is no apparent difference in the foot-to-wing surface area data for the available species; in all cases the ratio of foot area to wing area ranges from about 9 to 10 percent, in spite of the differing roles of the feet in these three groups. On the other hand, the ratio between foot area and total body weight does suggest that auks have substantially smaller foot areas relative to body weight than do loons or grebes, and that grebes perhaps have the most efficient surface area to body weight ratios for paddling and diving when using the feet alone.

Differences in the hind limbs of these birds may also be found in the skeletal elements of their legs. Shufeldt (1904) described the anatomy of the hind limbs of loons and grebes in some detail. He stated that the pelvic limb of a grebe is "altogether one of the most beautiful

Table 9: Speeds and Wingbeat Rates during Flight and Diving in Loons, Grebes, and Auks

Species	Speed (km/hr)	Wingbeat Rate (beats/min)	References
Loons			
Arctic (flight)	71 ± 6.8 (air)	—	Davis 1971
Arctic (diving)	2.5–5[a]	—	Lehtonen 1970
Red-throated	75–78 (air)	—	Davis 1971
Yellow-billed	64 (ground)	—	Davis 1971
Common	—	256–62	Meinertzhagen 1955
Common	74.9 (air)	—	Kerlinger 1982
Grebes			
Great crested (flight)	—	360–400	Meinertzhagen 1955
Great crested (diving)	4.3	—	Bolam 1921
Auks (flight)			
Common murre	129–42 (ground)	270–348	Meinertzhagen 1955
Razorbill	127–40 (ground)	—	Meinertzhagen 1955
Black guillemot	—	324–482	Meinertzhagen 1955
Atlantic puffin	62–132 (ground)	320–400	Meinertzhagen 1955
Dovekie	—	720–1080	Rüppell 1969
Auks (diving)[b]			
Common murre	3.2–6.1	—	Stettenheim 1959
Common murre	Avg. 4.07 (n = 12)	109	Kenneth Fink and author
Pigeon guillemot	—	104	Kenneth Fink and author
Rhinoceros auklet	Avg. 3.96 (n = 4)	82	Kenneth Fink and author
Tufted puffin	Avg. 3.84 (n = 14)	92	Kenneth Fink and author
Horned puffin	Avg. 4.07 (n = 19)	—	Kenneth Fink (pers. comm.)
Atlantic puffin	Avg. 3.60 (n = 2)	—	Kenneth Fink (pers. comm.)

[a]For dives lasting 5 seconds or more.
[b]In confinement at Sea World, San Diego.

adapted structures," which by appropriate articulations is "an avian oar." Its tarsus is compressed to the maximum degree, so that when it and the bladelike toes are brought forward they offer minimum resistance, but during the backward stroke the articulation permits the spread and expanded toes to be turned, and to some extent also the tarsus, thus exposing the maximum surface. The femur and tarsometatarsus are of about equal length, and the head of the femur is large. The tibiotarsus has a long cnemial crest, and this is supported posteriorly by a separate patella, providing substantial mechanical advantage for muscle contraction.

In loons the femur is short and thick and lies perpendicular to the sagittal plane of the body, causing the legs to be strongly splayed and essentially preventing walking on land. It has a large and globular head and is distinctly shorter than the tarsometatarsus, which as in grebes is extremely flattened. The tibiotarsus is marked by a greatly developed cnemial process but no separate patella, the latter apparently having become fused with the cnemial process in earlier evolutionary development.

By comparison, the hind-limb structures of the alcids are relatively simple, and they have been analyzed by

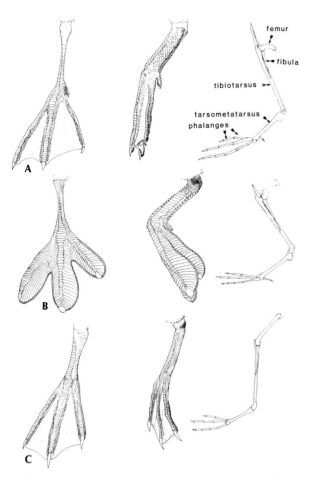

6. Comparison of foot structures and relative lengths of leg and foot bones: A, loons (*Gavia*); B, grebes (*Podiceps*); C, auks (*Uria*).

two categories, the burrowing forms (*Synthliboramphus*) and the surface-nesting forms (*Brachyramphus*). The burrowing forms tend to have heavy femurs and heavy claws, but in general the murrelets have short femurs and tarsometatarsi, suggestive of selection for swimming and diving rather than burrowing. However, in *Brachyramphus* the legs are usually short and the pelvis is especially broad, in conjunction with surface-nesting behavior.

It is impossible to discuss the hind limb without reference to the pelvic structure as well as the leg structure, since the two skeletal groups are functionally coevolved. Storer (1945) states that although the pelvises of the alcids average narrower than those of gulls and typical shorebirds, significant narrowing only occurs in a few forms such as in *Uria* and *Alca*. Kuroda (1954) reported a narrow pelvis in *Synthliboramphus* as well as in *Uria* and an unusually flat and broad pelvis in *Brachyramphus*, as may be seen in figure 7. In this fig-

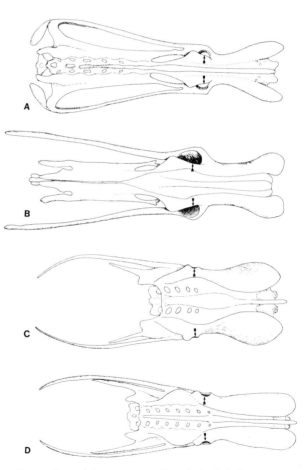

7. Comparison of the synsacrum (dorsal view): A, loons (*Gavia*); B, grebes (*Podilymbus*); C, murrelets (*Brachyramphus*); D, murres (*Uria*). In part after Cracraft 1982.

Storer (1945) and Kuroda (1954) in some detail. Storer noted that the leg length is greatest in the puffins, auklets, and *Cepphus* and is intermediate in the "auks" (great auk and razorbill), murres, dovekie, and murrelets except for *Brachyramphus*, in which it is shortest. Storer noted that in the genus *Cepphus* the legs are long and have heavy joints, a feature that in connection with a broad pelvis is associated with extensive walking. In puffins the long legs and heavy joints are also associated with ambulatory and fossorial adaptations; these are the only alcids that regularly stand with the tarsometatarsus raised above the ground (digitigrade posture). Fossorial adaptations of the puffins are evidenced in the thickness of the tarsometatarsus and the long, deep and curved claws, which are apparently used in burrow excavation. Auklets tend to have long legs but short pelvic bones, and the dovekie also has short, weak legs, so all these species are rather poor walkers. The murrelets fall into

ure it is apparent that the alcids have appreciably broader pelvises than do either loons or grebes. Thus the ratio of total pelvis length to width at the acetabulum (as shown by the arrows) is 9.6 for the illustrated example of *Gavia immer*, 6.8 for *Podilymbus*, 5.9 for *Uria*, and 3.5 for *Brachyramphus*. By this measure the loon would easily be the most specialized diver and *Brachyramphus* the least.

Some similar relative proportional relationships of the pelvic girdle and leg are shown in figure 8, which utilizes numerical and graphic information presented by Storer (1945) for a representative loon (*G. stellata*) and grebe (*P. grisegena*) and calculated means for eighteen species of Alcidae. Besides having the narrowest relative pelvis, the loon also has a remarkably short preacetabular component and an extremely long postacetabular length for all three of the pelvic bone components. In *Podiceps* the postacetabular pelvic component is essentially as long as that of *Gavia*, suggesting a similarly important muscular role for the leg muscles originating on the posterior pelvis, while the preacetabular pelvic

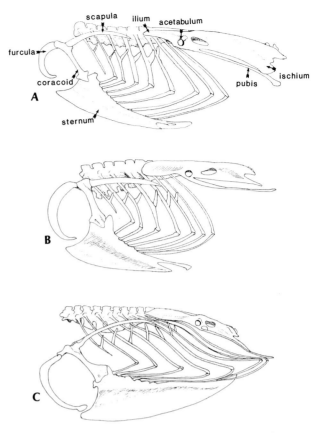

9. Comparison of trunk and sternal characteristics: A, loons (*Gavia*); B, grebes (*Podiceps*); C, auks (*Cepphus*). In part after Kartashev 1960.

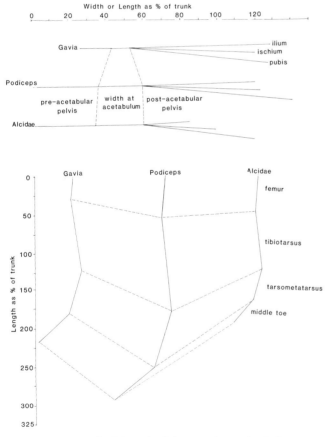

8. Comparison of pelvis and pelvic appendage ratios in loons (*Gavia stellata*), grebes (*Podiceps grisegena*), and auks (mean of eighteen spp.), based on data provided by Storer 1945.

length is substantially longer than in loons and approximately the same as that of the mean alcid length. Finally, in the alcids, the postacetabular ilium and ischium are both extremely short compared with those of the foot-propelled divers. The width of the pelvis at the acetabulum is also the greatest in the Alcidae, as indicated earlier.

Similar data summarized by Storer can be used to show the proportional relationships of the leg, foot, and toe bones (fig. 8). In this diagrammatic representation the extremely long legs and feet of loons and grebes as compared with the average alcid condition become evident. It may also be seen that in grebes the proportional lengths are greater throughout than in loons, and this is especially evident in the elongation of the tibiotarsus. Combined with the large surface area of the foot of a grebe relative to body weight, the very long feet are extremely effective propulsive mechanisms.

The relative roles of the muscles associated with the pelvic girdle as compared with those of the sternum in foot-propelled and wing-propelled divers becomes quite clear when the body skeletons of these groups are examined (fig. 9). Thus in alcids the sternum is extremely large, in association with the important wing muscles used for flight and diving, whereas in grebes and loons it is as small as or smaller than the pelvic girdle. In all three groups a substantial amount of "streamlining" is evident in the torso, the ribs are to varying degrees expanded, the coracoid is rigidly fused to the sternum, the furcula is enlarged and U-shaped, and the scapula is highly developed. In alcids the pectoralis major muscle is highly developed and constitutes 7–9 percent of the total body weight, whereas in the great crested grebe, for example, it makes up only 4 percent of the weight (Kartashev 1960). On the other hand, in grebes the muscles of the lower extremities may compose about 15–19 percent of the weight (Hartman 1961). Similarly, in the alcids the supracoracoideus muscle is unusually well developed, in conjunction with its role in underwater propulsion. Thus in alcids the weight ratio of the pectoralis major muscle to the supracoracoideus ranges from 3.04:1 to 3.65:1, while in the arctic loon it is 14.96:1 and in the eared grebe it is 7.48:1 (Kozlova 1961).

3. Comparative Egocentric and Locomotory Behaviors

As used here, "egocentric behavior" means those categories of individual survival and maintenance behaviors that are exclusive of such social interactions as aggressive, sexual, and parental behaviors, which will be considered in detail in the individual species accounts and also will be separately summarized in chapter 5. A discussion of the ecological aspects of foraging behavior will also be deferred until chapter 4, though behavioral aspects of underwater locomotion and prey catching will be considered here.

Comfort and Maintenance Behaviors

Self-directed maintenance behaviors relate to basic individual survival needs such as thermal regulation, drinking, food ingestion, and waste elimination. Other more periodic and generally leisurely types of egocentric behavior concern apparent comfort and body care, such as preening, oiling, bathing, sunbathing, stretching, shaking, and the like. These behaviors are fundamental to all birds and do not tend to differ greatly among rather distantly related groups, at least beyond the constraints set by anatomy and proportions.

The loons, auks, and grebes all spend considerable time preening and oiling their plumage. In all these groups the birds usually perform these behaviors while on the water, frequently rolling over on their sides to varying degrees while preening their breast and underpart feathers. Loons may remain in this partially inverted position for several minutes, waving the upper foot slowly in the air (Olson and Marshall 1952). Grebes sometimes also slowly wave one foot in the air well above the water surface while maintaining a normal swimming position (fig. 10C); the purpose of such behavior is uncertain but may relate to drying or warming. McIntyre (1975) described a probably comparable "foot waggling" in common loons, during which the

foot is extended and shaken and then placed under a wing. This behavior sometimes occurs when the bird is maintaining its position with the aid of only one foot. Preening in grebes often also occurs in a social situation during display, during which a ritualized version of preening, called "habit preening," often occurs. Similarly, preening of another individual, or "allopreening," is an extremely important social and sexual activity in various auks, especially the murres and the razorbill.

Stretching behavior in these birds takes two common forms. One is the wing and leg stretch, during which one wing and the corresponding leg are stretched laterally and posteriorly to the maximum degree (fig. 10A,D). The other common type of stretching motion is a simultaneous stretching of both wings above the back, while the neck is stretched forward to varying degrees (fig. 10B). Probably all three groups perform these behaviors in much the same manner, usually while on the water, though close comparisons remain to be made.

Wing flapping is similarly performed in much the same manner among all three groups. Among loons it is usually performed while the bird is swimming on the surface (fig. 10G), with the body often only partially raised from the water and the primaries sometimes striking the water at the bottom of the downstroke. A similar but much more erect wing flapping occurs during threat display in loons, providing an example of a ritualized form of behavior that coexists with its non-display precursor. Wing flapping in grebes also usually occurs on water, but among auks it can often be observed among birds resting on land or perched on rocky sites (fig. 10H). Wing flapping has not been obviously ritualized into a display function in either grebes or auks.

General rotary shaking movements, involving only the head and neck (fig. 10F) or at other times the entire body, are frequent in all species and perhaps serve to

10. Some general behavioral traits of loons, grebes, and auks: A, wing and leg stretching, horned grebe; B, wing and neck stretching, horned grebe; C, foot shaking, western grebe; D, wing and leg stretching, horned grebe chick; E, resting in "pork pie" posture, western grebe; F, drinking, western grebe; G, wing flapping, arctic loon; H, wing flapping, rhinoceros auklet; I, yawning, tufted puffin. After various sources.

free the bird of loose feathers or possibly foreign materials attached to the feathers. The same also applies to various head shaking or bill flicking movements that occur in diverse forms.

Last, there are a few minor kinds of behavior such as the bill stretching or "yawning" movement (fig. 10I). These movements seem to be more prevalent in puffins than in other species of diving birds, and their function is rather obscure. Grebes are said to perform both jaw stretching and true yawning behavior, and they also perform direct head scratching, throat touching to drain water from the bill, and feather eating (Cramp and Simmons 1977), the last of which will be discussed under foraging behavior.

Sunbathing behavior is well developed in some grebes. The birds orient themselves away from the sun and both lift their folded wings and tilt them somewhat forward so that the fluffy rump feathers are exposed to the sun. While resting or sleeping, grebes typically keep the head directed forward and lay the neck and nape back on the scapular feathers, assuming a flattened posture that has been called the "pork pie" position (fig. 10E). On the other hand, loons twist the head around and tuck the beak into the scapulars while resting or sleeping, as do the auks.

Terrestrial and Aerial Locomotion

The ability to stand on dry land is very differently developed among these groups of birds, as a reflection of their very different pelvic and leg structures. For loons standing appears to be extremely difficult, and I believe the birds rarely if ever are able to attain a digitigrade posture by lifting their tarsi above the ground. Instead they adopt a partially raised body posture (fig. 11A), and even this is assumed only when absolutely necessary, as when an incubating bird rises in the nest to turn or inspect its eggs. A very similar posture is assumed by

27

11. Some postural traits of loons, grebes, and auks: A, standing, common loon; B, standing, western grebe; C, resting on ground, arctic loon; D, resting on ground, eared grebe; E, walking, common murre; F, flying, common loon; G, flying, western grebe. After various sources.

grebes in the same circumstances (fig. 11B), though it is clear from observations of nesting western grebes that the birds can assume a digitigrade posture and can even run if necessary (Nero, Lahrman, and Bard 1958). When resting on dry land, loons and grebes typically let their long legs and feet splay out to the front or back (fig. 11C,D) while resting all their body weight on the substrate. Auks stand relatively erect much of the time, although when resting for extended periods many of them rest on their breasts. Auklets and murrelets typically stand in a plantigrade manner, as do the larger alcids with the exception of the puffins, which are more inclined to stand erect in a digitigrade manner. Puffins have the relatively longest legs of any of the alcids and are adapted not only for efficient walking and running but also for burrow-digging (Kozlova 1961). However, digitigrade walking is possible even for the relatively large murres, especially the common murre (fig. 11E), when an erect, penguinlike posture is assumed. I have seen no evidence that walking is regular in loons, though it has been noted that for short distances the red-throated loon can assume a semierect posture and

walk forward with its neck bent forward and its bill near the ground (van Oordt and Huxley 1922). However, in my experience loons move about on flat ground by pushing themselves forward with both feet simultaneously (fig. 12A), lifting the breast from the substrate only temporarily at the start of each propulsive effort.

Flight in loons is swift and powerful (see table 9), with the head, neck, and body all held in essentially the same plane (fig. 11F) or the head and neck slightly lower, resulting in a slightly humpbacked body outline. During landing the wings are stiffly outstretched or may be raised in a sharp dihedral as the bird loses altitude, drops its feet, and skids into the water. At times a gliding flight, with the wings held in a V above the back, has also been observed in loons, and these flights may represent a kind of aerial display (Olson and Marshall 1952).

Flight in grebes is infrequent and very poorly studied. The flight posture assumed by the western grebe (fig. 11G, after a photo by Gary Nuechterlein) is much like that of loons, but most observers have noted that the birds rarely attain any great altitude during flight, and

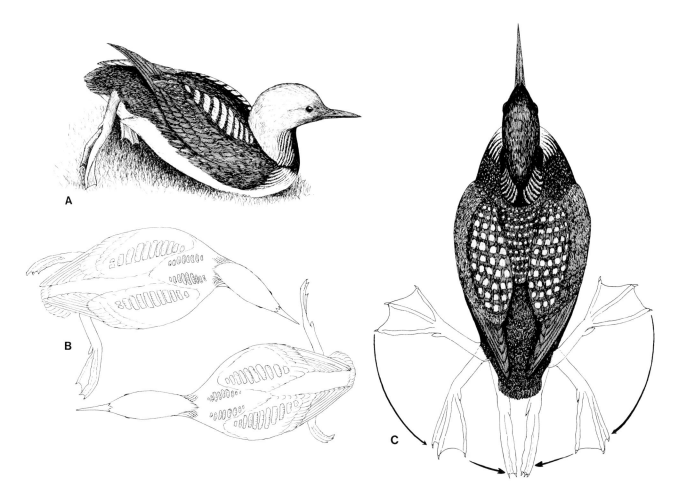

12. Walking and swimming behavior of loons: A, "walking," arctic loon; B, underwater swimming, arctic loon; C, swimming, common loon. After photos by author except for C, which is in part after Heilmann 1927.

Nuechterlein (1982) suspected that during its nesting season the western grebe may become incapable of flight because the breast muscles atrophy. I have seen flight in pied-billed, western, eared, and red-necked grebes but have been unable to estimate flight speed in any. Storer (1971) noted that a flying New Zealand dabchick was soon outdistanced by a group of New Zealand scaup (*Aythya novaeseelandiae*) and that it remained within a yard or so of the water most of the time it was airborne.

Takeoff in grebes and loons is preceded by a long running takeoff along the surface of the water. Olson and Marshall (1952) estimated that for the common loon the distance of this takeoff varies from 20 yards to as much as a quarter of a mile, and such long takeoff requirements are no doubt a severe handicap for loons and grebes when they are forced down on small water areas or, worse yet, on land. The red-throated loon is the only loon that has so far been reported as being able to take off from land (Harle 1952), and it also typically takes

flight from water after skittering 15 to 40 meters (Norberg and Norberg 1971).

Many of the auks regularly nest on cliff sites, and thus landing and takeoff require different capabilities from those of loons or grebes. In particular an effective braking ability, often facilitated by an upward stall just before landing, is needed for cliff landings. Takeoff from such sites is much easier and simply involves falling forward as the wings are spread. For water takeoffs or land takeoffs by the burrow nesters and surface nesters these tactics clearly are inappropriate, and as with the loons and grebes, rapid running over the substrate is required until flight speed is reached. When murres are taking off from water they may run along the water for as little as 6 feet or as much as 50 feet or more, depending on wind conditions; takeoffs are against the wind whenever possible. Airspeeds in murres probably range from 60 to 100 kilometers per hour, and the wingbeat rate may approach 500 per minute (Stettenheim 1959). Similar rates seem to be typical of other alcids (table 9).

Swimming and Diving

Swimming behavior in auks, loons, and grebes is surprisingly poorly documented, and except under special conditions of captivity it is rather difficult to observe. When grebes are resting on the water, their feet are oriented almost directly below them (fig. 13A). Slow paddling is done with alternating foot movements and varying degrees of lateral spreading of the legs. However, when underwater grebes exhibit an extreme degree of lateral foot orientation, with the legs being lifted to the midplane of the body or even above it, and locomotion is attained by simultaneous or alternate foot thrusts (fig. 13B,C), the former being associated with more rapid swimming (Fjeldså 1973a).

In loons the legs are more or less permanently splayed to the side, and during surface swimming they are moved in laterally, typically more or less simultaneously (fig. 12C). Such lateral orientation probably allows the feet to serve better as a rudder (Höhn 1982), and perhaps a maximum separation of the two feet also is more efficient in reducing drag and turbulence. Similarly, when underwater the feet also move in a lateral orientation, and in this case they may be moved either

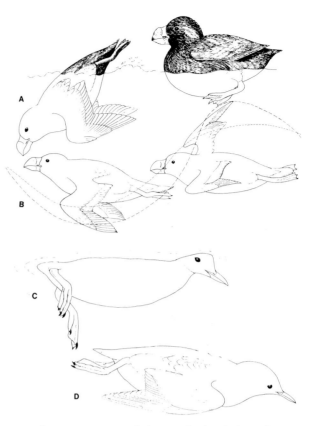

14. Underwater swimming behavior of auks: A, diving by tufted puffin; B, underwater swimming by tufted puffin, with paths of wingtips and wrist indicated by dotted lines; C, peering by pigeon guillemot; D, underwater "gliding" by pigeon guillemot. After photos by author and David Rimlinger.

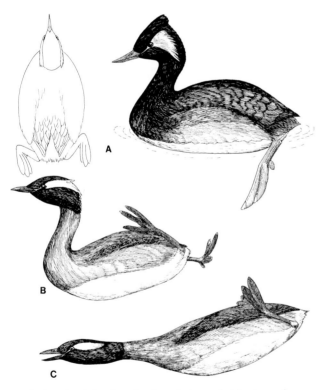

13. Swimming behavior of grebes: A, top and side views of surface swimming, eared grebe; B, slow and C, rapid swimming under water, horned grebe. After Fjeldså 1973c and photos by author.

simultaneously or alternately. A lateral extension of one foot also seems to be employed as a turning mechanism (fig. 12B).

In the auks, paddling while surface swimming is ducklike (fig. 14) with the legs oriented directly downward and moved alternately, simultaneously, or sometimes singly when the bird is slightly adjusting its orientation in the water. Additionally, paddling may occur while alcids are underwater. However, in the cases I have observed such paddling is used (by guillemots) not for propulsion but simply to hold the bird more or less motionless just above the substrate as it probes the bottom for food. Stettenheim (1959) observed similar "hovering" behavior in the common murre, with the two wings used independently, supplementing the propulsive effects of foot paddling. At the end of such a feeding session the paddling stops and the bird quickly returns almost vertically to the surface, its wings partly spread and its feet trailing behind. Likewise in grebes the return to the surface is done passively, in a posture similar to that shown in figure 13B (Fjeldså 1973a).

Often during such surface swimming the bill and head are partially immersed, the eyes slightly below the waterline (figs. 14C, 15D). This "peering" behavior is particularly prevalent in the alcids, but a comparable behavior occurs in loons (Olson and Marshall 1952). Peering also is common in grebes, and Fjeldså (1973a) illustrates a similar posture in a horned grebe capturing submergent prey while surface swimming.

These birds dive in a variety of ways. A typical manner of diving in grebes is shown in figure 15A, in which first the head and then the rest of the body slips under the water, with little apparent effort expended. Nuechterlein (1981a) calls such dives "level dives," compared with the more energetic "springing dives" in

which the body is almost lifted out of the water as the bird dives in a more nearly vertical fashion. Similar springing dives are typical of loons (fig. 15C), and I have not seen the wings used during the dives of either loons or grebes. However, Olson and Marshall (1952) stated that during rapid escape dives (as when shot at) the wings are used to strike the water and propel the loon down out of sight. A similar kind of emergency dive, or "crash dive," also occurs in grebes, especially among the species of *Podilymbus* and *Poliocephalus* (Fjeldså 1983b). In this behavior the bird suddenly sinks directly downward by rapid vertical propulsion with the feet (fig. 15E). Apparently least and pied-billed grebes do not typically swim great lateral distances during escape dives, but in the closely related giant pied-billed grebe a lateral movement of up to 90 meters has been reported, and

15. Diving behavior of loons, grebes, and auks: A, low and B, high dives of western grebe (after Nuechterlein 1981a); C, diving by common loon (after photo by author); D, diving by horned puffin (after photos by author); E, diving by common murre (after photo by David Rimlinger); F, crash dive by hoary-headed grebe (after Fjeldså 1983b).

red-necked grebes have moved underwater as far as 60 meters (Bleich 1975). Underwater speeds of up to 2.2 meters per second have been estimated for grebes (Bowes 1965). In the arctic loon lateral movements of from 400 to 600 meters have been reported, and underwater swimming rates of 2.5 to 5 meters per second appear to be common (Lehtonen 1970). Underwater movements by several species of alcids observed in captivity at Sea World had average speeds of from 0.96 to 1.13 meters per second, with little interspecies variation among puffins, rhinoceros auklet, and murres (Kenneth Fink, pers. comm.; see table 9).

When diving, the alcids hardly spring upward at all; instead, the wings are quickly opened and swept downward on the first power stroke as the bird submerges. The feet likewise go backward but sometimes seem to contribute little if any propulsive force (figs. 14 and 15). The angle of the dive is steep, which places the wings under the water surface immediately, and the partially folded wings are then quickly swept downward and backward, which together with a simultaneous backward kick of both feet pulls the bird completely underwater (Stettenheim 1959).

Underwater propulsion of the alcids is extremely interesting and warrants special consideration. When swimming underwater, alcids move their partially folded wings in a vertical action quite different from the wing action used during flight (fig. 14B). The wings are scarcely raised above the level of the back, and during the downstroke they are distinctly down-tilted, producing a hydrofoil effect that propels the bird forward. During this phase the wing is rotated as a single unit, with the wrist turning nearly 180° by the action of the humerus, while the rest of the wing components maintain a constant relation to one another (fig. 16A; 0.22 to 0.30 seconds). Then the humerus is swung posteriorly in a horizontal plane, pushing the forearm and hand posteromedially (fig. 16A; 0.30 to 0.48 seconds) and providing the second propulsion phase. Finally, the forearm and hand swing back medially, and the wing is raised for another power stroke (Spring 1971). Veering can be achieved by asynchronous wing movements, and turning is performed with the aid of the feet, by extending the foot on the side of the body toward which the bird is going to turn and using its extended webs as a brake. Turning may also be assisted by paddling with the opposite foot. Spring (1971) described three types of underwater turning in murres and noted that the thick-billed murre exhibited only two of the three types, suggesting greater dexterity in the common murre than in the thick-billed. However, he judged on the basis of anatomical differences that in the air the thick-billed murre should be an energetically more efficient flier

16. Swimming behavior of auks: A, turning sequence, common murre (after Spring 1971, with intervals in seconds); B, turning in common murre (after photo by author); C, comparison of underwater profile of common murre with that of a penguin and diving petrel (after photos by author except for diving petrel, which is based on a museum specimen).

and should also be capable of moving over greater underwater distances than the common murre.

When swimming underwater, alcids assume a fusiform body shape that is strongly similar to that of a penguin underwater (fig. 16D). Comparable photographs of the underwater posture and wing movements of diving petrels are not available, but judging from their anatomy it seems likely that these birds are virtually identical in shape and behavior to such alcid species as the ancient murrelet (Kuroda 1967).

When "flying" underwater, alcids do not move their wings continuously, but often "glide" for a time with their wings held close to their bodies (fig. 16D). This tends to reduce the average wingbeat rate while submerged, which in general tends to be only about 25 or 30 percent of the rate attained in aerial flight. Perhaps because of the varied rate of wing flapping underwater,

17. Foraging and parental feeding behaviors of loons, grebes, and auks: A, spearing by western grebe; B, surface and C, aerial catching of insects, horned grebe; and parental feeding of young in D, crested grebe; E, arctic loon; F, ancient murrelet; G, black guillemot; H, Atlantic puffin. Mostly after drawings by Fjeldså 1973c, 1975.

there is no clear relation between body size and wing-beat rate while underwater (see table 9).

Although wing movement and overcoming relative environmental resistance during flying and underwater swimming must be very difficult in alcids, I have seen murres plunge directly underwater from flight and emerge from dives directly into the air. Kenneth Fink (pers. comm.) once saw a horned puffin perform "porpoising" behavior, alternating dives of 3 to 6 feet horizontal distance with short flights 18 to 24 inches above the water and repeating the sequence several times in a few minutes, apparently out of sheer exhilaration. Direct diving into the water from flight has also been observed in Atlantic puffins, razorbills, and guillemots (Stettenheim 1959).

Foraging Behavior

Loons, grebes, and auks capture prey in varying ways, depending on its speed, abundance, and elusiveness. There do not appear to be any good descriptions of prey catching in loons, but there can be little doubt that it is chased down visually and caught in open-field chases, judging from the kinds of fish that usually make up their diet. Fjeldså (1973a, 1975) has described and illus-

trated the underwater behavior of various grebes, especially the horned grebe. The horned grebe feeds mainly by diving and underwater pursuit, though this may be supplemented by such techniques as picking up individual floating items from the water surface (fig. 17B), skimming the surface for masses of floating invertebrates, or even snatching insects from the air (fig. 17C). In the western grebe the prey may initially be speared with the needlelike bill (fig. 17A), but probably most grebes capture their prey by grasping it between the upper and lower mandibles.

There are few observations on prey catching in auks, but at least in captivity it is clear that such species as murres regularly obtain their prey by visual pursuit. In puffins it is quite regular for the birds to capture additional prey while still holding crosswise in their beaks with the aid of their fleshy tongues fish that they have captured earlier. At times a half-dozen or more fish may accumulate in the beak of a puffin (fig. 17H), often neatly organized in an efficient alternating left- and right-handed manner. Apparently this rather remarkable arrangement is made possible by the bird's swimming in a school of fish and alternately making prey-grasping movements to the right and left, automatically "stacking" prey in its bill. Razorbills likewise often carry two or more fish crosswise in their beaks, but with no special orientation. Fish carrying is done differently by the murres, however, with a single fish typically carried head-inward in the beak, with only the tail dangling out near the end of the bill. Similarly, guillemots carry single prey items back to their young by holding them crosswise in the bill, grasped at the head end. Curiously, at least in the black guillemot, individual birds show a preference for carrying their prey with the head held to either the left or the right side of the beak, though there is no obvious advantage to this kind of preference (Slater 1974). The dovekie and all five species of auklets have special gular pouches that facilitate the capture and carrying of large loads especially of plankton-sized prey, and at least in some species prey size varies with the reproductive state of the bird (Speich and Manuwal 1974). Neck pouches are not

known to occur in any of the murrelets, but the young have unusually well developed feet, probably allowing them to begin diving and feeding on their own well before their wings are sufficiently developed to aid much in diving. However, one or both parents tend to remain with their still-flightless chicks for some time after they go to sea and continue to feed them there (fig. 17F). In grebes and loons there are no neck pouches, and the adults directly feed their young individual bits of food by offering it bill to bill (fig. 17D,E).

Depths and Durations of Dives

As predatory birds, loons, grebes, and auks spend a good deal of time and energy capturing their prey while underwater, and it is of some interest to compare their relative diving efficiencies, at least indirectly. This

Table 10: Relative Depths (as Percentages) of Foraging Dives in Loons, Grebes, and Auks

Water Depth (ft)	Common Loon (N = 226)	Red-throated Loon (N = 207)	Great Crested Grebe (N = 282)	Horned Grebe (N = 156)	Little Grebe (N = 102)	Razorbill (N = 453)	Common Murre (N = 208)	Dovekie (N = 58)
1–6	0%	16%	29%	71%	100%	36%	24%	95%
6–12	8	54	48	29	0	52	25	5
12–18	46	24	20	0	0	11	37	0
18–24	36	4	20	0	0	1	12	0
24–30	9	2	3	0	0	0	2	0
30–36	1	0	0	0	0	0	0	0
Maximum depth (ft)	33.5	29	21	12	6	24	28	8

SOURCE: Data of Dewar 1924.   NOTE: Depths are of waters in which diving occurred rather than known diving depths.

Table 11: Average Durations (in Seconds) of Observed Foraging Dives in Loons, Grebes, and Auks

Water Depth (ft)	Common Loon	Red-throated Loon	Great Crested Grebe	Horned Grebe	Razorbill	Common Murre
1–6	—	18.2	22.1	14.9	18.9	15.0
6–12	28.2	27.6	26.6	25.2	24.8	24.9
12–18	35.5	27.6	33.3	—	34.2	36.8
18–24	49.3	35.2	45.6	—	48.7	49.1
24–30	60.2	47.8	—	—	—	61.0
30–36	68.0	—	—	—	—	—
Maximum depth (ft)	69	67	50	41	52	68

SOURCE: Data of Dewar 1924.   NOTE: Sample size as in table 10.

question was first addressed seriously by Dewar (1924), who spent several years gathering data on the behavior of not only these three groups of birds but other divers such as cormorants and diving ducks. He accumulated a great deal of interesting comparative data, some of which are summarized in tables 10 and 11 for the species of loons, grebes, and auks that he was able to observe. In general he found that loons tend to dive deeper than auks or grebes and that the durations of dives of representative species of all three groups are very similar in water of comparable depth, but that the maximum duration of observed loon dives tends to average somewhat greater than for auks or grebes.

Using a variety of criteria, including the longest dive

Table 12: Reported Maximum Diving Records for Loons, Grebes, and Auks

Species	Record	Reference
Loons	*Duration of Dive*	
Common loon	180 seconds[a]	Palmer 1962
Arctic loon	302 seconds	Lehtonen 1970
Red-throated loon	90 seconds	Höhn 1982
Grebes		
Horned grebe	180 seconds	Eaton 1910
Red-necked grebe	60 seconds	Cramp and Simmons 1977
Great crested grebe	56 seconds	Cramp and Simmons 1977
Eared grebe	50 seconds	Cramp and Simmons 1977
Auks		
Cassin auklet	120 seconds	Dewar 1924
Thick-billed murre	98 seconds	Glutz and Bauer 1982
Black guillemot	78 seconds	Glutz and Bauer 1982
Common murre	74 seconds	Glutz and Bauer 1982
Razorbill	74 seconds	Glutz and Bauer 1982
Dovekie	71 seconds	Glutz and Bauer 1982
Tufted puffin	61 seconds	Personal observations
Loons	*Depth of Dive*	
Common loon	60–70 meters	Höhn 1982
Arctic loon	46 meters	Höhn 1982
Grebes		
Great crested grebe	30 meters	Cramp and Simmons 1977
Horned grebe	25 meters	Palmer 1962
Auks		
Common murre	180 meters	Piatt and Nettleship 1985
Razorbill	120 meters	Piatt and Nettleship 1985
Black guillemot	50 meters	Piatt and Nettleship 1985
Rhinoceros auklet	31 meters	Stettenheim 1959
Atlantic puffin	60 meters	Piatt and Nettleship 1985

[a]Submersion to 15 minutes has been reported in a wounded bird (Schorger 1947).

duration, the "dive/pause ratio" (the average duration of dive per unit of resting between dives), and the rate of increase in pause time per fathom of diving depth, Dewar concluded that the auks are the most efficient of the divers, followed sequentially by loons, cormorants, grebes, and diving ducks.

Since Dewar's studies, many additional observations on diving depths and durations have been made by various observers. In general they prove difficult to compare directly, since such variables as water depth, food abundance, speed of swimming, and perhaps other factors enter into the question of diving efficiency. Thus, table 12 presents various observations on maximum observed dive durations and depths of dives, suggesting that maximum dive durations are sometimes much greater than those Dewar reported for the same species and frequently exceed a minute in species of all three groups. Diving depths also have been reported to exceed those Dewar observed, and almost incredible depths of more than 500 feet have been reported for the common murre. Olson and Marshall (1952) have mentioned that for a diving bird to reach a depth of 200 feet it must be able to withstand a water pressure of 86.7 pounds per square inch as well as deal with the problems of oxygen availability. It is well known that most of the needed oxygen comes from stored oxyhemoglobin in the blood and muscles (Schorger 1947), and additionally the metabolic rate during diving is probably sharply decreased by a drastic reduction in heart rate and the temperature of peripheral organs, especially the feet, judging from stud-

ies of diving ducks. Höhn (1982) has reviewed the major physiological aspects of diving in loons, and I will not repeat them here.

Storer (1945, 1952) has suggested that alcids such as murres and razorbills are considerably better adapted for diving than such genera as *Cepphus,* for example, which is anatomically more generalized. The limited data in table 12 suggest that maximum depths and durations of dive in murres average somewhat greater than those of guillemots, though the differences are not marked. Similarly, in a common display tank at Sea World where both species were maintained in captivity, the average foraging dive duration of common murres was only slightly longer (mean 27 seconds, range 17–39 seconds, sample of 10 dives) than those of pigeon guillemots (mean 21.7 seconds, range 4–36 seconds, sample of 36 dives). However, the dive/pause ratio of the common murre was 5.46:1, while that of the guillemot was 2.07:1, suggestive indeed of a considerably greater degree of diving efficiency in the murre than in the guillemot. Typical dive/pause ratios in grebes appear to be in the range of 1.5 to 2.7:1, at least for foraging dives (Bleich 1975; Storer 1971). Dewar's (1924) data on dive/pause ratios in loons indicated a ratio range of from 1.6 to 3.9:1, with the higher ratios generally associated with deeper rather than shallower dives, as might have been imagined. This might lead one to question the value of dive/pause ratios as a direct measure of diving stress.

4. Comparative Diets and Foraging Ecologies

Diets

Auks, loons, and grebes are birds whose anatomies and behaviors have been sharply influenced by their foraging niche adaptations. In each species these have evolved through natural selection over extended periods of geologic time as a reflection of available food resources, the presence of competing species, and the limitations on innate variations in anatomy, physiology, and behavior imposed by available genetic mutations and recombinations. To a very considerable degree the auks, loons, and grebes seem to have adjusted to the effects of interspecific competition by evolving differences in bill shape and body size that sometimes open specific new foraging niche opportunities to them and thus reduce direct competition with other species of their group. In the loons, for example, all four species of which overlap rather extensively in at least their wintering and sometimes also their breeding distributions, there is a rather marked stepwise gradation of body size and bill length but no major changes in basic bill shape throughout the series (fig. 18). Within each species, males tend to be slightly larger than females, and there is some evidence that, at least in the common loon, males tend to take more large prey than do females (Barr 1973).

All four species of loons are evidently almost exclusively piscivorous in both breeding and wintering areas. However, so far no studies have directly addressed possible interspecific differences in the diets of loons. A tabular summary (table 13) of prey reported from the digestive tracts of the four loon species suggests that certain families of fish (clupeids, salmonids, gadids, gasterosteids, cottids, ammodytids, and gobiids) are probably important prey items for most if not all species. Data for the common loon suggest that freshwater fish of such families as the sucker and catfish groups may be more important in this species than in the other more generally arctic-nesting forms. So far too few specimens of the yellow-billed loon have been examined to make any firm statements about its preferred diet, but it seems reasonably clear that the red-throated and arctic loons are very similar to one another in their general dietary intakes.

Other than fish, the diets of loons include varying amounts of crustaceans, mollusks, aquatic insects, and

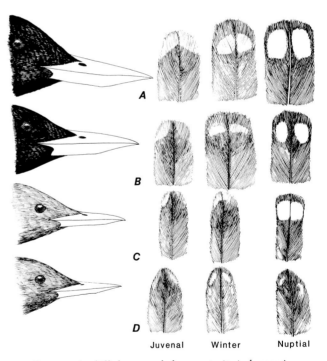

18. Comparative bill shapes and plumage traits in loons: A, yellow-billed; B, common; C, arctic; D, red-throated. Adapted from Bauer and Glutz 1966.

37

Table 13: Reported Prey of the North American Loons

Food Type	Red-throated	Arctic	Common	Yellow-billed
Fish				
Clupeidae (herrings)				
Brevoortia (menhaden)	—	—	X	—
Clupea (herring)	XX	XX	X	—
Dorosoma (gizzard shad)	—	—	X	—
Sardinops (sardine)	—	—	X	—
Sprattus (sprat)	XX	XX	X	—
Anguillidae (eels)				
Anguilla (eel)	—	—	X	—
Esocidae (pikes)				
Esox (pike)	—	—	X	—
Salmonidae (salmonids)				
Coregonus (whitefish)	XX	—	—	—
Leucichthyes (cisco)	—	—	X	—
Salmo (trout)	X	X	—	—
Salvelinus (char)	X	X	—	—
Thymallus (grayling)	X	X		
Osmeridae (smelts)				
Mallotus (capelin)	X	—	X	—
Osmerus (smelt)	—	—	X	—
Cyprinidae (cyprinids)				
Alburnoides (bleak)	X	—	—	—
Alburnus (bleak)	X	X	—	—
Cyprinus (carp)	—	X	—	—
Leuciscus (dace)	X	X	X	—
Phoxinus (minnow)				
Catostomidae (suckers)	—	—	XX	—
Ictaluridae (catfish)	—	—	XX	—
Gadidae (cods)				
Boreogadus (polar cod)	X	—	—	—
Gadus (cod)	XX	XX	—	X
Melanogrammus (haddock)	—	X	X	—
Merlangus (whiting)	—	X	X	—
Microgadus (tomcod)	X	—	—	X
Zoarcidae (eelpouts)	X	X	X	—
Cyprinodontidae (toothcarps)				
Fundulus (killifish)	X	X	X	—

Food Type	Red-throated	Arctic	Common	Yellow-billed
Fish (*continued*)				
Atherinidae (silversides)				
Atherina (sand smelt)	—	XX	—	—
Syngamidae (pipefish)	—	—	X	—
Gasterosteidae (sticklebacks)	XX	XX	X	—
Cottidae (sculpins)	XX	X	X	XX
Percichthyidae (temperate bass)				
Morone (bass)	—	—	X	—
Centrarchidae (sunfish)				
Lepomis (sunfish)	—	—	X	—
Micropterus (black bass)	—	—	X	—
Poxomis (crappie)	—	—	X	—
Percidae (perches)				
Perca (perch)	X	X	XX	—
Stizostedion (pike perch)	—	—	X	—
Embiotocidae (surfperches)				
Cymatogaster (shiner perch)	XX	X	X	—
Stichaeidae (pricklebacks)				
Lumpenis (eelblenny)	X	—	—	—
Pholididae (gunnels)				
Pholis (butterfish)	X	—	—	—
Ammodytidae (sand eels)				
Ammodytes (launce)	XX	XX	X	—
Gobiidae (gobies)	XX	X	X	—
Pleuronectidae (righteye flounders)				
Amphibians (newts and frogs)	X	X	X	—
Cephalopod mollusks (squid)	X	XX	—	—
Other mollusks	X	X	X	X
Insecta	X	X	X	—
Crustacea	X	X	X	X
Annelida (polychaetes, leeches)	—	X	X	X

SOURCE: Summarized from available literature, especially Ainley and Sanger 1979.

NOTE: Prey that have been reported as regular or frequent components are shown as XX; other positive records are shown as X.

other prey, especially during the breeding season. Frogs, leeches, polychaetes, and other items have also been reported, although in some cases these trace items might simply reflect food materials in the stomachs of prey species. In general, plant materials are rarely eaten, but there have been a few cases of apparent consumption of mosses (Hypnaceae) and seaweeds in considerable quantities. Seeds and fibers of some freshwater plants such as pondweeds and bulrushes have also been reported at times.

Among the grebes, there are also substantial differences in body size, ranging from species such as the western grebe, which approximates the weight of the smallest loon, to the least grebe, which approaches the size of the smallest alcids (table 14). Within this size gradient the grebes exhibit a good deal more variation in bill shape than the loons, with the larger fish-eating grebes having rather loonlike bills and the smallest grebes having bill shapes not very different from those of murrelets, for example. Among the North American

Table 14: Size Categories and Usual Diets of North American Loons, Grebes, and Auks

Weight Category	Typical Diet	Representative Species		
		Loons	Grebes	Auks
Very large (over 2,000 g)	Fish (to ca. 30 cm)	Yellow-billed Common Arctic	—	Great auk
Large (1,200–2,000 g)	Fish (to ca. 25 cm)	Red-throated	Western	—
Medium large (500–1,200 g)	Fish (to ca. 17 cm), invertebrates	—	Red-necked	Common murre Thick-billed murre Tufted puffin Razorbill Rhinoceros auklet
Medium small (250–500 g)	Fish (to ca. 15 cm), invertebrates	—	Pied-billed Horned Eared	Horned puffin Atlantic puffin Pigeon guillemot Black guillemot Crested auklet Parakeet auklet
Small (100–250 g)	Invertebrates, Fish (to ca. 10 cm)	—	Least	Marbled murrelet Kittlitz murrelet Ancient murrelet Cassin auklet Dovekie Craveri murrelet Xantus murrelet Whiskered auklet
Very small (under 100 g)	Planktonic invertebrates (to ca. 1.5 cm)	—	—	Least auklet

NOTE: Organized by descending average weights within each size category.

grebes the most divergent of all bill shapes is to be found in the pied-billed grebe, which eats a widely diversified diet, including a considerable amount of crustaceans, for the capture and crushing of which its heavy and compressed bill seems to be admirably adapted (fig. 19).

A summary of prey types reported for the North American species of grebes (table 15) indicates considerable overlap among the larger species of grebes (red-necked and western) and loons in terms of fish families utilized, specifically the clupeids, gasterosteids, and cottids among marine forms and the centrarchids and percids among the freshwater families. Certain fish families, such as the anguillids, gasterosteids, and cottids, appear to be of rather general significance to several species of grebes, and amphibians appear to be of greater importance to grebes than to loons. Among the noninsect invertebrates, amphipod and decapod crustaceans, polychaete worms, and various mollusks (mainly bivalve and univalve types) seemingly are of general food value. However, it is the insects that clearly are of special significance to grebe species other than the two

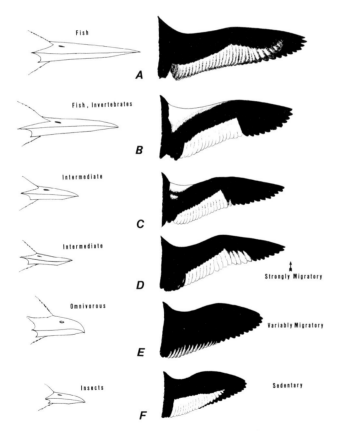

19. Comparative bill shapes and wing traits in grebes: A, western; B, red-necked; C, horned; D, eared; E, pied-billed; F, least. Adapted in part from Bauer and Glutz 1966.

largest and fish-adapted forms, with aquatic beetles, true bugs, and dragonflies being of particular importance.

Beyond these food types, grebes also have the unusual behavior trait of swallowing varying amounts of feathers. The function of such activity is still unproved, but it has generally been believed that feather swallowing may be related to fish consumption, and that feathers may enmesh swallowed fish bones that might be a potential danger to the bird. Feathers not only are swallowed by the older birds, mainly during self-preening, but they are often also fed to the young, sometimes within a day of hatching. These feathers soon decompose into a feltlike, amorphous mass, often forming a ball. Apparently all grebes except the two species of *Poliocephalus* swallow feathers, and in general the species of grebes that have diets rich in fish are more prone to feather eating. However, the two species of *Poliocephalus* are known to eat fish under some conditions, and so the apparent absence of feather eating in these forms is difficult to explain (Fjeldså 1983a).

The diets of the alcids are much more diverse than those of the loons and grebes, partly reflecting the considerably greater number of species involved, which exhibit a size range from larger than the largest loon to smaller than the smallest grebe (table 14). Throughout this range the larger species (guillemots and larger) eat mostly fish, while the smaller auklets and murrelets eat varying quantities of invertebrates, including those of planktonic size.

The diets of the North American alcids have not been well documented in some cases, especially those of several murrelets, but tables 16 and 17 give summaries of fifteen of the twenty-one North American species. It is clear from this summary that murres and puffins overlap with loons and the larger grebes in at least some aspects of their diets, showing an apparent dependence on such fish as clupeids, osmerids, gadids, scorpaenids, cottids, and ammodytids. The pigeon guillemot seems to have a considerably more diverse diet than these other fish eaters, and it specializes on bottom-dwelling fish that are associated with the intertidal and inshore coastal zones.

The bill shape, upper palate, and tongue characteristics of the alcids provide excellent clues to their diets (fig. 20), as has been amply demonstrated by Bedard (1969d). Bedard classified the alcids into plankton feeders (*Aethia, Alle,* and *Ptycoramphus*), fish feeders (*Uria, Alca,* and *Cepphus*), fish and plankton feeders (*Fratercula* and Cerorhinca), and a remaining group of little-studied and unclassified types (*Synthliboramphus* and *Brachyramphus*) that apparently feed on a diverse array of small fish and marine invertebrates. He

Table 15: Reported Prey of the North American Grebes

Food Type	Least	Pied-billed	Horned	Red-necked	Eared	Western
Fish						
Clupeidae (herrings)	—	—	X	XX	—	XX
Anguillidae (eels)	—	X	X	X	X	X
Engraulidae (anchovies)	—	—	X	—	—	—
Osmeridae (smelts)	—	—	—	—	—	X
Cyprinidae (cyprinids)	—	XX	X	—	—	—
Catostomidae (suckers)	—	X	—	—	—	—
Ictaluridae (catfish)	—	X	—	—	—	—
Atherinidae (silversides)	—	—	—	—	—	X
Gadidae (cods)	—	—	—	—	—	X
Cyprinodontidae (toothcarps)	—	X	—	X	—	—
Poeciliidae (live-bearers)	X	X	—	—	—	—
Gasterosteidae (sticklebacks)	—	X	XX	XX	—	—
Scorpaenidae (rockfish)	—	—	X	—	—	—
Cottidae (sculpins)	—	XX	X	XX	X	XX
Percidae (perches)	—	X	—	—	—	XX
Embiotocidae (surfperch)	—	—	X	—	—	XX
Stichaeidae (pricklebacks)	—	—	—	—	—	X
Gobiidae (gobies)	—	—	—	—	X	—
Amphibians	X	X	—	X	X	X
Crustaceans						
Euphausiacea	—	—	X	—	—	—
Mysidae	—	—	X	—	XX	—
Amphipoda	—	X	XX	X	XX	—
Decapoda	X	XX	X	X	—	X
Annelida						
Polychaeta	—	—	X	X	X	X
Hirudinea	—	XX	—	—	—	—
Mollusca	—	X	X	X	X	X
Insecta						
Coleoptera	XX	XX	XX	X	XX	X
Hemiptera	XX	XX	XX	X	X	X
Odonata	XX	X	X	X	XX	X
Other orders	X	X	X	X	X	X

NOTE: Symbols as in table 13.

Table 16: Reported Prey of Primarily Fish-Eating Species of North American Alcids

Food Type	Murres		Pigeon Guillemot	Rhinoceros Auklet	Puffins		
	Common	Thick-billed			Tufted	Horned	Atlantic
Fish							
Petromyzontidae (lampreys)	—	—	X	X	—	—	—
Chimaeridae (chimaeras)	—	—	X	—	X	—	—
Clupeidae (herrings)							
Clupea (herring)	XX	—	—	XX	XX	—	XX
Sardinops (sardine)	—	—	—	—	—	—	XX
Sprattus (sprat)	—	—	—	—	—	—	XX
Engraulidae (anchovies)	XX	—	—	XX	—	—	—
Salmonidae (salmonids)							
Salmo (trout)	X	—	—	—	—	—	—
Onchorhynchus (salmon)	—	—	—	X	—	—	—
Osmeridae (smelts)							
Allosmerus (smelt)	—	—	—	X	—	—	—
Hypomesus (smelt)	XX	—	X	XX	XX	—	—
Mallotus (capelin)	XX	X	—	XX	XX	X	X
Spirinchus (smelt)	—	—	—	XX	—	—	—
Thaleichthya (eulachon)	X	—	—	—	—	—	—
Bathylagidae (deep-sea smelt)							
Nansenia (argentines)	—	—	—	XX	—	—	—
Myctophidae (lanternfish)	X	X	—	X	—	—	—
Paralepididae (barracudinas)	—	—	—	X	—	—	—
Gadidae (cods)							
Boreogadus (polar cod)	XX	XX	X	—	XX	XX	—
Ciliata (rockling)	—	—	—	—	—	—	X
Eleginus (saffron cod)	—	—	—	—	XX	—	—
Gadus (cod)	—	—	—	—	—	—	XX
Gaidropsarus (rockling)	—	—	—	—	—	—	X
Melanogrammus (haddock)	X	X	—	—	—	—	—
Merlangus (whiting)	—	—	—	—	—	—	X
Microgadus (tomcod)	X	—	—	—	XX	—	—
Pollachius (pollack)	X	—	—	—	—	—	X
Theragra (walleye pollack)	XX	XX	—	X	XX	XX	—
Ophididae (cusk eels)	—	—	—	X	—	—	—
Zoarchidae (eelpouts)							
Gymnelis (ocean pout)	—	X	—	—	—	—	—
Lycodes (eelpout)	—	X	—	—	—	—	—
Scomberesocidae (sauries)							
Cololabris (saury)	—	—	—	XX	—	—	—

(*continued*)

43

| | Murres | | Pigeon Guillemot | Rhinoceros Auklet | Puffins | | |
Food Type	Common	Thick-billed			Tufted	Horned	Atlantic
Fish *(continued)*							
Gasterosteidae (sticklebacks)	—	—	X	X	—	—	—
Scorpaenidae (rockfish)	XX	X	XX	XX	XX	—	—
Anoplopomatidae (sablefish)	—	—	—	XX	—	—	—
Hexagrammidae (greenlings)							
Hexagrammas (greenling)	—	—	—	XX	—	—	—
Pleurogrammus (Atka mackerel)	—	—	—	X	XX	XX	—
Cottidae (sculpins)							
Gymnocanthus (sculpin)	—	X	—	—	—	—	—
Hemilepidotus (lordfish)	—	—	XX	—	X	—	—
Icelus (sculpin)	—	X	X	—	—	—	—
Myoxocephalus (sculpin)	X	X	XX	—	—	—	—
Triglops (sculpin)	X	XX	XX	—	—	X	—
Seven additional genera	—	—	X	X	—	—	—
Agonidae (poachers)	—	—	X	—	X	—	—
Liparidae (snailfish)	—	X	X	—	—	—	—
Embiotocidae (surfperch)							
Cymatogaster (shiner perch)	X	—	X	—	—	—	—
Kyphosidae (sea chubs)	—	—	—	X	—	—	—
Trichodontidae (sandfish)	—	—	—	—	—	X	—
Bathymasteridae (ronquils)	—	—	X	—	—	—	—
Clinidae (clinids)	—	—	X	—	—	—	—
Stichaeidae (pricklebacks)							
Cebidichthys (monkeyface eel)	—	—	X	—	—	—	—
Chirolophus (blenny)	X	X	—	—	—	—	—
Lumpenus (blenny)	—	XX	XX	—	—	—	—
Xiphister (blenny)	—	—	X	—	—	—	—
Pholidae (gunnels)	X	—	X	—	—	—	—
Cryptacanthodidae (wrymouths)	—	—	X	—	—	—	—
Zaproridae (prowfish)	—	—	—	—	X	—	—
Ammodytidae (sand eels)	XX	XX	XX	XX	X	XX	XX
Centrolophidae (medusafish)	—	—	—	X	—	—	—
Stromateidae (butterfish)	—	—	—	X	—	—	—
Bothidae (lefteye flounders)	—	—	X	X	—	—	—
Pleuronectidae (righteye flounders)							
Hoppoglossoides (sole)	—	—	—	X	—	—	—
Lipidosetta (sole)	—	—	X	—	—	—	—
Reinhartius (halibut)	—	X	—	—	—	—	—

(continued)

Table 16: (*Continued*)

Food Type	Murres		Pigeon Guillemot	Rhinoceros Auklet	Puffins		
	Common	Thick-billed			Tufted	Horned	Atlantic
Crustaceans							
Copepods	—	X	—	—	—	—	—
Euphausiacea	XX	XX	—	X	X	—	—
Amphipods	X	XX	X	—	X	XX	—
Isopods	X	—	X	—	—	—	—
Decapods	X	X	X	—	—	—	—
Polychaete annelids	X	XX	—	—	XX	X	XX
Cephalopod mollusks	X	XX	X	XX	XX	X	X

NOTE: Symbols as in table 13.

observed that the ratio of bill width to gape length provides a useful index to the species' diet, with plankton feeders having ratios of 0.3 or more, fish feeders ratios of less than 0.2, and intermediate types ratios of between 0.2 and 0.3. He also observed that the species that eat considerable amounts of plankton have a large number of cornaceous papillae (denticles) in the anterior palate region, while in fish feeders the number of denticles is greatly reduced and the individual papillae are more sharply pointed. The tongues of such fish eaters as murres are long and slender, with a rigid horny shield at the tip, apparently adapted to "locking" prey against the palatal denticles. In the plankton eaters the tongue is much less cornified and tends to be short and wide. In the puffins the tongue is of an intermediate type, with a cornified tip but a generally fleshy upper surface. This adaptation may help in holding several prey items simultaneously and also may be related to an increased proportion of invertebrates in the diet.

Bedard made the important point that in the alcids the bill not only serves as a food-getting device but also is important as a visual releaser in social interactions, which probably also influences the degree of interspecific variability in bill shape and appearance. He also stated that the fish-feeding alcids have evolved toward an optimum size that appears to approach the upper threshold of body weight compatible with both aerial and underwater flight. The smallest of the fish feeders, the murrelets, are so small that it is doubtful they rely entirely on fish, and it also is questionable whether they can effectively carry fish back to their nestlings. The plankton feeders of about the same general size have evolved gular pouches for carrying food back to their young, but the murrelets seem to have dealt with

this problem by reducing the nestling period. Thus in the marbled and Kittlitz murrelets the nestling period is probably less than a month, while in the genus *Synthliboramphus* the nestling period has been reduced to only a few days, during which the young are apparently not fed. These two murrelets have seemingly modified this important aspect of their reproductive biology as a result of dietary considerations.

Similarly, the plankton feeders have evolved body sizes that presumably cannot exceed the upper limits that are probably set by their prey size, while lower size limits are presumably set by physiological factors such as surface/volume ratios, in Bedard's view. Although Storer (1945) considered the *"Endomychura"* (marbled and Kittlitz) murrelets relatively primitive, Bedard concluded that they are actually specialists, particularly insofar as their modified nesting biology is concerned.

Foraging Ecologies

The ecological aspects of foraging similarities and differences in the loons, grebes, and alcids are of great interest and have only recently begun to receive the attention of ornithologists. There are as yet no good studies on the comparative foraging ecologies of the rather widely sympatric red-throated and arctic loons, though some fairly extensive samples of winter foods of these two species are now available from Danish waters (table 17). Thus Madsen (1957) found that cod (*Gadus morhua*) made up over 50 percent of the total volume of foods found in 173 samples of red-throated loons and also composed about a third of the diet of arctic loons, based on an analysis of 123 samples. Cod remains were found in 71 percent of the red-throated loon samples

Table 17: Reported Prey of Primarily Plankton-Eating Species of North American Alcids

Food Types	Dovekie	Murrelets		Auklets				
		Ancient	Marbled	Cassin	Parakeet	Least	Whiskered	Crested
Copepoda								
Calanoidea	XX	—	—	—	XX	XX	—	XX
Malacostraca								
Euphausiacea	XX	XX	XX	XX	XX	X	—	XX
Mysidacea	XX	XX	XX	—	—	X	—	XX
Amphipoda	XX	X	—	XX	XX	XX	XX	XX
Gammaridea								
Gammaridae	—	XX	—	—	—	—	XX	—
Hyperiidae	XX	—	—	—	—	—	—	—
Decapoda								
Caridea	—	—	X	—	X	XX	X	X
Decapod larvae	X	X	—	—	—	—	—	—
Polychaetes	X	—	—	—	X	—	—	—
Cephalopod larvae	X	—	X	X	X	—	—	X
Fish								
Engraulidae (anchovies)	—	—	XX	—	—	—	—	—
Osmeridae (smelts)	—	—	X	—	—	—	—	—
Gadidae (cods)	—	—	—	—	XX	X	—	X
Scorpaenidae (rockfish)	—	X	X	—	—	—	—	—
Cottidae (sculpins)	—	—	—	—	X	—	—	—
Stichaeidae (pricklebacks)	—	—	XX	—	—	—	—	—
Ammodytidae (sand eels)	—	XX	XX	—	—	—	—	—

NOTE: Symbols as in table 13.

and were the only food present in 38 percent, while in the arctic loon a combination of cod, gobies (mainly *Pomatoschistus* and *Chaparrudo*), and sticklebacks (*Gasterosteus*) made up 90 percent of the total food and were the only fish present in 80 percent of the total sample. Gobies and sticklebacks were also frequently found in the samples from red-throated loons but collectively made up only about 25 percent of the total sample. It thus seems that, at least during winter, there are rather marked similarities in the diets of arctic and red-throated loons in Danish waters. Along the Pacific coast of North America arctic and red-throated loons have broadly overlapping winter distributions, although the red-throated loon tends to winter much more along the Atlantic coast than does the arctic loon, which may help to reduce foraging competition during that time of year.

McIntyre (1975) studied the winter feeding behavior of common and red-throated loons along the coast of Virginia and noted that typically the red-throated loons foraged in small groups in areas where the tidal currents were swift but only occasionally were seen in bays and coves. However, common loons were regularly found feeding singly in the quiet waters of bays and coves, suggesting that these two species might utilize quite different foraging strategies. She estimated that each common loon used an average of 10 to 20 acres for its foraging area, which she believed to represent typical

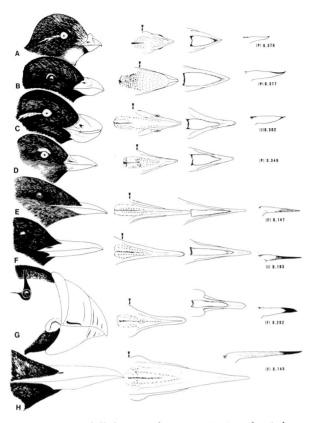

20. Comparative bill shapes and tongue traits in auks: A, least auklet; B, dovekie; C, parakeet auklet; D, Cassin auklet; E, marbled murrelet; F, pigeon guillemot; G, horned puffin; H, common murre. The palate surface, lower mandible and tongue, and tongue profile are shown, with shading of the tongue indicating relative cornification. The arrows indicate the commisural point, the numbers indicate the ratio of bill width to length, and the letters indicate primary foods (P = plankton, I = intermediate, F = fish). Adapted from Bedard 1969a.

wintering loon density in optimum habitat. The apparently greater sociability of red-throated loons in winter compared with common loons should be investigated in terms of the possible role of social rather than individual foraging tactics. Although detailed information is lacking, the arctic loon also appears to be less social in winter than the red-throated loon. The yellow-billed loon also reportedly migrates and winters singly or in small parties that may be family groupings.

The foraging ecologies of the grebes have received substantially more attention than those of loons and offer several points of interest. The skull and bill anatomy of such fish-catching grebes as the great crested grebe is remarkably streamlined and highly adapted as a fish-getting device (fig. 21) and shows certain convergent

similarities to the skulls of loons and fish-catching alcids. Of the North American grebes, only the western (including *clarkii*) appears to be almost exclusively a fish eater (table 18), but the red-necked grebe probably takes most of its foods from this type of resource, at least in the case of the North American race.

In other parts of the world, as in North America, most of the grebe species appear to have foraging ecologies that are predominantly dependent upon aquatic invertebrates (table 19), with the smallest species largely or exclusively insect eaters, and only the largest species that have bill lengths of more than 30 millimeters being essentially fish dependent. North America, South America, and Eurasia each support two fish-dependent species. South America supports an additional seven species of grebes, and North America and Eurasia have four and three more respectively, making South America the most grebe-rich area in the world. There is a rather striking similarity between the grebe fauna of North and South America, in that beyond the commonly occurring least and eared grebes (the South American population of eared grebes is considered by some to be a distinct species), the remaining four species of North American grebes have close replacement counterparts in South America, at least in terms of their

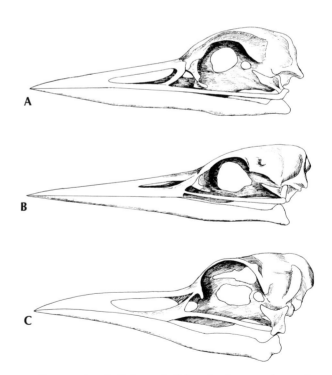

21. Comparative skull shapes in fish-eating loons, grebes, and auks: common loon (*top*); western grebe, female (*middle*); common murre (*bottom*). After museum specimens.

Table 18: Reported Percentages of Various Components in the Diets
of North American Loons, Grebes, and Selected Alcids

Species	Sample Size	Fish	Crustaceans	Insects	Polychaetes	Mollusks	References
Loons							
Red-throated	173	100.0	—	—	—	—	Madsen 1957
Arctic	123	100.0	tr[a]	—	—	tr	Madsen 1957
Common	27	100.0	tr	tr	—	tr	Olson and Marshall 1952
Yellow-billed	4	100.0	tr	—	—	tr	Cottam and Knappen 1939
Grebes							
Least	6	—	—	100	—	—	Cottam and Knappen 1939
Pied-billed	174	24.2	31.1	46.3	—	tr	Wetmore 1924
Horned	122	34.6	17.9	46.0	—	tr	Wetmore 1924
Red-necked	46	55.5	20.0	21.5	—	tr	Wetmore 1924
Eared	27	9.8	tr	84.2	—	—	Wetmore 1924
Western	19	100.0	—	—	—	—	Wetmore 1924
Alcids							
Razorbill	71	90.0	10.0	—	tr	tr	Madsen 1957
Common murre	117	95.2	3.6	—	—	1.2	Hunt, Burgeson, and Sanger 1981
Thick-billed murre	233	76.0	17.1	—	0.1	5.3	Hunt, Burgeson, and Sanger 1981
Black guillemot	26	67.0	33.0	—	tr	—	Madsen 1957
Least auklet	258	0.7	92.7	—	—	—	Hunt, Burgeson, and Sanger 1981
Crested auklet	107	tr	100.0	—	—	—	Bedard 1969a
Parakeet auklet	55	26.6	48.5	—	23.5	0.4	Hunt, Burgeson, and Sanger 1981
Tufted puffin	23	79.7	3.4	—	11.9	1.7	Hunt, Burgeson, and Sanger 1981
Horned puffin	39	81.4	11.1	—	3.9	0.7	Hunt, Burgeson, and Sanger 1981
Atlantic puffin	117	83.1	11.8	—	11.8	—	Wehle 1980

NOTE: Calculated from volumetric percentages except for Atlantic puffin, which is based on frequency-of-occurrence data. For unknown reasons, data of Hunt et al. 1981 do not approach 100 percent in some cases.

[a]tr = trace.

bill shape and general head plumage characteristics (fig. 22).

By far the best discussion of the foraging ecologies of grebes is the review by Fjeldså (1983a), based on studies of nearly three thousand museum specimens and extensive fieldwork in Europe, South America, and Australia. He has noted that in all the observed cases where two closely related species overlap locally, either one or both of these species exhibit indications of divergent bill morphology, or "character displacement." In at least three of these cases there was evidence that these morphological changes were associated with dietary differences that reduced the degree of interspecific food overlap. He suggested that such ecological foraging displacement is most likely to occur in stable environments utilized by species showing K-strategy reproductive characteristics (deferred reproductive maturity, longer reproductive lives, extended parental care, etc.). In isolated areas supporting only a single species of grebe there is a tendency for that species to evolve an "all purpose" bill that permits opportunistic fish catching without loss of the ability to forage efficiently on small aquatic arthropods. Furthermore, grebes that live under relatively poor foraging conditions tend to exploit

Table 19: Distribution of Grebe Species by Diet and Bill-Length Categories

Usual Foods	Average Bill Length	Central and/or South America	North America	Eurasia	Africa	New Zealand	Australia
Fish and invertebrates	Over 30 mm	Great grebe Titicaca grebe	Western grebe Red-necked grebe ---------------------	Crested grebe --			
Primarily invertebrates	20–30 mm	Puna grebe[a]				New Zealand dabchick	Hoary-headed grebe
						---------------------Australian dabchick	
						Madagascar Aloatra grebe Madagascan dabchick	
		Hooded grebe	Horned grebe -------------------------				
		---------------------------Eared grebe-------------------------------------					
		White-tufted grebe		Little grebe -----------------------------------			
		Giant pied-billed[a]					
		---------------------------Pied-billed grebe					
Insect eaters	Under 20 mm	Silvery grebe					
		Least grebe -----------------------------------					

[a]May be primarily fish eating.

all the available potential foods, whereas specialization on optimal foods tends to occur when foods are easy to find.

Part of Fjeldså's evidence for character displacement came from his study of the red-necked grebe, which has a relatively broad geographic distribution in Eurasia and North America. In Europe the species forages largely on arthropods, with fishes eaten only locally or temporarily. In this way it apparently attains an efficient ecological isolation from the fish-adapted great crested grebe of Eurasia. However, in eastern Siberia and North America the red-necked grebe is represented by a large and long-billed race that in some respects matches that of the great crested grebe, and fish eating appears to be a general characteristic of red-necked grebes in North America. Similarly, in northern Norway and Iceland, where the horned grebe does not encounter competition from several other grebe species (as is true farther south in Europe), the birds have larger and deeper bills and are

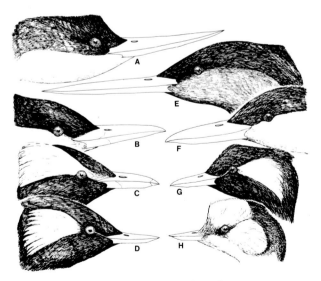

22. Convergent evolution in North and South American grebes: A, western; B, red-necked; C, horned; D, eared; E, great; F, Titicaca; G, white-tufted; H, silvery.

more opportunistic foragers, using a wider array of habitats and eating a more flexible diet. Fjeldså now considers this a probable case of character release in the nonsympatric populations rather than of character displacement, which was his earlier view.

The alcids offer an even greater number of closely related and sympatric species to investigate for foraging niche differences, and many such studies have been carried out over the years. Thus Hunt, Burgeson, and Sanger (1981) investigated the feeding ecologies of common and thick-billed murres, three species of auklets, and two species of puffins that breed in the eastern Bering Sea. Bedard (1969a) also compared three auklet species in the vicinity of Saint Lawrence Island. Pearson (1968) investigated the comparative foraging ecologies of nine

species of seabirds of the Farne Islands, including the Atlantic puffin and the black guillemot, and Cody (1973) attempted to analyze the ecological isolating mechanisms of six alcid species found along the Pacific coast of Washington.

With regard to the comparative ecologies of the common and thick-billed murres, it is now evident from a variety of studies that these two outwardly similar species have some marked morphological differences associated with locomotion (Spring 1971). They also show marked dietary differences, with the thick-billed murre exhibiting a considerably greater reliance on invertebrate foods (Schwartz 1966; Hunt, Burgeson, and Sanger 1981).

Studies of the three widely sympatric auklets (least,

Table 20: Reported Prey Differences in Some Syntopic Alcid Species

| Wild-Caught Prey (length) | Least Auklet | | Crested Auklet | | Parakeet Auklet | | |
	Crustaceans	Fish	Crustaceans	Fish	Crustaceans	Fish	References
to 7.0 mm	3,169 (3.7%)	187 (87.4%)	82 (0.4%)	0 (0%)	7 (0.1%)	0 (0%)	Bedard 1969a[a]
7.1–15.0 mm	81,986 (96.0%)	21 (9.8%)	9,698 (46.5%)	5 (8.3%)	4,566 (60.7%)	35 (15.4%)	Bedard 1969a[a]
over 15.0 mm	257 (0.3%)	6 (2.8%)	11,057 (52.1%)	55 (92.7%)	2,944 (39.2%)	192 (84.6%)	Bedard 1969a[a]

Wild-Caught Prey	Common Murre	Atlantic Puffin	References
Length (mm) of *Ammodytes*			
Length range	50–175	50–100	Pearson 1968
Commonest length	100–125	75–100	Pearson 1968
Weight (g) of all prey			
Range of weights	1–32	0–32	Pearson 1968
Average weight	8	2	Pearson 1968

Captive-Fed Birds	Common Murre	Razorbill	Atlantic Puffin	References
Weight of prey (g) (preferred/maximum)				
Clupea	14/96	4/18	4/18	Swennen and Duiven 1977
Trisopterus	16/62	6/16	6/16	Swennen and Duiven 1977
Height of prey (mm) (preferred/maximum)				
Clupea	23/44	15/26	15/26	Swennen and Duiven 1977
Trisopterus	23/41	15/23	15/23	Swennen and Duiven 1977

[a]Total quantities present in gullet samples during chick-rearing period as determined from table 1 of Bedard 1969a.

crested, and parakeet) likewise indicate some important foraging niche differences among them. Bedard (1969a) initially reported that the least auklet consumes the smallest prey items, especially small crustaceans, the crested auklet eats prey of intermediate size, again primarily crustaceans, and the parakeet auklet takes the largest prey (table 20). Additionally, the least and crested auklets are essentially zooplankton specialists, foraging in middle and surface depths, while the parakeet auklet takes a much wider variety of invertebrates and fishes, at least some in near-bottom (demersal or epibenthic) zones. Hunt, Burgeson, and Sanger (1981) confirmed these differences and pointed out that these dietary differences may have important implications in determining local distribution patterns, with crested and least auklets largely restricted to islands having large shelf-edge zooplankton populations while the parakeet auklet occurs more widely in coastal waters supporting diverse demersal and epibenthic prey. Further, these food preference patterns appeared to be stable over several years of study, though they varied most obviously in the more generalized parakeet auklet, which is the most opportunistic of the three auklet species.

Studies by Pearson (1968) of seabirds breeding on the Farne Islands indicated a substantial overlap in the size and species of fish taken by each of the nine species of seabirds breeding there, though the birds differed considerably in the average distance flown in search of food and the depth at which food was obtained. Of the two alcid species, common murre and Atlantic puffin, the larger common murre tended to select longer prey fish (*Ammodytes*) and heavier prey than did the Atlantic puffin, though the degree of overlap was substantial. Later studies with captive birds by Swennen and Duiven (1977) have confirmed these differences between the common murre and Atlantic puffin (table 20). The razorbill, also included in this study, took foods of essentially the same weight and height as did the Atlantic puffin. These authors concluded that the maximum size of prey fish in these three species of alcids is determined not by length but rather by diameter, and that the preferred prey size is approximately half of the maximum that the bird can swallow. This prey-size selection is evidently made visually.

In an extensive review of foraging relationships of seventy seabird species breeding in the Bering Sea and northeastern Pacific Ocean, Ainley and Sanger (1979) concluded that fewer than 7 percent feed on a single type of prey, about 60 percent feed on two or three types, and the rest feed on four or more prey types. Where dietary overlap exists, foraging partition is done by different feeding methods, selection of different-sized prey, and zonation of foraging habitats. Some of these

interrelationships are evident in figure 23, which attempts to summarize some aspects of prey choice and horizontal foraging zonation tendencies (during winter), based largely on a similar diagram by Tuck (1960) for Newfoundland. Also shown are varied patterns of diurnal activity for these or related species, based on Sealy's (1972) summary, which suggests there may be significant differences in diurnal foraging intensities, at least during the summer breeding period.

Cody (1973) emphasized the possible significance of differential foraging zones in the six species of alcids that he studied off the coast of Washington, suggesting that these six species all have similar diets and breed at the same time of year and that differences in bill shape, foraging depths, and other possible differences are less important than the zonation of foraging areas in reducing interspecific competition. Bedard (1976) has strongly criticized these conclusions and in particular has illustrated how foraging zonation patterns can be locally affected by such factors as coastline and slope configuration, water circulation patterns, and oceanographic conditions. Bedard emphasized that both data from Cody's study and other data from the Atlantic Ocean tend to show considerable overlap in foraging

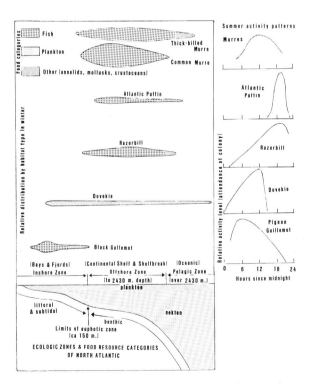

23. Comparative foraging ecologies of North Atlantic auks. Adapted in part from Tuck 1960.

Table 21: Major Foraging Habitats and Foods of Loons, Grebes, and Fish-Adapted Alcids

Habitats and Prey Types	Loons	Small Grebes	Large Grebes	Razorbill	Murres	Guillemots	Puffins
Saltwater areas							
Surface-dwelling fish	x	—	x	x	—	—	x
Ammodytidae (juveniles)							
Atherinidae							
Blenniidae (juveniles)							
Clupeidae (juveniles)							
Gadidae (juveniles)							
Mid-depth fish	X	x	X	X	X	x	X
Clupeidae							
Engraulidae							
Gadidae (some)							
Osmeridae							
Salmonidae							
Benthic and littoral forms	x	x	x	x	x	X	x
Fish							
Agonidae							
Ammodytidae							
Bathymasteridae							
Blenniidae							
Bothidae							
Clinidae							
Cottidae							
Cryptacanthodidae							
Embiotocidae							
Gadidae (some)							
Hexagrammidae							
Liparidae							
Pholidae							
Pleuronectidae							
Stichaeidae							
Scorpaenidae							
Trichodontidae							
Zoarchidae							
Invertebrates							
Crustaceans	x	X	X	X	X	X	X
Annelids	—	—	x	—	—	—	X
Brackish and fresh waters							
Fish	X	x	X	—	—	—	—
Anguillidae							
Catostomidae							
Centrarchidae							
Cyprinidae							
Cyprinodontidae							
Esocidae							
Gasterosteidae							
Ictaluridae							
Percidae							
Petromyzontidae							
Invertebrates	x	X	x	—	—	—	—

[a]Organized in part after Pearson 1968; X indicates major food sources; x denotes an apparently minor food source from indicated habitats of prey types.

zonation rather than spatial segregation among the species.

By way of summary, table 21 lists major prey types of loons, grebes, and the fish-adapted alcids of North America, organized by habitat and water depth. All three groups of birds tend to forage on mid-depth fishes, with more limited use of surface-dwelling and bottom-inhabiting forms, and all except loons also eat crustaceans to a considerable degree. Freshwater fish are important prey items of loons and the larger grebes, while freshwater invertebrates are major food sources for the smaller grebes. Annelids appear to be of minor importance in all groups except puffins, which sometimes eat polychaetes in substantial numbers.

5. Comparative Pair-Forming and Copulatory Behaviors

All species of loons, grebes, and auks are monogamous, with adults forming strong pair bonds that are established or reestablished each year, probably during the spring prenesting period. The extent of remating by birds mated the previous year in these groups is still largely undocumented, though at least in alcids it is fairly high, given the relatively long life-spans and the tendencies of the birds to return year after year to essentially the same territory and sometimes to the same nest site. Such conditions would promote reestablishment of contracts between previously paired birds, since it is not believed that in these birds pair or family units normally remain intact through the winter. Although in some grebes such as the horned grebe breeding site tenacity and mate fidelity may be very strong (Fjeldså 1973c), in others such as the colonial-nesting eared grebe there is no evidence of this and mate fidelity is poorly developed (Fjeldså 1982a).

Pairing behavior in the loons is certainly understood the least well of all these three groups, and as such is very difficult to summarize. It is apparent (table 22) that all loons are highly vocal and that vocalizations are extremely important social signals, especially in territorial advertisement. It is also evident that much of the described "courtship" behavior of loons is little more than variably ritualized hostile behavior, since it is often extremely difficult to distinguish between hostile territorial encounters and actual pair-forming displays between potential mates. Indeed the most complex and spectacular displays of loons, like those of some grebes, seem little more than highly ritualized aggressive behavior that probably serves to avoid serious fighting and promote social bonding between these sexually monomorphic species. It is thus not surprising that, as in some grebes, "racing" ceremonies are often present, as are erect "penguin" postures that involve treading water while maintaining an essentially threatening posture. The variable visibility of the species-specific throat and neck markings of loons is associated with differential amounts of bill tilting, neck stretching, and sometimes also head turning; in general the degree of exposure of these throat and neck patches seems to be directly related to the degree of hostility or sexual intensity of the display. The conspicuous back patterning of loons is not obviously utilized during display, and it seems likely that their white underparts are related to countershading requirements rather than being important as social display features.

Copulatory behavior has been described for all four species of loons, and the general pattern seems to be highly stereotyped and very similar in all (fig. 24). Thus, unless visual aspects of species-specific plumages are significant, copulatory behavior in loons is unlikely to serve as an effective reproductive isolating mechanism. McIntyre's (1975) observations on the common loon are probably representative of loons in general and involve a large sample of twelve observed copulations. She found that either sex might initiate copulation, which always occurred on land but not on a specific "copulation platform" as has at times been alleged. If the male precedes the female to land he typically utters a "soft" call while waiting for her to approach the copulation site, which often later becomes the nest site. The female's receptive posture is one of lowered head and bill (fig. 24A,D), and when the male mounts he stands on the female's back or scapular area with his head directly over hers (fig. 24B,E). After treading the male walks over whichever one of the female's shoulders is nearer to water (fig. 24C). No specific postcopulatory displays occur in loons, though preening and bathing behavior often are performed by both sexes. Some nest-building behavior may also follow copulation.

Pair-forming behavior in grebes, by contrast with that of loons, is extremely conspicuous and highly ster-

Table 22: Social Behavior Patterns and Calls of Loons

Behavior	Red-throated	Arctic	Common	Yellow-billed
Calls				
Croaking	X	X		
Wailing	X	X	X	X
Moaning	X	X	X	X
Long call/yodeling			X	X
Tremolo calling			X	X
Short call		X		
Posturing				
Antagonistic (appeasement/alarm)				
Neck stretching/alert	X	X	X	X
Alarm/prone	X	X	X	X
Antagonistic (threat/attack)				
Hunched/forward	X	X	X	X
Bill dipping	X	X	X	X
Splash diving	X	X	X	X
Fencing/penguin/bow jump	X	X	X	X
Surface rushing	X	X	X	X
Mutual or group displays				
Plesiosaur race/surfing	X			
Snake ceremony	X			
Circle dance/pivoting		X	X	X

NOTE: See species accounts for descriptions and possible functions.

1969, 1971, 1982) have contributed greatly to our understanding of grebes and their displays. Simmons has divided crested grebe courtship into two categories, water courtship and platform behavior. The first of these is not confined to pair formation but occurs both during and after pairing, though it is largely limited to the pre-egg stage of the reproductive cycle. Platform behavior includes soliciting, copulation, and sometimes ceremonial nest building. Such platform behavior may precede nesting by several months. It probably serves both in establishing pair bonds and later also as a mechanism for fertilization of the female, and as such it may appropriately be called "platform courtship."

Solitary grebes of at least nearly all *Podiceps* species as well as the western grebe and the white-tufted grebe exhibit advertising behavior. This calling, performed by solitary birds of either sex, occurs in unpaired birds seeking a mate, in paired birds visually separated from their mates, and in parents that have lost contact with their chicks. The advertising calls are typically species specific but may also have sufficient individual variability to permit individual recognition. Mutual calling, or "duetting," has also been recorded in all six of the

eotyped and exhibits a high degree of species specificity (table 23). Why such differences exist between loons and grebes is not at all clear, since in general there seem to be no differences in relative need for reproductive isolation among sympatric and congeneric forms in the two groups. As with the loons, most of the species-specific nuptial plumage characteristics of grebes are concentrated in the head and neck region, and furthermore in both groups the displays and calls are performed mutually and apparently identically between the sexes. Indeed, one might wonder how sexual recognition is attained in these groups; possibly it is achieved by relative dominance and submissive behavior.

Most of the classic observations on grebe pair-forming and pair-maintaining behavior have been performed on the great crested grebe, beginning with the studies of Huxley (1914) and continuing through to much more recent work by Simmons (1955, 1975). A variety of papers by Fjeldså (1973c, 1975, 1982a) and Storer (1962, 1967,

24. Comparative sexual behaviors of loons: A, soliciting, B, copulation; and C, postcopulatory behavior of arctic loon (after photos in Höhn 1982); D, soliciting and precopulatory approach of red-throated loon (after Cramp and Simmons 1977); E, copulation in yellow-billed loon (after Cramp and Simmons 1977).

Table 23: Interspecific Distribution of Some Sexual Behavior Patterns in North American Grebes

Behavior	Least	Pied-billed	Eared	Red-necked	Horned	Western
Courtship behavior						
Duetting	X	X	X	X	X	X
Crest erection	—	—	X	X	X	X
Head turning/shaking/waggling	—	X	X	X	X	X
Discovery ceremony (cat—ghostly penguin)	—	—	X	X	X	—
Penguin dance	—	—	X	X	X	—
Weed ceremonies	—?	X?	X	X	X	X
Triumph ceremonies	X	X	X	X	X	—
Barging or parallel swimming	X	X	X	X	X	X
Rushing ("racing")	—	—	—	—	—	X
Habit preening	—	—	X	—?	X	X
Threat pointing	—	—	—	—	—	X
Platform behavior						
Female behavior						
Inviting	X	X	X	X	X	X
Rearing	—	X?	X	X	X	—
Breast stroking	X	X	—	—	—	—
Male behavior						
Copulation call	X	X	X	X	X	X
Water treading	—	X?	X	X	X	?

NOTE: X indicates presence of behavior; X? indicates uncertain ritualization; — indicates apparent absence.

North American grebe species (table 23). In some species this behavior is called a "triumph ceremony," and at least in the horned grebe it is the only mutual display that is frequently produced by well-established pairs (Fjeldså 1973c). In that species it often occurs after territorial combat or when a pair meets after a temporary separation. A triumph ceremony without calling occurs in the red-necked grebe.

Crest erection, often in conjunction with various kinds of head turning or head shaking, is common in many grebes and occurs in at least five of the North American species. Frequently it is incorporated into more elaborate ceremonies such as the "discovery ceremony" or "penguin dance" or, as in the crested grebe, may form a conspicuous part of an elaborate head-shaking ceremony.

A major ceremony of grebes during the period of initial pairing, and also later among paired birds on territory, especially as a greeting ceremony after brief separations, is the "discovery ceremony." In this cere-

mony one of the birds (the searcher or "ghost diver") takes the active searching role while the other waits to be "discovered." In the crested grebe the approach by the "searching" bird resembles the threatening approach made during antagonistic encounters, while in other species such as the red-necked grebe, horned grebe, and eared grebe the searching bird periodically exposes itself in a "bouncy" posture, with the plumage depressed and the breast well puffed out. When within about a meter of the waiting bird, the searching bird suddenly rises vertically out of the water in a "ghostly penguin" posture. The waiting bird, which in some species such as the horned grebe and great crested grebe has assumed a raised wing and expanded tippet "cat posture," now rises in the water in synchrony with the other, and the two perform a mutual "penguin dance." A penguin dance has been described for nearly all of the typical *Podiceps* species, although in the great crested grebe it is replaced by the head shaking ceremony. Most probably the discovery ceremony is ritualized or symbolic attack and the cat display is similarly a ritualized defensive posture.

In some species the penguin posture is also assumed while both birds are holding aquatic vegetation in the bill, a variation Simmons (1975) called the "weed dance" to distinguish it from those species that do not use weeds and thus perform typical "penguin dance" ceremonies. Simmons has pointed out that, unlike the discovery ceremony, in the weed ceremony of the crested grebe the roles of male and female are identical throughout, and during this posture both birds rise breast-to-breast in penguin postures and perform a weed dance. This kind of ceremony has been observed in both red-necked and western grebes and has also been reported in some populations of horned grebe but not others.

A somewhat similar ceremony of some grebes is the "weed rush." In the horned grebe this occurs when, after the discovery ceremony, the two birds appear at the water surface with plant material in their bills. They thus emerge from the water in straight-necked attitudes and swim toward one another. As they collide they rise in the water into penguin attitudes and swim parallel for a distance in a weed rush ceremony. Although red-necked and western grebes have essentially stationary weed dance ceremonies, only the horned grebe among the North American grebes performs the weed rush. However, the rushing display of the western grebe is distinctly similar, and the slower "barging" of the eared grebe may also be related. Likewise, the least grebe performs parallel swimming displays, accompanied by loud calling. In other species of grebes, parallel swimming or barging often terminates with unison diving, as is also the case with the western grebe's rushing ceremony (Fjeldså 1973c). Fjeldså suggested that in the more primitive grebes such as *Tachybaptus* and *Podilymbus* there is much weed-presenting behavior but that such weed ceremonies are not incorporated into breast-to-breast or parallel swimming displays. He thus suggested that the weed rush ceremony, like the penguin dance ceremony, originated from attack behavior. He further suggested that the discovery ceremony allows potential mates to attain close contact, which might be necessary for sex recognition in grebes. More recently (1982a) he has reaffirmed his position that the discovery ceremony probably plays a central initiating role in pair formation of several *Podiceps* species and that this and other complex rituals have most probably evolved from antagonistic responses. Furthermore, he believes that no pair bonding develops (at least in *Podiceps*) unless the birds perform another ceremony, usually with weeds, after the discovery ceremony. In gregarious, colony-nesting grebes incipient partnerships apparently change often,

and promiscuous platform behavior may be an important aspect of attaining mutual recognition.

Beyond these major ceremonies, several less complex displays occur in various species of North American grebes, such as "habit preening" and "threat pointing," which are both highly developed aspects of courtship behavior in western grebes but seem to have no counterpart in other North American grebes, except possibly for the less conspicuously ritualized preening that occurs in the eared grebe.

Copulation in grebes invariably occurs on floating vegetation or land. Usually it is done on platforms that initially are built for this specific purpose and later are often modified to serve as nests. However, at times these preliminary platforms are too exposed or otherwise prove unsuitable for nesting, and thus a new and better site may be selected for the nest. In apparently all grebes, not only do both sexes solicit copulation, but either one can take the active role during treading. Such "reversed mounting" was initially reported for the crested grebe (Simmons 1975) but has been observed in several other species. However, during the period immediately before and during egg laying nearly all copulations are of the normal type. Simmons used the term "mating ceremony" for the solicitation, copulation, and postcopulatory sequence and "platform behavior" for all the calls and behavior associated with copulation as well as with collecting materials and building mating structures. Although an "inviting on the water" behavior has been seen in crested grebes, soliciting typically occurs on the mating platform, when the inviting bird assumes an immobile, nearly prone posture (fig. 25A,D). At times, especially when the other bird approaches, it may also assume a "rearing" posture with the body raised and neck arched downward, at times performing wing quivering. Treading involves a rather erect posture by the active bird (fig. 25E). Its bill may be open or closed, and it typically utters a trilling or rattling call. It does not hold the nape of the passive partner, which in *Podiceps* and *Aechmophorus* holds its head as low as or even lower than during inviting, though its bill may be tilted upward. However, in *Podilymbus* and *Tachybaptus* the passive bird raises its head and rubs it against the other's breast (fig. 25E). Apparently intermediate behavior occurs in *Rollandia* (Fjeldså 1982a). Dismounting in grebes is often accompanied by water treading by the active partner (fig. 25F), while the passive bird simply raises its head. In some species such as the silver grebe and red-necked grebe a rather stereotyped postcopulatory posturing is also present.

Copulatory behavior in the grebes thus appears to be extremely conservative and similar throughout the en-

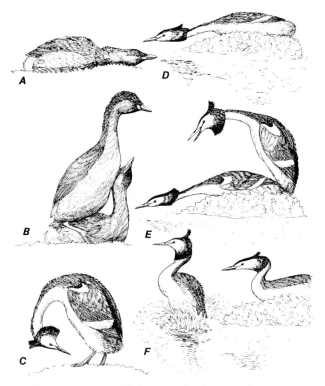

25. Comparative sexual behaviors of grebes: A, soliciting and B, copulation in least grebe (after Cramp and Simmons 1977); C, rearing, D, solicitation, E, copulation, and F, postcopulatory display of great crested grebe (primarily after Simmons 1955).

crested and least auklets (Sealy 1975a), rhinoceros auklet (Leschner 1976), and Atlantic puffin (Ashcroft 1976, 1979). These monogamy-promoting behaviors may ease the need for highly species-specific signals in alcids, which nonetheless do have specific bill shapes or colors and sometimes distinctive head plumages and which also tend to have distinctive vocal signals associated with pair interactions. Yet by comparison with the grebes, the posturing tends to be relatively simple and frequently is very similar between closely related species (table 24).

Even in the most fully studied alcids, the murres, there is still a good deal of uncertainty about the timing and mechanisms of pair bonding. Tuck (1960) noted that there is no conclusive evidence that murres are mated before they return to their nesting sites, though some authors have claimed this is so. He mentioned "joy flights" and "water dances" as important social flock activities associated with the birds' return to their nesting colonies and judged that such communal displays may help stimulate and maintain their reproductive condition. He reported a high level of individuals' remating with previous mates and judged that perhaps pairing normally occurs among birds that choose potential nesting sites close to one another. Thus if males (which are normally the first to arrive) gradually come to tolerate females that attach themselves to the particular site already occupied by the male, pair bonding may easily develop or redevelop. In any event, at least among murres and razorbills, mutual billing and preening behavior is apparently a major pair-maintaining mechanism that occurs throughout the entire breeding season, even after the young are hatched. Mutual billing and preening or its solicitation, and mutual billing in the absence of preening as occurs in the puffins, certainly place maximum visual and tactile significance on the bill and facial region and thus might help explain the bright colors and head plumages typical of so many alcids. Among species where mutual billing or preening does not seem to be present, as in the murrelets, the facial plumage and the bill shape and coloration seem undeveloped compared with that of other alcids.

In contrast to the loons and grebes, copulation in the alcids may occur either in water or on land, depending upon the species (fig. 26 and table 24). In the murres, razorbill, guillemots, and dovekie copulation occurs on a solid substrate, often but not necessarily the nesting site itself. In the murres at least it may also rarely occur in water, and thus there may be no hard and fast taxonomic distinctions among the alcids as to the distribution of this kind of behavior. During copulation solicitation the female leans forward and raises her

tire group, and as such it is less likely to offer opportunities for reproductive isolation than is aquatic courtship, which is far more species specific. Testing of behavioral reproductive isolating mechanisms has scarcely begun for grebes, although Nuechterlein (1981a) concluded that they are attained in the two morphs of the western grebe by differences in the advertising call rather than by any postural differences, which appear to be lacking altogether. Fjeldså (1982a) was unable to find any evidence of ethological character displacement among sympatric versus allopatric populations of species in the genus *Podiceps.*

The alcids, like the loons, are often extremely long-lived, and not only do most species return to the same breeding colony year after year, but the birds typically return to the same burrow, cliff ledge, or other specific nesting site. Such nest site tenacity occurs in at least eleven species of alcids (Leschner 1976), and successive-year mate retention has been reported for the dovekie (Norderhaug 1967), razorbill (Lloyd 1979), thick-billed and common murres (Tuck 1960), pigeon guillemot (Drent 1965), black guillemot (Preston 1968), ancient murrelet (Sealy 1975a), Cassin auklet (Manuwal 1974a),

Table 24: Distribution of Various Structural and Behavioral Traits Associated with Reproduction in Auks

Trait	Dovekie	Razorbill	Murres	Guillemots	Murrelets	Small Auklets	Rhinoceros Auklet	Puffins
General nuptial traits								
Mutual billing	X?	X	X	X	?	X	X	X
Mutual preening	—?	X	X	—	?	—?	—?	—
Bill enlarged and colorful	—	—	—	—	—	X	x	X
Head plumes or crests	—	—	—	—	—	X	x	x (1 sp.)
Colorful feet	—	—	—	X	—	—	—	X
Primarily nocturnal	—	—	—	—	X	x (1 sp.)	X	—
Food presentation display	—	—	X	?	?	?	?	—
Duetting	—	—	—	—	X	X	?	—
Copulatory behavior								
Land copulation typical	X	X	X	X	?	X?	?	—
Water copulation typical	—	—	—	—	?	X?	?	X
Precopulatory circling	—	—	—	X	?	?	?	—

NOTE: X indicates trait well developed; x indicates variable or limited development; X? indicates probable presence; — indicates apparent absence.

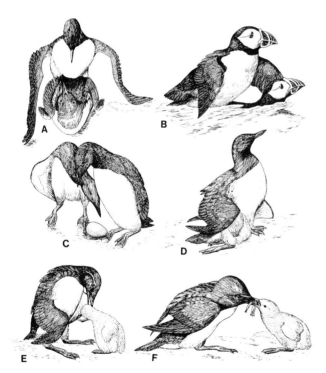

26. Comparative sexual behaviors of auks: A, copulation in common murre; B, copulation in Atlantic puffin; C, nest relief behavior of common murre; D, brooding, and E, parental feeding by common murre; and F, parental feeding by razorbill. After Glutz and Bauer 1982.

rump. The male mounts from the side, either drooping his wings over her or at times flapping his wings to maintain balance. The female then opens her bill and utters a hoarse call. As treading is completed the female usually rises, causing the male to slide off her back.

On the other hand, copulation in some alcids, including at least all three typical puffins, normally occurs on water. The situation in the rhinoceros auklet is still unknown, but copulations or at least attempted copulations have been observed among swimming crested and least auklets, although nest site copulations have also been observed in these two species (Sealy 1968; Thoreson, in press). Atlantic puffins sometimes attempt to copulate on land, but these attempts are usually not successful. In the puffins, copulation is typically preceded by a male's following a female and performing head jerking and probably also uttering vocalizations. If the female is receptive she will allow the male to approach from behind, alight on her back, and cause her to become completely submerged except for her head. During copulation the male flaps his wings to maintain his balance, and copulation is usually terminated by the female's diving and resurfacing some distance away (Wehle 1980). Unlike copulation, billing in puffins may occur either in the water or on land and is usually initiated when one bird nuzzles the other's throat and breast feathers. Billing occurs in all four puffins and often serves as a greeting, in a "triumph ceremony" after an aggressive encounter by one member of the pair with another bird, during courtship, or in other situations. In guillemots billing typically initiates a copulatory sequence; in the pigeon guillemot billing is a certain sign that the birds are a mated pair (Drent 1965), and billing with associated vocalizations is apparently of great importance in establishing and maintaining the pair bond. Like the murre's mutual preening behavior, billing in guillemots is performed by pairs whenever they meet throughout the breeding season. According to Drent (1965), "twitter billing" in the pigeon guillemot is functionally comparable to silent mutual billing in the Atlantic puffin and to the combination of billing and allopreening behavior typical of common murres and razorbills. Mutual billing also occurs in at least some of the auklets, but it does not seem to have been described yet for the murrelets. However, food presentation behavior, which is a frequent part of precopulatory ceremonies in the murres, is apparently absent in razorbills and puffins and is of questionable occurrence in guillemots. It has not yet been observed, but might occur, in the auklets, given the adults' ability to carry substantial amounts of food back to the nesting burrow in special throat pouches. Thus billing seems to be the most universal form of pair-maintaining behavior in the alcids and billing invitation behavior apparently plays an important role in the earlier stages of pair formation, while mutual billing is a regular part of alcid pair-maintaining behavior. In murres and razorbills, preening has supplemented and to some degree replaced billing in fulfilling these roles, but the two behaviors are closely related in occurrence and function. Like the discovery ceremony of the grebes, billing is probably derived from hostile behavior, and it too may well have evolved as a mechanism for allowing two potential antagonists to approach one another and establish individual contact. Vocalizations may play a larger role in individual recognition in the alcids than in the grebes, and this is likely to be especially true in the nocturnal forms such as the murrelets and nocturnal auklets.

6. Comparative Life Histories and Reproductive Success Rates

Life Histories

It is now well recognized that, like behavior, a species' life history characteristics, such as age at sexual maturity, clutch sizes, and incubation, brooding, and fledging patterns, are evolved traits that may be strongly influenced by a variety of ecological factors (Lack 1968). Within the auks, loons, and grebes one can find variations in the age at sexual maturity and time of first breeding ranging from as little as 1 to as many as 5 or 6 years, average clutch sizes that range from 1 to 4 eggs, and substantial variations in adult survival rates and maximum longevity (table 25). However, compared with such similar-sized aquatic birds as ducks and geese, all three groups tend toward relatively small clutches, a greater tendency to defer sexual maturity and breeding, and substantially higher adult survivorship rates and thus potentially greater maximum longevity. In general such characteristics are associated with groups of birds that breed under relatively difficult conditions, where age and experience probably are significant factors influencing breeding success and for which, therefore, it is advantageous to the species to limit reproductive efforts to older individuals, who are most likely to be successful. In these three groups all the species are exclusively monogamous, with both sexes participating in incubation, brooding, and to varying degrees in feeding the young, which though generally precocial cannot capture prey on their own until they are relatively well developed (see figs. 17, 26, and 27).

Clutch sizes in loons are highly uniform, with two-egg clutches typical. However, single-egg clutches are not uncommon, usually because one of the eggs is lost from the platformlike nest. Olson and Marshall (1952) reported an average clutch of 1.55 for 47 common loon nests, and McIntyre (1975) found a mean of 1.67 eggs in 51 nests of this species. Bergman and Derksen (1977) reported average clutch sizes of 2.0 eggs for arctic loon nests and 1.86 eggs for 21 red-throated loon nests. Peterson (1979) found an average clutch of 1.93 for 43 arctic loon nests, and Lehtonen (1970) found that 78 first clutches of arctic loons had an average of 1.88 eggs. A collective total of 39 first clutches of red-throated loons averaged 1.79 eggs (Cramp and Simmons 1977). The available records of yellow-billed loons clutches are nearly all of 2 eggs; Bailey (1948) reported 22 such clutches.

There are apparently as yet no definitely established records on the age of initial breeding in loons, though the definitive adult plumage is not attained until the third year of life, by which time some believe breeding occurs (Palmer 1960). However, Lehtonen (1970) has postulated that in the arctic loon initial breeding may not occur until the birds are 6–7 years old, which would require a very high adult survival rate. Indirect evidence supporting the probability of such high adult survival rates are records of individual banded arctic and red-throated loons that lived over 20 years, though no similar records have yet accrued for common loons. In North America there are still regrettably few loon banding results. As of 1979, only 5 recoveries had been obtained for red-throated loons and 2 for arctic loons (Jonkel 1979). As of 1981, 624 common loons had been banded, with 82 recoveries (Clapp et al. 1982). This 13 percent recovery rate is appreciably higher than might have been expected for a protected species that has little contact with humans.

For grebes the information is little better. Fjeldså's (1973b) rather indirect calculation of an approximate 50 percent adult mortality (and recruitment) rate in the horned grebe seems to be the only available estimate of adult survivorship for grebes. Horned grebes and eared grebes account for the largest numbers of grebes banded

Table 25: Life History Data for Representative Loons, Grebes, and Auks

Species	Annual Adult Survival (%)	Maximum Known Longevity (yr)	Age at First Breeding (yr)	Modal Clutch	References
Loons					
Common	? > 80[a]	7	2–3 (?)	2	Clapp et al. 1982
Arctic	89	28	2–3 (?)	2	Nilsson 1977
Red-throated	? > 85[a]	23	2–3 (?)	2	Bauer and Glutz 1966
Grebes					
Eared	?	5	1	3	Clapp et al. 1982
Horned	ca. 50	6	1	4	Fjeldså 1973b
Great crested	? 45–50[a]	9	2 (?)	4	Cramp and Simmons 1977
Auks					
Cassin auklet	83	5	2–3	1	Speich and Manuwal 1974
Black guillemot	80–85	17	2–3	2	Glutz and Bauer 1982
Common murre	87.0–93.7	32	4–5	1	Glutz and Bauer 1982
Razorbill	89–92	25	4–6	1	Glutz and Bauer 1982
Atlantic puffin	95.0–95.2	18	4–6	1	Glutz and Bauer 1982

[a]See figure 29.

in North America; as of November 1, 1984, there had been 69 recoveries of banded horned grebes and 57 recoveries of eared grebes. Maximum survival after banding was about 5 years for the horned grebe and 6 years, 1 month for the eared grebe. Additionally, there had been 55 recoveries of western grebes, with a maximum survival after banding of almost 8 years. There have been 8 recoveries of red-necked grebes, with a maximum survival of 4 years after banding (Bird Banding Laboratory records, Patuxent, Maryland). Compared with the loons, where the recovery rates have averaged more than 10 percent, grebe recovery rates in North America of about 2.5 percent are relatively low. Although inadequate to construct life tables, these data do not suggest high survival rates for the grebes, and the maximum record of survival in grebes seems to be one of 9 years, 8 months for a great crested grebe (Rydzewski, cited in Cramp and Simmons 1977).

As with loons, there is still some uncertainty about the age of initial reproduction in grebes. In the case of the great crested grebe, which is certainly the best-studied grebe species, it is known that young birds acquire their nuptial plumage in their first year. They may also pair and seem to be capable of breeding by then, at least in years when the population has been depressed following a severe winter (Simmons 1955, 1974). However,

first-year birds may be unsuccessful at breeding because of competition with older and more experienced birds. Fjeldså (1973c) thus observed that some presumed "first-year" horned grebes gave up sustained attempts to establish territories and finally moved elsewhere. On the other hand, those "first-year" birds that were members of a flock before the breakup of ice at the nesting sites exhibited a productivity level almost as high as in "old" birds.

Clutch sizes in grebes tend to vary considerably around a mean rather than being somewhat rigidly fixed as in loons and alcids. In the great crested grebe the usual clutch is of 3 or 4 eggs in Britain, although clutches of 2 or 5 are not infrequent (Simmons 1974). A similar finding seems typical of Europe and southern Africa, though there may be significant local differences in average clutch size from site to site or year to year in the same general area (Cramp and Simmons 1977; Dean 1977). Fjeldså (1973c) found that clutch sizes of horned grebes in favorable (fertile) habitats were larger than those in food-poor and infertile lakes and ponds, and he also reported that "old" pairs of horned grebes nesting on traditional sites had an average clutch size of 4.89 eggs, while the mean clutch of "first-year" birds was only 3.62 eggs. There may also be a gradual reduction in average clutch size through the breeding season, which

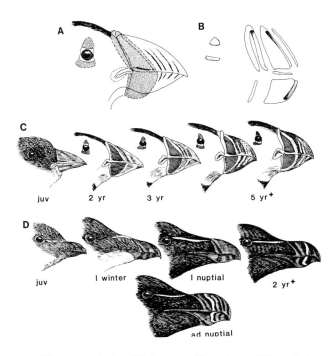

27. Bill characteristics of Atlantic puffin and razorbill: A, B, deciduous portions of the rhamphotheca; changes in bill with age in C, the Atlantic puffin and D, razorbill. After Glutz and Bauer 1982.

may well reflect smaller clutches in second breeding cycles. Second broods constitute about 1–5 percent of the total broods in Great Britain and Lower Saxony, and such second broods average considerably smaller than first broods (Zang 1977).

Grebes vary greatly in their relative coloniality, with eared and western grebes being examples of highly colonial species and the pied-billed grebe an example of a highly territorial species. Fjeldså (1973c) reported that in Iceland the horned grebe's relative territoriality was inversely correlated with abundance of food; thus congregations of grebes in infertile and food-poor lakes resulted in such adverse effects of high territorial conflict as highly asynchronous egg laying and correspondingly reduced fecundity.

These two major variations, of clutch size and of relative territoriality, may be important devices for grebes in terms of their reproductive adaptions. Other important mechanisms that might also be important in Simmons's (1974) view include a limited degree of asynchronous hatching of the young so that the first- and second-hatched chicks have improved chances of survival, brood division of the young by the two adults, with consequent reduced food competition within the family, and a capacity for renesting and sometimes also for double brooding. There may also be a favoring of

specific chicks by each parent for receiving limited food, while others may receive little or none, and asynchronous hatching may give adults the option of taking a reduced brood away from the nest and abandoning some viable eggs if food is limited. Some additional reproductive adaptations, such as the prolonged dependence of young grebes on their parents, no doubt help to increase chances of survival while the young are learning to forage for themselves.

Life history data are generally far better for the alcids than for the loons and grebes, largely because they are much easier to capture on their nesting sites. As a result, a considerable amount of survival and longevity information is available (table 25), which strongly suggests that alcids are among the longest-lived of any North American birds, sometimes attaining adult mortality rates of less than 10 percent per year. Lloyd (1974) noted that British razorbills had an adult annual mortality rate of only 11 percent, and the sample of 626 birds included an individual with a maximum longevity of 20 years following banding as an adult. A very similar annual adult mortality rate of 12 percent was calculated by Birkhead (1974) for British common murres. Even lower adult mortality rates of about 5 percent were estimated by Ashcroft (1979) for British puffins, although she also estimated a very high overall mortality rate of about 85–90 percent between fledging and probable initial breeding at 4 years of age.

The relatively long 4 to 6-year period to reproductive maturity in such species as puffins, razorbills, and

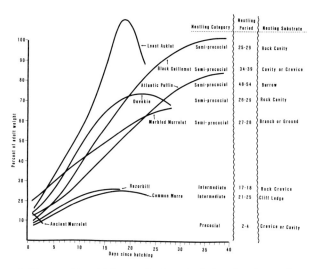

28. Comparative chick-growth characteristics of various auks. Mainly after Glutz and Bauer 1982, with addition of least auklet (Byrd and Knudtson 1978), ancient murrelet (Sealy 1976), and marbled murrelet (Simons 1980).

murres is of special interest, and at least in the case of puffins and razorbills it is accompanied by fairly conspicuous age-related differences in the size and appearance of the bill. These differences might be important bases for age estimation and individual recognition between mates or potential mates, and they also have relevance in human estimates of age in these species. Bill features in puffins are fairly indicative of age up to the fifth or sixth year of life (Peterson 1976), and similar progressive changes are evident in the razorbill for the first several years (fig. 27).

In all the alcids the newly hatched young are very similar in appearance and slightly resemble downy loons (see fig. 49). In contrast, grebe chicks usually are patterned with stripes and spots, especially on the head region. They typically also have bare spots on the head that are variably colorful and quite possibly serve as important signals between parents and offspring (see plate 11). One of the major ways the alcids differ among themselves is in the length of the nesting period, which ranges from a minimum of 2 or 3 days in some murrelets to 50 days or more in the puffins. On this basis the alcids can be classified as precocial (the nidifugous species whose young leave the nest almost immediately) to semiprecocial (the species whose young remain in the nest until they are nearly ready to fly). Intermediate sit-

Table 26: Biological Traits of the North American Auks

Species	Relative Hatching Weight (% of adults)	Age of Initial Thermoregulation (days)	"Fledging" Weight[a] (g)	Age of Sexual Maturity (yr)	Maximum Known Longevity[b]
Dovekie	13.4	?	114.3 (71.4%)	2	9 yr
Common murre	9.6	9–10	200 (19%)	4–5	32 yr
Thick-billed murre	11.0	9–10	260 (26%)	4–5 (?)	22 yr, 8 mo
Razorbill	9	3–4	140 (20.4%)	4–6	25 yr
Black guillemot	9.4	3–4 (brooded 5–6)	384 (89%)	2–3	17 yr
Pigeon guillemot	9.2	3–5	411 (91%)	2–3 (?)	9 yr, 3 mo
Marbled murrelet	11.7	1–2	157 (70%)	?	—
Kittlitz murrelet	11	?	?	?	—
Xantus murrelet	17.3	2	?	?	—
Craveri murrelet	?	2	?	?	—
Ancient murrelt	15	2	26 (12.6%)	?	5 yr
Cassin auklet	10.6	5–6	150 (90%)	2–3	5 yr, 8 mo
Parakeet auklet	9.8	ca. 3 (brooded to 6)	223 (78%)	3	—
Least auklet	ca. 9	ca. 5	81 (88%)	2–3	—
Whiskered auklet	16.5	?	?	?	—
Crested auklet	10.2	ca. 3	228 (80%)	2–3	—
Rhinoceros auklet	11.5	ca. 4 (range 0–9)	357 (69%)	?	6 yr, 2 mo.
Tufted puffin	7.8	5–6	561 (70.4%)	4–5	—
Atlantic puffin	9.4	6–7	283 (59%)	4–6	20 yr, 2 mo
Horned puffin	7.9	6	407 (68%)	4–5	—

NOTE: Adapted primarily from Thoresen, in press, and Sealy 1972, 1973b.

[a]Percentages of average adult weight in parentheses. "Fledging" refers to nest departure rather than flight attainment.

[b]Noncaptives. After Glutz and Bauer 1982 and Clapp et al. 1982.

uations also exist, as in the murres and razorbill. In the ancient murrelet, and probably also in the Xantus and Craveri murrelets, the newly hatched young are not fed at all by the adults, and the chicks lose weight until they leave the nest and make their way to sea (fig. 28). Young murres and razorbills are fed by both parents until they reach approximately one-fourth of the adult weight, at which time they too flutter down from their nesting ledges and take to the ocean, approximately 3 weeks after hatching. In the largest number of species, including the puffins, guillemots, auklets, dovekie, and marbled and Kittlitz murrelets, the nestling period is typically at least 4 weeks, and the young attain a weight ranging in various species from about half of adult weight to full adult weight before leaving the nest site. Such a prolonged nestling period obviously is advantageous to the young in keeping them safe from predators, chilling, and other hazards associated with early sea life, but it places considerable strain on the parents to carry enough food to maintain them. Thus among the small murrelets only the species that have precocial young regularly have clutches of two eggs; among the alcids with semiprecocial young only the guillemot species normally have two eggs, with one-egg clutches the norm for all the rest. The evolution of special throat pouches for carrying food has apparently promoted the evolution of a prolonged nestling period in the small auklets, whereas the precocial murrelet species tend to attain thermal regulation remarkably soon after hatching (table 26), in conjunction with their extremely short nestling periods.

The relative length of the nestling period in alcids is highly variable (tables 27 and 28) and is probably influenced by a wide variety of ecological factors, including

Table 27: Nest, Egg, and Nestling Traits of the North American Auks

Species	Nesting Site	Usual Clutch Size	Relative Egg Weight (%)[a]	Incubation Period (days)	Nestling Type	Nestling Period (days)
Dovekie	Cavity, crevice	1	15.0	29	Semiprecocial	26–29
Common murre	Ledge	1	10.8	33	Intermediate	25
Thick-billed murre	Ledge	1	11.3	32	Intermediate	23
Razorbill	Ledge, crevice, cavity	1	12.5	37	Intermediate	18
Black guillemot	Cavity, burrow, surface	2	12.5	30	Semiprecocial	34–39
Pigeon guillemot	Cavity, burrow, surface	2	12.5	30	Semiprecocial	35
Marbled murrelet	Tree crotch, ground	1	16.2	ca. 30	Semiprecocial	28
Kittlitz murrelet	Rocky tundra	1	15.2	?	Semiprecocial	ca. 21?
Xantus murrelet	Crevice, cavity, ground	2	23.7	31–33	Precocial	3–4
Craveri murrelet	Crevice, cavity	2	23.7	ca. 32	Precocial	2–4
Ancient murrelet	Burrow, crevice	2	22.0	34	Precocial	2–3
Cassin auklet	Burrow	1	16.8	38	Semiprecocial	41
Parakeet auklet	Cavity, crevice	1	13.0	35	Semiprecocial	35
Least auklet	Cavity, crevice	1	19.0	31–36	Semiprecocial	29
Whiskered auklet	Cavity, crevice	1	21.4	35–36	Semiprecocial	30+
Crested auklet	Cavity, crevice	1	14.2	36–41	Semiprecocial	34
Rhinoceros auklet	Burrow	1	14.8	45	Semiprecocial	42–60
Tufted puffin	Burrow, crevice	1	11.4	45	Semiprecocial	42–50
Atlantic puffin	Burrow, crevice	1	11.8	46	Semiprecocial	48–52
Horned puffin	Usually crevice	1	9.5	42	Semiprecocial	36–42

NOTE: Adapted in part from Sealy 1972, 1973b and Thoresen, in press.

[a]Expressed in percentages of adult weight as estimated by Thoresen, in press.

Table 28: Reproductive Traits of Alcids That May Influence the Length of the Fledging Period

Coloniality	Clutch Size	Adult Foraging Ecology	Foods and Feeding of Chicks	Diurnality; Nest Visibility	Chick Feeding Rates	Ratio of Fledging to Incubation	Species
		Larger offshore fish eaters	*Long foraging flights, one fish/load*	*Diurnal; exposed*	*3–8 food flights/day*	ca. 70	Murres
				Diurnal; exposed	*2–5 food flights/day*	ca. 50	Razorbill
	One-egg clutches		Mostly shorter foraging flights; Several fish/load	Diurnal; hidden	5–6 food flights/day	ca. 100	Puffins
				Nocturnal; hidden	*1–2 food flights/night*	120–50	Rhinoceros auklet
Coastal; colonial		Smaller offshore plankton eaters	Plankton carried to young in throat pouches	Variable; hidden	Feeding rates Variable	ca. 100	Dovekie, small auklets
		Inshore foragers; eat diverse bottom fauna	Short foraging flights; one fish/load	Diurnal; hidden	Up to 16 food flights/day	115–70	Guillemots
	Two-egg clutches	*Offshore foragers; mostly small shoal fish eaten*	Chick diet unknown	*Nocturnal; hidden*	*Chicks not fed at nest*	ca. 10	Ancient, Craveri, Xantus murrelets
Inland; solitary	One-egg clutches	Inshore foragers; mostly small shoal fish eaten	Chick diet unknown	*Nocturnal; exposed*	*1–2 food flights/night*	ca. 100	Marbled and Kittlitz murrelets

NOTE: Traits in italics are those that are likely to favor increased precocity (reduced fledging/incubation period ratio) in species.

the number and length of food-carrying trips adults can make each day (or night) to feed their young, the amount of food they can carry on each trip, the relative vulnerability of the chicks at the nest to predators, the total length of the season suitable for rearing young at the nest, and the competition between chicks (in cases where two are hatched) for being fed. Some species such as the rhinoceros auklet have seemingly extended the nestling period far longer than might be expected, in conjunction with relatively slow growth rates in the chicks, while others such as several murrelets have evolved an absolute minimum nestling period in favor of taking the young to water and to begin feeding them at offshore foraging areas.

Sealy (1972) summarized data supporting the view that nest site tenacity, which is typical of alcids, is probably adaptive inasmuch as it allows experienced birds to occupy known nest sites as early as possible in the breeding season. He listed known cases of nest site retention for seven species of alcids, including instances when the same site was used for as many as four successive seasons. He also pointed out that available sex-ratio data suggest that adult sex ratios in alcids approach equality, thus reducing competition for mates. He further noted that diurnality and nocturnality differences in alcid attendance at the colonies are marked, but that their biological significance is unknown though certainly complex. For some species, such as plankton feeders, it may be related to the diel cycles of prey availability rather than being solely a means of diurnal predator avoidance, which has often been offered as an explanation. However, predation certainly plays a role in the diel cycles of some species, and it is likely that the nocturnal egress of alcid chicks to sea is adaptively related to avoidance of such diurnal predators as gulls.

Sealy also pointed out that in alcids there is a correlation between breeding dispersion and food supplies, in that those species (such as guillemots and the two *Brachyramphus* species) that nest solitarily, or nearly so, tend to be inshore feeders, whereas colonial nesters typically forage far from shore. However, the presence of safe nesting sites is probably the major factor affecting relative coloniality, and limitation of acceptable nesting spaces (narrow cliff ledges, talus slopes, soil suitable for burrowing) is also likely to have profound effects on local nesting distribution. Obviously, broad controls on distribution and abundance of alcids are determined by general availability of food resources; this is evident from the apparent relation between oceanic productivity levels and the locations of extremely large alcid colonies, as shown in chapter 1.

Average durations of incubation between nest reliefs by the mate, and of intervals between successive visits to the nest with food for the young, are related to the distance the adults must fly for food and the ease with which it can be gathered. Sealy has pointed out that these intervals are shortest in the diurnal, inshore-foraging guillemots and longest in the nocturnal, offshore-foraging ancient murrelet. In most other species the incubation shifts usually occur at 24-hour intervals, although during brooding of the young this pattern is markedly altered and feeding is much more frequent.

The egg characteristics of the alcids also exhibit a number of interesting attributes (table 29). As might be expected, cliff-nesting and highly colonial species of alcids lay heavily spotted eggs that are probably less visible than unspotted eggs and may also aid individual egg recognition by adults (Tschanz 1959). However, spotted eggs also rather inexplicably (perhaps atavistically) are laid by such crevice nesters as Xantus, Craveri, and ancient murrelets as well as by the more diversely nesting guillemots. These three murrelets also lay the relatively heaviest eggs of all the alcids, and their newly hatched young are likewise among the heaviest relative to adult weight (tables 26 and 27). This is of course related to the precocity of their young, and the fact that they are apparently not fed by their parents before leaving the nest site. The stress of laying such large eggs is magnified because in these species the usual clutch size is two eggs laid, at least in the case of the ancient murrelet, about 7 days apart (Sealy 1976). In *Cepphus* the rather typical occurrence of two-egg clutches is probably related to the inshore foraging behavior of guillemots and their corresponding ability to make numerous foraging trips each day for their young. Single-egg clutches apparently are typical of younger birds, probably those breeding for the first time, and it is probably adaptive that such inexperienced birds have only one offspring per pair to brood and feed efficiently. Egg

Table 29: Typical Nesting Sites and Egg Patterns of Alcids in North America

Nesting Habitat	Breeding Species		Egg Pattern or Color
	Pacific Coast	Atlantic Coast	
Mature forests	Marbled murrelet (N)[a]		Spotted
Rocky tundra	Kittlitz murrelet		Spotted
Cliff ledges	Common murre	Common murre	Spotted
	Thick-billed murre	Thick-billed murre	Spotted
Ledges or crevices		Razorbill	Spotted
Soil burrows	Tufted puffin	Atlantic puffin	White
	Rhinoceros auklet (N)		White
Rock crevices	Horned puffin		White
	Xantus murrelet (N)		Spotted
	Craveri murrelet (N)		Spotted
Talus cavities	Parakeet auklet	Dovekie	Bluish white
	Cassin auklet (N)		White
	Whiskered auklet (N)		White
	Crested auklet		White
	Least auklet		White
	Ancient murrelet (N)		Spotted
Cavities or burrows	Pigeon guillemot	Black guillemot	Spotted

[a](N) indicates a nocturnally active species.

shape in alcids also varies considerably and is obviously related to the danger of rolling.

Reproductive Success

Information on relative breeding success under varying ecological conditions and at different times provides important clues to optimum breeding conditions and maximum tolerance limits of various species, and it may also allow for determination of recruitment rates for a species, which can be compared against survival rates and thus used to project population trends.

Reproductive success in at least three loon species seems to be positively correlated with lake size (table 30). It appears that the red-throated loon is highly adapted for breeding on small water areas of generally

Table 30: Loon Breeding Densities and Breeding Success Rates Reported for Various-Sized Lakes

Species and Lake Area (ha)	Total Pairs	Hectares per Pair	Fledged Young	Nesting Success[a]	Young per Pair	Fledging Success
Red-throated loon[b]						
1	193	2.1	79	—	0.41	—
1–5	59	7.7	23	—	0.39	—
5	35	—	4	—	0.11	—
Total or average	278	—	106	47.8%	0.37	63.4%
Arctic loon[c]						
200	385	96[e]	106	—	0.28	—
200–1,000	105	—	53	—	0.50	—
1,000	213	160[f]	60	—	0.28	—
Total or average	703	—	219	—	0.31	—
Common loon[d]						
500	9	35	2	—	0.22	—
500–4,000	88	53	43	—	0.49	—
4,000	187	111	71	—	0.76	—
Total or average	284	—	116[g]	57.7%	0.41	94.4%

[a]Percentage of all nesting pairs having young, including possible renesting efforts.

[b]Adapted from Bundy 1976, 1978 for Shetland Islands, 1973, 1974, and 1976.

[c]Adapted from Andersson et al. 1980 for Sweden, 1971–73.

[d]Adapted from McIntyre 1975 for various areas in Minnesota (tables 15 and 23).

[e]Based on reported usage of available waters of that size range; midpoint of range assumed as average water area.

[f]Based on a sample of eight lakes for varying numbers of years.

[g]Data did not distinguish young according to age; number thus includes unfledged young.

less than 5 hectares, and these birds also seem to attain their highest average number of fledged young per pair on such small waters. Bundy (1978) estimated that the average territory size of red-throated loons on larger waters is only about 0.5 hectare, while on smaller waters of up to 1.0 hectare the entire area is defended. On the other hand, the highest nesting success in the arctic loon, based on Swedish data, seems to be on lakes ranging in size from 200 to 1,000 hectares. Territories in this species tend to be considerably larger than in red-throated loons. They were estimated by Lehtonen (1970) to range from 100 to 150 hectares in Finland, and they ranged from 43 to 96 hectares in Norway according to Dunker (1974). In the common loon territories are extremely large and may range from areas less than 10 hectares in the bays of some lakes to entire lakes of 40 hectares or more (Olson and Marshall 1952). Reproductive success as well as overall usage by this species ap-

parently increases with increasing lake size (tables 30 and 31). Sawyer (1979) found that in Maine only 8 percent of the loon usage was on lakes up to 40 hectares, while 56 percent of the total loon usage was on lakes larger than 200 hectares (500 acres). Similarly, lakes up to 10 hectares had only 18.8 percent utilization, those of 40 to 100 hectares had over 80 percent utilization, and those over 100 hectares had over 90 percent utilization (Cross 1979). Likewise, McIntyre found that about half of all Minnesota lakes surveyed that were larger than 4 hectares had resident pairs, but almost 80 percent of those larger than 20 hectares had resident pairs. She judged that from 40 to 80 hectares are typically needed per resident pair, with birds on smaller lakes using adjacent water areas for supplemental feeding.

The average number of young fledged per breeding pair of loons generally ranges from as little as 0.3 in some years and areas to as high as 0.8 in others, but in

Table 31: Reported Loon Densities in Various Regions

Species	Density (ha/pair)	Reference
Red-throated loon		
Shetland Islands (3 areas)	118–28 (avg. 124)[a]	Merrie 1978
Northern Alaska (5 years)	125–67 (avg. 146)[a]	Bergman and Derksen 1977
Arctic loon		
Western Alaska (2 years)	20–65[a] (avg. 42.5)[a]	Petersen 1979
Finland	100[a]	Lehtonen 1970
Northern Alaska (2 years)	125[a]	Bergman and Derksen 1977
Sweden (8 lakes >10 km²)	100–500 (avg. 160)[b]	Andersson et al. 1980
Arctic and red-throated		
Scotland and Hebrides (5 areas)	275–3,033 (avg. 1,268)[a]	Merrie 1978
Common loon		
North America (5 areas)	39–502 (avg. 196)[b]	Table 37
Yellow-billed loon		
Northern Alaska	2,000[a]	Derksen, Roth, and Eldridge 1981

[a]Includes both water and associated land areas.
[b]Includes water areas only.

general seems to average about 0.4–0.5 per nesting pair (table 30). This means an average egg or chick mortality between laying and fledging of 75–80 percent, much of which seems to occur before hatching (table 32). Loon eggs are often lost because of water fluctuation or wave action, and predation by mammalian and avian predators is frequently significant. Losses during the egg stage are partially compensated for by replacement clutches in loons, at least in more southerly populations. In Minnesota up to two replacement clutches have been reported for common loons (Olson and Marshall 1952; McIntyre 1975). Bundy (1976) estimated that 14 of 22 pairs of red-throated loons he studied in the Shetland Islands laid replacement clutches, and Lehtonen (1970) stated that 7 of 85 Finnish clutches he observed were replacement clutches. However, the incidence of renesting in these species in high-arctic areas appears to be very low or nil (Bergman and Derksen

1977), and very probably the same is true of the yellow-billed loon, which is exclusively a high-arctic breeder.

Compared with loons, there is relatively little information on reproductive success rates in grebes. Some of the best data are on the great crested grebe, as summarized by Cramp and Simmons (1977). One national sample of 431 pairs raised an average of 1.3 young per pair, while another British sample of 169 pairs produced an average of 1.5 reared young per pair. This compares with a general European average of 2.1 young raised per pair. A breeding success of 1.3 young per pair would mean a recruitment rate of 39 percent (assuming all pairs attempted to breed), while an average fledged brood size of 2.1 per pair would represent a recruitment rate of 68 percent. An average of these two extremes would be a 53 percent recruitment rate, which perhaps provides a very rough approximation of productivity (and mortality) rates in this species. Simmons (1974) judged that great crested grebes normally attempt to raise only a single brood, but that something less than 5 percent of the pairs raised second broods, which would tend to increase overall recruitment rates slightly. Similar rates of double brooding have been found in Germany (Zang 1977). Reproductive success data for other species of grebes, including the North American species, are equally scanty and difficult to interpret. Probably the best available data are those of Fjeldså (1973b) for the horned grebe, summarized in part in table 33. He studied the success of 721 nests, established the exact fate of some 1,332 eggs in 339 nests, and estimated an overall hatching success of 63.2 percent of all eggs laid, plus a 75.5 percent nesting-success rate. Nesting success was higher for "old" females than for "first-year" birds and was also greatly affected by nest placement, with nests around sedge-fringed islets much more successful than nests close to shore. Nest desertion in dense colonies was also an important aspect of nesting success. Survival of young to 60 days after hatching (or to fledging) averaged 1.93 young per successful pair, or an approximate 49 percent recruitment rate, assuming a nest failure rate of 25 percent. This approximates Fjeldså's estimate of an approximate 50 percent annual mortality rate, which he based on an estimated 4.1 percent monthly disappearance rate of adults during his period of observation. Fjeldså judged that second broods were too rare to affect overall productivity, since the few cases he observed (9 clutches) were probably all too late to produce any fledged young. Apparently most of the chick mortality in Fjeldså's study area occurred during the first 20 days following hatching, after which there were no more significant losses before fledging.

An approximate 50 percent hatching-to-fledging survival rate seems to be fairly typical of other grebes as

Table 32: Sources of Nest or Egg Failure in Loons

	Common Loon			Arctic Loon
	Maine[a]	New Hampshire[b]	Saskatchewan[c]	Finland[d]
Total nests (or eggs)	51 (nests)	130 (nests)	424 (eggs)	159 (eggs)
Total losses	18 (35.3%)	83 (63.8%)	261 (61.7%)	72 (45.3%)
Predation losses				
Mammals (raccoons)	1	35	—	—
Birds (gulls, crows, ravens)	1	4	27	31
Unknown or other sources	—	—	119	6
Eggs lost or abandoned				
Water levels or waves	8	12	59	1
Other weather effects	2	—	—	—
Eggs lost from nest	—	—	15	—
Human or dog harassment	5	6	—	26
Unknown or other causes	1	17	18	—
Eggs inviable or infertile	—	9	23	10
Cause of loss unreported	—	—	—	14

[a]Sawyer 1979.   [b]Sutcliffe 1979b.   [c]Fox, Young, and Sealy 1980.   [d]Lehtonen 1970.

well, judging from information summarized by Fjeldså for the little grebe, red-necked grebe, and eared grebe over various parts of their ranges. This seems appreciably lower than the brood-rearing efficiency of loons. However, because of their larger average clutch sizes, the number of grebe chicks fledged per successful pair usually approximates two, while in loons the average brood size of fledged young is about half that.

There also seems to be a fairly high incidence of nest failure in grebes (table 33), though the data supporting this view are still rather limited. There may be a somewhat higher incidence of renesting following nest failure in grebes than in loons; Fjeldså (1973c) reported that some pairs of horned grebes he studied laid as many as four clutches, and the same may also be true of red-necked grebes (Palmer 1960). Replacement clutches have also been reported for eared, pied-billed, and western grebes, and multiple broods are a regular feature of least grebes (Palmer 1960).

Reproductive success rates in the alcids (tables 34–36) do not seem to differ much between the cliff-nesting forms and those that nest in burrows or rock cavities. For the cliff-nesting murres and razorbill, all of which

have one-egg clutches, there appear to be rather substantial variations in hatching success rates but relatively high fledging success for those chicks that do hatch successfully. Egg replacement is certainly regular in these ledge-nesting forms; Tuck (1960) estimated that in a population sample of about 400 pairs of thick-billed murres at Cape Hay egg loss was continuous during the entire egg-laying period, with some 44 percent of the sample losing at least one egg during a 32 day period. Thirty percent of the pairs laid one replacement egg, 11 percent laid two replacements, and the remaining 3 percent deserted or did not lay again. Approximately 1.6 eggs were produced per nesting pair. Gaston and Nettleship (1981) reported a much lower rate of re-laying and a considerably higher nesting success at Prince Leopold Island. However, studies of the cliff-nesting murres and razorbill generally support the view that about 0.5–0.7 young per pair are typically fledged by these species. Fledging in these species occurs only about 3 weeks after hatching, however, and undoubtedly much additional juvenile mortality occurs in the first summer of life. Thus estimated breeding success rates are certainly several times higher than actual fall recruitment rates,

Table 33: Nesting Losses in Various North American Grebes

	Pied-billed Grebe[a]	Western Grebe[b]	Western Grebe[c]	Horned Grebe[d]	Horned Grebe[e]
Total nests	138	516	—	—	721
Total eggs	—	—	224	637	1,332
Hatched					
Nests	97 (70.4%)	353 (68.4%)	47 (10.9%)	—	75.5%
Eggs	—	—	—	30.3%	850 (63.2%)
Mortality sources					
Eggs inviable	—	—	—	—	
Deserted	—	17 (3%)	10 (4.5%)	2%	27 (2.0%)
Depredation					
Humans	—	—	—	—	181 (13.6%)
Predators	ca. 7.5%	31 (6%)	89 (39.7%)	38%	93 (7.7%)
Other Losses					
Fell into water	—	—	—	—	17 (1.3%)
Disappeared	—	—	—	18%	20 (1.5%)
Wave action or water levels	ca.15%	111 (22%)	67 (29.9%)	12%	95 (7.1%)
Miscellaneous	—	4 (1%)	11 (4.9%)	—	9 (0.6%)

[a]Glover 1953.   [b]Nuechterlein 1975.   [c]Lindvall and Low 1982 (known-fate nests only).   [d]Ferguson and Sealy 1983.
[e]Fjeldså 1973b.

judging from known adult mortality rates in murres and razorbills.

For the cavity- and burrow-nesting puffins, rather similar results seem to be typical. Single-egg clutches are likewise the rule, and probably a low rate of egg replacement is required in these species because of their protected nest sites. Hatching success rates seem in general to exceed 50 percent, but fledging success rates seem to average somewhat lower than in the cliff nesters. Perhaps this is a reflection of their generally longer fledging periods, which are approximately twice as long as in the ledge nesters. The productivity of successful adults is remarkably similar in all four species and approximates 0.3 fledged young per nesting pair.

The data for the guillemots are of special interest, since they typically have two-egg clutches rather than single-egg clutches as do most alcids (table 36). Studies of three species in North America suggest a rather con-sistently high hatching success of about 50 percent (lower in one area of high disturbance in Quebec) and a fledging success rate of from about 50 to 90 percent, resulting in a typical overall reproductive success of about 0.4–0.8 fledged young per nesting pair. Preston (1968) estimated the adult survival rate of banded birds in the population he studied at about 80 percent, which would obviously require an annual recruitment rate of 20 percent to maintain. Even with a substantial postfledging mortality this should be quite feasible, assuming (as Preston estimated) that the actual breeding population represents about half of the total summering population. Indeed, with an average annual production of 0.73 fledged young per nesting pair and a nonbreeding component of 50 percent, the recruitment rate would be approximately 18 percent, very close to Preston's estimated annual mortality rate for this population.

There is still relatively little information on the re-

Table 34: Average Breeding Success Rates of Dovekie, Razorbill, and Murres

	Dovekie[a]	Razorbill[b]	Thick-billed Murre[c]	Common Murre[d]
Total number of pairs	—	—	—	486
Total number of eggs	98	170	2,015	—
Eggs produced per pair	—	—	1.05	—
Number of eggs hatched	64	143	1,587	392
Hatching success	65.3%	84%	78.8%	80.7%
Number of chicks fledged	50–62	138	1,420	349
Percentage fledging success	80–95%	96.5%	89.5%	89.0%
Estimated breeding success[e]	32–62%	81.1%	70.5%	71.8%
Fledged young per nesting pair	est 0.3–0.6	est. 0.8	0.75	0.72

[a]Data from Spitsbergen, 1974–75 (Stempniewicz 1981a).
[b]Data from Kandalaksha Bay, 1957–59 (Bianki 1977).
[c]Data from Prince Leopold Island, 1975–77 (Gaston and Nettleship 1981).
[d]Data from Skomer Island, 1973–75 (Glutz and Bauer 1982).
[e]Hatching success x fledging success.

productive success rates of the small auklets, but what is available indicates rather marked locational or yearly variations in both hatching and fledging success rates (table 36). It is obviously dangerous to generalize much from these limited data, but they suggest a rather low rate of overall reproductive success that perhaps ranges about from 0.3 to 0.5 fledged young per nesting pair. Except for the Cassin auklet, there are probably no opportunities for renesting in these primarily high-arctic forms. Further, at least in the Cassin auklet there is also a substantial "floater" population that is unable to breed because of territorial defense of limited suitable nesting sites (Manuwal 1974b). In this species such "floaters" compose about half the total nonbreeding population, representing a productivity loss roughly comparable to 15 percent of that of the total breeding population. The estimated annual adult mortality rate of this population is 17 percent. Replacing these losses would require an annual productivity of 0.5 fledged young per pair, assuming a total nonbreeding (floater and immature) component of about 30 percent of the total population and postfledging mortality rates approx-

Table 35: Average Breeding Success Rates of Puffins and Rhinoceros Auklet

	Rhinoceros Auklet N (# studies)	Tufted Puffin N (# studies)	Atlantic Puffin N (# studies)	Horned Puffin N (# studies)
Number of burrows with eggs	—	723 (13)	552 (4)	222 (9)
Percentage of burrows with eggs	63.6%	57.3%	—	—
Number of eggs hatched	132 (4)	367 (12)	347 (4)	168 (12)
Percentage hatching success	71–91% (5)	55.9%	62.8%	75.7%
Number of chicks fledged	82 (4)	181 (12)	180 (4)	60–61 (6)
Percentage fledging success	62.1%	48.4%	51.9%	46%
Fledged young per pair	0.38 (2)	0.29 (11)	0.33 (4)	0.31 (5)
Estimated breeding success[a]	44–56%	27.1%	32.6%	34.8%

NOTE: Adapted from tables 49–52 in Wehle 1980, using comparable numerical data from various studies.

[a]Hatching success x fledging success (should approximate percentage of nesting pairs fledging young).

Table 36: Average Breeding Success Rates for Guillemots and Small Auklets

Species and Location	Hatching Success	Fledging Success	Overall Success	References
Black guillemot				
Kent Island, Maine	52.4% (44/84)	50% (22/44)	26.2% (22/84)[a]	Winn 1950
	(1.83 eggs/nest)		(0.47 young/pair)	
Kent Island, Maine	55.8% (353/633)	88.7% (142/160)	42.1% (142/337)[a]	Preston 1968
	(1.56 eggs/nest)		(0.73 young/pair)	
Quebec (3 areas)	32–66%	59–71%	19.5–46.9%[b]	Cairns 1980
	(1.88 eggs/nest)		(0.38–0.97 young/pair)	
Pigeon guillemot				
Washington	53.8% (42/78)	86% (50/58)	46.3%[b]	Thoresen and Booth 1958
Cassin auklet				
Farallon Islands	—	—	69%	Manuwal 1972
			(0.69 young/pair)	
Farallon Islands	33% (25/75)	64.5%	26.6% (20/75)[a]	Thoresen 1964
			(0.27 young/pair)	
Parakeet auklet				
Saint Lawrence Island	67.7% (21/31)	76.2% (16/21)	51.6%[b]	Sealy and Bedard 1973
			(0.52 young/pair)	
Least auklet				
Buldir Island	67.8% (19/28)	75% (9/12)	43% (9/21)	Byrd and Knudtson 1978
			(0.43 young/pair)	
Saint Lawrence Island	60%	34%	20.4%[b]	Searing 1977
			(0.2 young/pair)	
Whiskered auklet				
Buldir Island	86% (6/7)	100% (3/3)	75% (3/4)[a]	Byrd and Knudtson 1978
Crested auklet				
Buldir Island	76% (45/59)	66.7% (14/21)	40% (14/35)[a]	Byrd and Knudtson 1978
Saint Lawrence Island	30%	—	—	Searing 1977

[a]Percentage of known-fate eggs producing fledged chicks.

[b]Estimated breeding success (hatching success x fledging success), assuming no renesting efforts.

imating those of adults. Thus, estimates of 0.3 to 0.5 fledged young per pair may not be too far from the mark for the small cavity-nesting auklets, depending on actual adult mortality rates and the incidence of nonbreeders in the population.

In summary, it seems that rather different reproductive strategies have evolved among the loons, grebes, and auks (fig. 29). Most alcids have evolved a one-egg clutch, a highly variable nestling period depending largely upon the ease with which food can be brought to the nestling for feeding, and a relatively high overall breeding success rate that reflects intense biparental care of a single egg or nestling. Guillemots (and a few murrelets) have evolved two-egg clutches. In the murre-

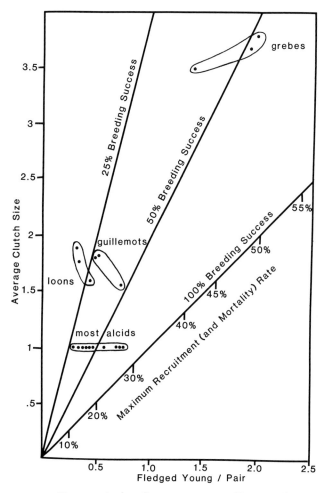

29. Comparative breeding success rates of loons, grebes, and auks, based on various literature sources as cited in the text.

lets this is related to the fact that the young are highly precocial and leave the nest before any parental feeding is required. In the guillemots the larger clutch size seems to occur because adult foraging is exclusively inshore, usually within a few kilometers of the nest, and the young are fed on fairly easily captured bottom-dwelling prey. These conditions allow the guillemots to "gamble" on a slightly larger clutch and the possibility of at least occasionally raising two young per season rather than a single chick. Loons similarly have two-egg clutches, but their breeding success averages about half that of guillemots, with most losses occurring during the egg stage, since the eggs are greatly exposed to environmental dangers. Only with high adult survival rates in excess of 80 percent can the loons "afford" such a breeding strategy. Finally, the grebes have the relative advantages of fairly large clutch sizes, repeated renesting efforts, and occasional double or multiple brooding, the last depending on the length of the breeding season. With all these advantages, they enjoy the highest recruitment rates of any of the three groups considered here and thus can sustain populations even with substantial adult mortality rates of about 50 percent annually. It seems quite possible that the northern breeding limits of grebes may well be affected by such considerations as opportunities for renesting or double brooding, whereas southern breeding limits in loons seem to be influenced by the distribution of large oligotrophic lakes offering adequate food supplies. Such lakes are virtually all of glacial origin (Dunker 1974), thus restricting the loons to northerly distributions and renesting opportunities that range from limited to nil.

Species Accounts

Loons (*Gaviidae*)

Red-throated Loon (Red-throated Diver)

Gavia stellata (Pontoppidan)

OTHER VERNACULAR NAMES; Røstrubet lom (Danish); plongeon catmarin (French); Sterntaucher (German); lomur (Icelandic); abi (Japanese); krasnozobaya gagara (Russian); smålom (Swedish).

Distribution of Species (see map 1)

BREEDS in North America on arctic coasts and islands from Alaska to Greenland, south along the Pacific coast through the Aleutian Islands to the Queen Charlotte Islands and (formerly) Vancouver Island, in the interior of the continent to central Yukon, southern Mackenzie, northern Saskatchewan, northern Manitoba, James Bay, and (formerly) the north shore of Lake Superior, and along the Atlantic coast to southeastern Quebec (including Anticosti Island), Miquelon Island, and northern Newfoundland (Ball Island); and in Eurasia from Iceland and arctic islands and coasts south to the British Isles, southern Scandinavia, northern Russia, Lake Baikal, Sakhalin, the Kurile Islands, Kamchatka, and the Commander Islands. Recorded in summer (and probably breeding) in northeastern Alberta and Newfoundland.

WINTERS in North America primarily from the Aleutian Islands south to northern Baja California and northwestern Sonora, and on the Atlantic coast south to Florida, ranging regularly to the Gulf coast of Florida; and in Eurasia south to the Mediterranean, Black, and Caspian seas, and along the western Pacific coast to China and Taiwan.

Description (after Witherby et al. 1941)

ADULTS IN BREEDING PLUMAGE (sexes alike). Crown somewhat glossy dark gray, streaked blackish, each feather having an even white margin on each side of web; upper mantle blackish brown, washed gray and spotted grayish white, each feather having at tip a pair of small grayish white spots as in winter but smaller and less pure white, a few similar feathers on sides of mantle, otherwise rest of upperparts usually unspotted glossy blackish brown; lores, sides of head, chin, sides of throat, and sides of neck dark ash gray; down middle of throat a long patch of dark chestnut (feathers with gray bases), narrow on upper part and increasing to full width of throat at base; immediately below this and on sides of upper breast feathers narrowly streaked blackish brown; rest of breast and belly white; feathers of flanks blackish brown more or less edged white and sometimes with a few small gray spots; under tail coverts as winter; tail feathers blackish brown, very narrowly tipped paler brown; wing feathers as winter; wing coverts much more narrowly and less distinctly fringed grayish white rather than white, and lesser coverts with small grayish spots or fringes and sometimes uniform blackish brown.

WINTER PLUMAGE. Whole crown and back of neck blackish gray, finely streaked white (each feather with blackish center and gray edges with short narrow white streak on each side at tip); rest of upperparts brownish black closely spotted white (each feather having at tip a pair of small white spots well separated and slightly oblong, on scapulars almost rectangular and set obliquely with usually an additional pair of spots proximal to them); underparts white, including lower part of lores, under eye, and side of neck, feathers joining back of neck tipped blackish and giving mottled appearance that occasionally extends onto throat; sides of upper breast and flanks more or less streaked black and white and also spotted; from side to side across vent narrow and somewhat ill-defined dark line; under tail coverts white, lower ones with basal halves dark brown; axillaries white with blackish brown median streaks; under wing coverts white, those covering primaries with pale brown shaft streaks; tail feathers brownish black, narrowly and evenly tipped white; primaries brownish

1. Current North American distribution of the red-throated loon, showing breeding areas (*hatched*), wintering areas (*shaded*), and southern limits of major migratory corridors (*wavy line*). The Eurasian range is shown on the inset map.

black on outer webs and tips, paler brown inner webs and whitish at extreme base; secondaries blackish brown, inner ones with narrow white fringes at each side of tip; primary coverts as primaries; wing coverts brownish black, feathers narrowly fringed white on each side of tip giving appearance of diagonal short streaks on greater and median but smaller and more spotlike on lesser.

JUVENILES. Like winter adult but crown and back of neck more uniform ash gray with narrow inconspicuous blackish brown streaks; rest of upperparts much browner, not so black as adult winter and spots more grayish, not so pure white, and smaller, being narrower and longer and forming narrow triangular edges to feathers, this being especially noticeable on scapulars and wing coverts; underparts white but cheeks, sides of neck, and throat thickly speckled brown; flanks browner than in adults and feathers edged white, not spotted, vent mottled brown; under tail coverts white with very narrow brown margins; lower ones brown with narrow white tips; tail feathers brown narrowly tipped grayish, becoming whitish when worn; primaries and secondaries as adult but all latter narrowly tipped white.

DOWNY YOUNG. Variable in color but usually warmer in hue than in the arctic loon, with a drab gray belly rather than white as in other loons (Fjeldså 1977).

Measurements and Weights

MEASUREMENTS. Wing (unflattened): males 272.5–292.5 mm (average of 10, 180.2); females 259–81 mm (average of 10, 270.4). Culmen: males 48.2–57.1 mm (average of 10, 52.4); females 46.4–54.6 mm (average of 10, 51). Eggs: average of 20, 73.6 x 45.1 (Palmer 1962).

WEIGHTS. In breeding season, males 1,526–2,265 g (average of 17, 1,833); females 1,356–1,887 g (average of 18, 1,543) (various sources). In winter, 6 males averaged 1,341 g and 9 females 1,144 g (Cramp and Simmons 1977). Estimated egg weight 83 g (Schönwetter 1967). Newly hatched young weigh about 65 g (Fjeldså 1977).

Identification

IN THE FIELD. This is the smallest of the loons, and its short neck feathers give it the slimmest conformation. Its bill is also the weakest and most stilettolike but has a distinct uptilt that is lacking in the arctic loon. In breeding plumage it is the only loon that lacks distinct white back patterning, has a reddish brown foreneck, and has black-and-white striping extending up the

hindneck and nape. In nonbreeding plumage the red-throated loon may be recognized by its small size, slim conformation, and unusually pale head and neck, with only the crown and hindneck relatively dark. It is thus almost as pale as is the yellow-billed loon's winter plumage, but unlike that species it has a dark-colored bill and lacks a dark auricular patch, and it is much smaller.

IN THE HAND. This smallest loon has a wing length that is usually less than 285 mm, and its culmen length is no more than 57 mm. Its measurements overlap considerably with those of the arctic loon, but unlike that species the tarsus is about as long as is the middle toe exclusive of its claw, and the culmen is virtually straight, emphasizing the upward tilt of the lower mandible.

Ecology and Habitats

BREEDING AND NONBREEDING HABITATS. Breeding habitat requirements of this species have been analyzed in Alaska by Bergman and Derksen (1977), who concluded that it prefers to nest in wetland areas occurring in large, shallow, and partially drained lakes of the arctic coastal plain, particularly relatively small, shallow ponds having emergent stands of Carex aquatilis and a central open-water zone. Larger ponds or lakes without emergent center vegetation but with stands of Arctophila fulva near the shore were also used throughout the summer, though not to the degree typical of the arctic loon. The average wetland area used by red-throated loons covered 0.4 hectare (range 0.1 to 0.8), compared with 3.0 hectares (range 0.7 to 12.1) in the arctic loon. Davis (1972) made an exhaustive comparison of habitat use by these two species and found that pond size is the most critical difference separating their nesting pond habitats. The two species have similar preferences with respect to shoreline features, pond depth, bottom type, water clarity, and general visibility from the pond, and none of these factors limited the suitability of a pond for nesting. Both species preferred island-type nesting areas, and a food supply had to be present within several kilometers, but not necessarily in the nesting pond. The average size of 67 nesting ponds of the red-throated loon in the McConnell River area was 0.32 hectare, while for 46 arctic loons it was 2.22 hectares. Apparently the greater takeoff ability of red-throated loons is a major reason they can exploit smaller ponds than those used by arctic loons. The difference in these two species with respect to space requirements is further pointed out in table 30.

Nonbreeding habitats have not been nearly so well

documented, but in general this species frequents shallow nearshore coastal waters, occasionally even freshwater lakes, reservoirs, and larger rivers, including the Great Lakes. Shallow tidal areas in estuaries, with associated submerged mud flats, seem to be favored foraging areas. On migration the birds often move along rivers from lake to lake or lagoon to lagoon or follow the shoreline of coasts between ice floes and coastal lagoons, concentrating in estuarine inlets and avoiding the open sea (Portenko 1981). Densities in the Gulf of Alaska and eastern Bering Sea are apparently low, and the birds seem to be limited to bay habitats (table 5).

SOCIALITY AND DENSITIES. This is perhaps the most gregarious of the loons during nonbreeding periods, and loose flocks sometimes develop that have been reported to number as large as 1,200 birds during fall migration on the Great Lakes (Palmer 1962). McIntyre (1975) noted that on wintering areas of coastal Virginia red-throated loons tended to feed in small groups among channels where the water currents were swift, whereas common loons foraged singly and used quieter waters in the area. However, there does not appear to be any evidence that the birds forage cooperatively, by driving fish into shallower waters where they can more easily be captured, for example. It is likely that pairing occurs before or during spring migration, since pairs typically arrive on breeding areas simultaneously.

Densities of breeding birds have been estimated by various authors, including Bergman and Derksen (1977), who found that over five years the average breeding density ranged from 125 to 167 hectares per pair on sample 2.6 square kilometer plots. The average densities of loons observed in summer over an area of 42,000 square kilometers of coastal tundra ranged from 0.6 to 1.7 birds per square kilometer, while the entire area of 92,000 square kilometers surveyed in the National Petroleum Reserve supported 0.35–0.40 bird per square mile. Red-throated loons composed 5 percent of this total, suggesting a density of only about one pair per 60 square kilometers (King 1979). Locally in the Shetland Islands the densities of red-throated loons may be quite high, approaching one pair of birds per square kilometer of water and associated land, or up to as many as 11 pairs sharing only 18 hectares of actual water area (Merrie 1978). In this area much feeding of the young was apparently done in shallow coastal waters within a few kilometers of the nesting pond, and thus these extremely high nesting densities are somewhat misleading. If these nearby coastal waters are included in the estimates a mean breeding density of 2.5 square kilometers per pair was estimated for nine different Shetland Islands and Scotland study areas (Merrie 1978).

COMPETITORS AND PREDATORS. Of the other species of loons, only the arctic loon is likely to be a significant competitor with the red-throated loon. Davis (1972) analyzed the habitat requirements of both species and noted that in one area where only red-throated loons nested that species exhibited only a slight tendency to use bigger ponds in the absence of the larger arctic loon. In high-density offshore areas the birds exhibited a random selection of various-sized ponds, and only on the mainland did they tend to choose larger ponds. Davis attributed this to an antipredator response rather than to a lack of competition with arctic loons.

Predators are apparently a major cause of egg losses in this species. Davis (1972) reported predation as the primary cause of egg mortality in his study and listed the chief egg predators as the parasitic jaeger (*Stercorarius parasiticus*), herring gull (*Larus argentatus*), sandhill crane (*Grus canadensis*), and arctic fox (*Alopex lagopus*). Of these, the jaeger was responsible for more egg losses than the rest of the predators combined. Predators of young and eggs were listed as humans, gulls, and red foxes (*Vulpes fulva*). Bergman and Derksen (1977) found that nests of red-throated and arctic loons were probably destroyed by arctic foxes, jaegers, and glaucous gulls (*Larus hyperboreus*), although for most nests the exact cause of failure was not determined. Bundy (1976, 1978) reported a "considerable" loss of both chicks and eggs in his study area on the Shetland Islands and judged that hooded crows (*Corvus corone*), jaegers, great skuas (*Stercorarius skua*), or gulls including great black-backed gulls (*Larus marinus*) were most probably responsible, although he obtained little direct evidence of this. The great black-backed gull has also been implicated as a serious predator on chicks in eastern Canada (Johnson and Johnson 1935). Predation on young chicks by the bald eagle (*Haliaeetus leucocephalus*) has also been observed (Reimchen and Douglas 1984).

Like other loons, this species probably has few serious predators once fledging has occurred, but substantial losses result from entanglement in fishing nets. Parmelee, Stephens, and Smith (1967) noted that a relatively large number were found caught in such nets in spite of the birds' relative scarcity in southeastern Victoria Island.

General Biology

FOOD AND FORAGING BEHAVIOR. The best information on food intake by wintering birds comes from Madsen's (1957) study of 203 birds taken between October and February. These birds had eaten fish for the most part, if not exclusively, with common cod (*Gadus collarias*) composing over half the total food by volume and present in over 70 percent of the birds examined. These fish measured up to 25 centimeters long. Gobies (*Gobius* spp.) were second in frequency of occurrence, and sticklebacks (*Gasterosteus* spp.) were third. Herring (*Clupea harengus*) were fourth in frequency, and over 80 percent of the sample had fed exclusively on one or more of these four types of fish. More than half had eaten a single type of fish, and about a third contained only two types of fish. The only nonfish remains were traces of crustaceans, polychaetes, and bivalve mollusks, at least some of which no doubt originated from the stomachs of prey. The total size range of fish was from very small (under 3 centimeters) to at least 25 centimeters.

On the breeding grounds, foods taken are less well known, but Davis (1972) reported that adults fed their young a variety of fish including capelin (*Mallotus villosus*), slender eelblenny (*Lumpenus fabricii*), sand launce (*Ammodytes dubius*), arctic char (*Salvelinus alpinus*), grayling (*Thymallus thymallus*), and unidentified sculpins. Few of these fish were obtained on the nesting ponds, since most are marine species. Bergman and Derksen (1977) similarly found that the adults fed their young on fish obtained from adjacent coastal waters. Capture of two chicks confirmed that they had been fed arctic cod (*Boreogadus saida*), and observations of wild birds feeding their young suggested that this species is the major food of young birds in that area. Crustaceans and aquatic insects may also be fed to the young during the first few days after hatching (Palmer 1962). The length of 11 fish carried to nearly fledged young birds in one study averaged 15.6 centimeters (range 9–20) and the estimated average weight was 50 grams (Norberg and Norberg 1976). In a more complete study, Reimchen and Douglas (1984) found that young were fed an average of eleven times a day, with a gradually increasing total amount of fish being fed through the fledging period as progressively large prey fish were brought to them. Major prey fish included shiner perch (*Cymatogaster aggregata*), unspecified osmerids, clupeids, and gadids, sand launce, gunnels (*Pholis*), and pricklebacks (*Lumpenus sagitta*), with average lengths ranging from 7.9 to 13.7 centimeters and average weights ranging from 1.5 to 22.5 grams. The young apparently are fed throughout the entire 7 week fledging period.

Most foraging seems to be done in relatively shallow waters less than 10 meters deep, with dives lasting no more than 90 seconds. These observed limits seem to be shallower and shorter than those reported for other loons (tables 10–12), although the data are still extremely limited in this regard. Reimchen and Douglas (1980) noted that average diving times of red-throated and common loons foraging in the same lake were very similar, although in red-throated loons 70 percent of all diving was in water less than 1 meter deep, whereas in the common loon only 15 percent of the dives were in such shallow water. In conjunction with this, the birds tend to winter on shallower waters than are typical of the other loons, though they may move to deeper waters during the night. Foraging flights of up to several kilometers from nesting ponds to sources of prey fish are regular in this species and are apparently unique among loons. Reimchen and Douglas (1984) concluded that breeding adults exhibit foraging adaptations that minimize the number of daily foraging flights, such as returning with prey nearly the optimal size for their variably grown chicks, but their preferential use of marine fish as prey, even when freshwater prey species are readily available, is of questionable adaptive value.

MOVEMENTS AND MIGRATIONS. A substantial coastal migration occurs on both the Atlantic and Pacific coasts of North America, with a very limited interior movement along the Mississippi River valley. As much as 12 weeks may elapse from the time of departure from southern wintering areas until arrival on tundra breeding grounds. Subadult nonbreeders also move north to coastal waters of the arctic or subarctic and remain there through the summer. Fall movements are also rather leisurely, with most migrants traveling coastally except for those passing through the Great Lakes, where large numbers of loosely associated birds sometimes assemble in October (Palmer 1962).

The red-throated loon is more maneuverable in flight than is the arctic loon as a result of its lower body weight relative to wing length and other mensural features (Davis 1972). Data on comparative flight speeds in these two species are not available, but it is known that the red-throated loon not only can take flight from very small water areas but can also, if pressed, take off from land. Normally, however, the birds skitter along the water surface for 15–40 meters and take off at about 35 kilometers per hour. The average flight speed during fishing flights is 60–64 kilometers per hour (Norberg and Norberg 1971).

Social Behavior

MATING SYSTEM AND TERRITORIALITY. Like the other loons this species is strongly monogamous, and pairs probably renew their pair bonds regularly each year as long as both are available. The high level of territorial stability Davis (1972) noted over a three-year period indicates such prolonged pair associations. Of 39 territories, 23 had nests all three years, 7 had nests during two years, and 2 had nests only one year. Seven had nests in two years and a resident pair was present the third year, and 2 had nests one year and a pair was present the next. Thus 30 territories were used all three years, a high number considering that in one of the three years only 32 total territories were occupied, as a result of severe spring flooding of breeding ponds. Davis concluded that the birds tended to modify territory size according to the number of ponds present, with from 1 to 4 ponds present in 36 territories, averaging 2.86 ponds. The actual ponds used for nesting often changed from year to year, especially if the previous year's effort had been unsuccessful. Although the data were not statistically significant, trends indicated that birds with a larger number of ponds on their territories had a higher chance of nesting success than did those with single-pond territories.

Davis believed that loons might select their breeding territories during the late summer before the breeding season in which they were to be used, partly because he observed that both arctic and red-throated loons spent a good deal of time on territory and in territorial defense even after their eggs or young had been lost. In red-throated loons the number of territorial encounters was about the same on territories with failed nests as during the prelaying period, but these encounters lasted longer and often included sustained aerial components. Thus red-throated loons were likely to call toward flying conspecifics intruding on their territories, causing them to change course and fly away. Since males sometimes were able to obtain new mates within a few days after the loss of their old ones, it is likely that pair bonding may also occur on the breeding grounds rather than wintering areas.

VOICE AND DISPLAY. Except for the long call, the vocalizations of this species are the same in both sexes. The long call is a harsh, loud, cooing call that is strongly pulsed, with the pulses varying in pitch and repeated up to ten times. The female's call is longer and softer; the call is often uttered as a duet during territorial encounters, contacts between mates, and "frog jumping" displays. A second major call is "wailing" or "mewing," a catlike meowing that lasts about a second and is of descending pitch. This call is apparently aggressive in function, and together with the long call is often uttered toward intruders flying over the territory. The croaking call is likewise aggressive and consists of short, barking crowlike notes repeated about four times per second. A moaning call is uttered as a contact call between mates and between parents and young; it is soft and low-pitched. Chicks utter chirping calls while begging and to attract parents, and a version of the long

call given in response to the same call by adults (Bylin 1971; Cramp and Simmons 1977).

Besides the classic descriptions of Huxley (1923), the displays of this species have been described by Bylin (1971) and Sjolander (in press; and cited in Cramp and Simmons 1977). Sjolander's observations seem especially useful, considering his wide comparative experience with other loon species. Initial pair bonding is probably achieved by young males' establishing territories and waiting for females to join them. Because of the small territories and frequent foraging outside their boundaries, territorial behavior is more complex than in the other loons, and because returning mates are often greeted by essentially agonistic mutual "triumph ceremony" displays (fig. 30A), such behavior has often been mistaken (as, for example, by Huxley) for courtship. When an intruder swims into or flies over a territory it is greeted by wailing calls and long calls by the resident birds. They will swim up to an intruder in a neck stretched and bill tilted (fig. 30B) "alert posture" ("start-

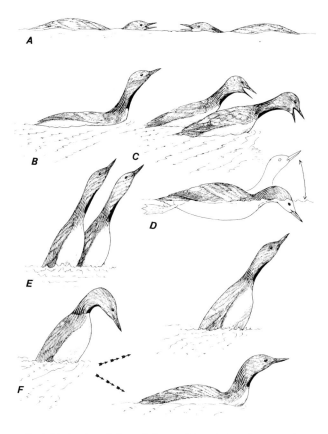

30. Social behavior of the red-throated loon (after Cramp and Simmons 1977 and Bylin 1971): A, mutual calling or "triumph ceremony"; B, alert posture; C, snake ceremony; D, bill dipping; E, plesiosaur race; F, penguin posture, followed by plesiosaur posture (above) or normal swimming (below).

ing position" of Bylin) and then often begin splash diving or bill dipping (fig. 30D). The resident pair may then perform the "snake ceremony," in which they swim in parallel with their necks outstretched diagonally while uttering long calls in duet (fig. 30C). Or they may form a silent but more intense "plesiosaur race" (fig. 30E), with their bodies partially raised from the water and their bills tilted upward as they swim in parallel. The intruding bird may participate in such races. A "penguin posture" (fig. 30F) similar to that assumed during the plesiosaur race may also be adopted by single birds and is probably the equivalent of the fencing posture of other loons. This posture may grade into a typical plesiosaur race posture or may terminate in a normal swimming posture. Rushing chases over the water are sometimes also performed, as in the other loons.

Precopulatory behavior in this species reportedly consists of mutual bill dipping and splash diving. Holding their necks diagonally outstretched, the birds begin to bill dip with increasing speed and sometimes dive past each other (Bylin 1971). Actual copulation is preceded by search swimming by the female, who moves along the shoreline in a crouched posture, eventually going ashore and inviting the male by remaining in this characteristic position with retracted neck and withdrawn bill. The male approaches, mounts, and copulates without specific displays except for sometimes uttering moaning calls. The female may remain on the shore for a few minutes afterward, sometimes making nest-building movements, but specific mating platforms are not needed for successful copulation (Cramp and Simmons 1977).

Reproductive Biology

BREEDING SEASON AND NESTING SUBSTRATE. In arctic Alaska the egg-laying period extends from June 8 to early July, and in the Canadian arctic it lasts from mid-June onward (Palmer 1962). In southern Canada egg records have been obtained as early as May 25 (Johnson and Johnson 1935), and on Victoria Island the egg-laying period is apparently from about June 27 to early July (Parmelee, Stephens, and Schmidt 1967). At even higher latitudes on northeastern Devon Island the observed egg-laying period was July 2–13, or about 2 weeks after the birds' initial arrival (Hussell and Holroyd 1974).

Nesting pond characteristics have been thoroughly studied by Davis (1972). He found that these birds choose small ponds (averaging about 0.25–0.40 hectare), regardless of whether arctic loons are nesting in the same area. The largest pond used in his study areas was 1.125 hectares, and the smallest was 0.01 hectare. Ponds with shorelines that were steep, rocky, or covered with thick and shrubby vegetation were not used. Ponds with islands or "marsh islands" (islands surrounded by marshy vegetation) were preferred over those with wet grassy substrates lacking these features, and dry shoreline was the least preferred nest substrate. Apparently ponds having a minimum amount of suitable shoreline for nest sites are as suitable as those having extensive areas of such shoreline. Although there is strong territorial fidelity from year to year, there is little fidelity to specific ponds for nesting, much less to nest sites. Pond depth is apparently more important than might be imagined, since loon chicks typically dive when frightened and escape by swimming along the bottom and stirring up mud, obscuring them from view. The minimum suitable depth for red-throated loons is probably 26–33 centimeters, and average depths are 39 centimeters. The bottom of the pond should preferably have a layer of loose sediment that can easily be stirred up by an escaping chick. Unlike arctic loons, red-throated loons do not require a food supply in the nesting pond. A final factor that might influence the choice of nesting substrate is general visibility from the nest site, particularly insofar as ease of rapid takeoff might be affected.

NEST BUILDING AND EGG LAYING. According to Davis (1972), this species constructs two types of nests. One is a rudimentary structure on dry shore, simply a hollowed area surrounded by small amounts of vegetation pulled out of nearby water. This nest type can be completed and ready for the first egg within a few hours. The second type of nest is constructed above a submerged site. It is built up of bottom mud and aquatic vegetation, and it apparently requires many hours of work before the first egg can be laid, possibly as long as several days. This type of nest may be somewhat safer from land-based predators and is apparently favored by loons if their ponds offer both dry and wet locations. Of 90 observed nesting attempts, 79 different sites were used, with only 9 cases of reuse of an earlier nest site. The eggs are laid at intervals of 2 days, but hatching typically occurs after a one-day interval, so incubation must begin almost immediately. One-egg clutches are not infrequent in this species; Bundy (1976) reported an average clutch size of 1.76 eggs for 39 clutches, 21 percent of the clutches being of single eggs. Davis (1972) found 12 one-egg clutches and 61 two-egg clutches. He noted that two-egg clutches were less likely to be completely lost than one-egg clutches, and that one-egg clutches had a lower hatching success than did the remaining eggs of two-egg clutches.

INCUBATION AND BROODING. Both sexes incubate, with the female probably doing the larger share. The incubation period is relatively short, 25–26 days being reported by Hussell and Holroyd (1974) for arctic Canada and 27.05 days by Cramp and Simmons (1977) for a sample of 19 eggs on the Shetland Islands. Johnson and Johnson (1935) reported that an abandoned egg was hatched under a chicken 38–40 days after the start of incubation, but this is certainly a case of unnaturally extended incubation.

GROWTH AND SURVIVAL OF YOUNG. Of a total of 134 eggs, Davis (1972) found that 91 hatched (68 percent) and that an average of 1.43 young hatched from two-egg clutches while 0.33 young per nest hatched from single-egg clutches. Overall brood survival for a three-year period was 39 chicks out of a total of 101, a 38.6 percent fledging success. Apparently the fledging success was not influenced by brood size at hatching; thus the presence of a second chick had no effect on the survival of the older one. However, the first-hatched young survived better than its younger sibling unless sufficient food was available. The second-hatched chick was fed only after the first had ceased to beg for food, which tends to result in maximum efficiency of brood-size regulation. The fledging period (hatching to initial flight) was determined to be 55–58 days (3 birds) by Peterson (1976), and Bundy (1976) estimated that juvenal plumage was attained in an average of 42 days for 27 chicks. In the early stages of chick growth one parent remains with the young while the other forages for food, and brooding is nearly continuous during the first week after hatching. After the young are 2 weeks old both adults begin to leave the territory simultaneously to forage for food for themselves and their young.

BREEDING SUCCESS AND RECRUITMENT RATES. Bundy (1976) estimated an overall breeding success rate of 0.41 young fledged per pair for 63 pairs in Scotland. Cyrus (1975) reported that 15 nesting pairs raised 5 young to fledging (0.33 young per pair), with most losses occurring in the egg stage. Davis (1972) reported an overall hatching success of 67.9 percent for 134 eggs and a survival rate of 38.6 percent for 101 hatched chicks, an approximate breeding success of 26.2 percent. With an average clutch size of 1.83 eggs (87 nests), the number of young raised per pair would be 0.47 chick. Because of the unknown percentage of nonbreeders in the population, recruitment rates cannot be derived from these figures.

Evolutionary History and Relationships

There seems little doubt that the nearest relative of the red-throated loon is the arctic loon, and Davis (1972) has postulated that these forms were isolated in

Pleistocene times into one occupying a colder arctic refugium and the other occurring south of the continental ice shelf. The former population, ancestral to the red-throated loon, became adapted to breeding on fast-thawing small ponds, which eventually placed the form at a competitive disadvantage with the larger arctic loon where the two met on larger and deeper waters.

Population Status and Conservation

There are no estimates available of North American or world populations of this species. King and Sanger (1979) assigned an oil vulnerability index of 49 to the red-throated loon, or somewhat less than the average (55) they calculated for the entire loon family.

Arctic Loon (Black-throated Diver)

Gavia arctica (Linnaeus)

OTHER VERNACULAR NAMES: Black-throated loon; Pacific loon; sortstrubet lom (Danish); Prachttaucher (German); glitbrusi (Icelandic); ohamu (Japanese); cheronozobaya gagara (Russian); storlom (Swedish).

Distribution of North American Subspecies (See Map 2)

Gavia arctica pacifica (Lawrence)[*]
BREEDS in North America from Cape Prince of Wales and Point Barrow, Alaska, east to Melville Peninsula and southern Baffin Island; south to the Alaska Peninsula, Kodiak Island, northern Saskatchewan, northern Manitoba, and northern Ontario; east to Labrador Peninsula (casually). Recorded in summer at the Carey Islands, near 77° N at the northern end of Baffin Bay, Devon Island, Hudson Bay, and James Bay, and on the Pacific coast from the Sitkan district of Alaska to the Queen Charlotte Islands, northwestern Washington, and Oregon; also breeds in Scotland, Norway, Finland, and northern Russia south to northern Germany, Baltic states, and to 55° N in Russia.

WINTERS mainly along the Pacific coast from the Aleutian Islands south to Guadalupe Island, southern Baja

---

[*]This form is sometimes considered a distinct species (Flint and Kishchinski 1982), since apparent breeding sympatry of *G. a. pacifica* with *G. a. viridigularis* has been reported. This taxonomic change was also adopted by the AOU (*Auk* 102, 680) after manuscript submission. The name Pacific loon thus applies to this form, while *viridigularis* is sometimes called the green-throated loon.

2. Current North American distribution of the arctic loon; symbols as in map 1.

California, the western side of the Gulf of California, and along the eastern side to southern Sonora; also on the coasts of Japan.

Gavia arctica viridigularis Dwight
BREEDS in northeastern Siberia, from the Khatanga River to Kamchatka and Sakhalin Island, and locally at Cape Prince of Wales, Alaska, where it is seemingly sympatric with *pacifica* (Bailey 1948).

WINTERS south to the Baltic Sea, Sakhalin Island, and Japan.

Description (Modified from Witherby et al. 1941)

ADULTS IN BREEDING PLUMAGE (sexes alike). Forehead and lores dark slate gray; crown and back of neck extending onto sides as nape ashy gray, base of back of neck merging into bluish black of mantle and upperparts; down each side of mantle a series of short rows of white spots, similar to but longer and broader than columns of larger spots on scapulars, rest of upperparts uniform bluish black but sides of back and rump tinged brown; sides of head as crown but below eye and lower ear coverts washed blackish brown merging into black of chin; throat black slightly glossed greenish purple; across upper throat a prominent line of short white parallel vertical streaks, down sides of neck long white streaks; flanks glossy black, and a blackish brown line across vent; rest of underparts white, but lower under tail coverts blackish brown, tipped with white; tail feathers brownish black with no white tips; wing coverts glossy black with twin oval white spots, but feathers along front edge of wing and primary coverts unspotted.

WINTER PLUMAGE. Forehead, crown, and back of neck brown washed grayish; base of back of neck and rest of upperparts blackish brown and somewhat glossy, some scapulars with a pair of small subterminal whitish spots; underparts white; white on sides of head extending from base of lower mandible under eye and slightly backward on sides of nape with slight mottling at juncture of white and brown; usually a narrow line of brown-tipped feathers across upper throat; flanks dark brown, a (usually incomplete) narrow dark brown line across vent; lower under tail coverts blackish brown tipped with white, upper ones white; axillaries white, sometimes with dark brown shafts and terminal shaft streaks; under wing coverts white, sometimes with pale brown median streaks; tail feathers black, tipped with white; wing feathers with glossy brownish black outer webs and tips, dark brown shafts and paler brown inner webs; upper wing coverts brown.

JUVENILES. Crown and back of neck like adult winter but browner and less grayish; rest of upperparts dark brown, feathers with ashy gray margins, prominent on mantle and scapulars, less conspicuous on rump and upper tail coverts; sides of neck adjoining white of throat more finely mottled brown than in adult and mottling sometimes extended onto throat; brown line across vent prominent; lower under tail coverts dark brown with narrow white tips; tail feathers dark brown with narrower white tips than in adult; wing feathers and primary coverts as adult but all wing coverts with grayish brown tips.

DOWNY YOUNG. First down rather short, thickly covering whole body and especially thick on underparts. Upperparts dark mouse brown, underparts paler, middle of breast and belly grayish, immediately around eye whitish. Second down like first but decidedly paler both above and below, center of breast and belly whitish. Iris reddish brown, legs and feet dark greenish gray, bill dark gray (Fjeldså 1977).

Measurements and Weights

MEASUREMENTS (of *pacifica*). Wing (flattened): males 285–307 mm (average of 10, 299.6); females 281–307 (average of 10, 295.7). Culmen: males 49.5–55.0 mm (average of 10, 51.9); females 49–54 mm (average of 10, 50.8). Eggs: average of 20, 76.0 x 46.7 mm (Palmer 1962). Measurements of *viridigularis* average about 5–10 percent larger.

WEIGHTS (of *pacifica*). In breeding season, males 1,382–2,409 g (average of 6, 2,025); females 1,616–2,152 g (average of 9, 1,823) (various sources). Estimated egg weight, 95 g (Schönwetter 1967). Weights for *viridigularis* average larger than for *pacifica*; Dementiev and Gladkov (1968) report males of that race weighing 3,280 and 3,792.5 g and females from 2,255 to 3,075 g (average of 3, 2,665 g). Newly hatched young, presumably of *arctica*, weigh about 75 g (Fjeldså 1977).

Identification

IN THE FIELD. This is a medium-sized loon, with a bill that is rather tapering and less uptilted than that of the red-throated loon. In breeding plumage the silvery gray nape and lower neck, with long lateral neck striping and a blackish foreneck patch, are distinctive. The back is extensively spotted with white patterning, including two large areas of white on the wings and two smaller areas on the upper back. In nonbreeding plumages the white areas on the mantle are lacking and the upperparts are uniformly dark, with white limited to the breast, foreneck, and lower half of the head. The bird is thus appreciably darker than the red-throated loon in corresponding plumage and is approximately the same darkness as the common loon. However, besides being smaller, it also lacks the pale "eyebrows" typically present in that species. Reportedly the best field marks for separating the arctic loon from the common loon in nonbreeding plumages include the arctic loon's tendency to exhibit a dark "chin strap," its more sharply demarcated patterning on the sides of the neck, and its more distinctly transversely barred back. Its nape is typically paler than that of the common loon, and a white flank spot is sometimes also visible at the waterline near the tail. The smaller bill size of the arctic loon is a less useful distinction from the common loon than might be imagined (*Loon* 52:59–61).

IN THE HAND. This is a medium-sized loon, with a wing length no greater than 315 mm, a culmen length no more than 70 mm, and the tarsus longer than the middle toe (including its claw). It is very similar in culmen length to the red-throated loon, but unlike that species the culmen is slightly convex rather than straight, producing a more tapering bill shape. Separation of the forms *pacifica* and *viridigularis* is rather difficult, but *viridigularis* has a bill that is longer (over 55 mm from forehead feathering, vs. under 55 mm for *pacifica*), and heavier (over 15 mm high from forehead feathers, vs. under 15 mm for *pacifica*) and is generally larger in other measurements. *Pacifica* also exhibits brown-tipped axillaries, a thicker-appearing neck, and a somewhat paler head and hindneck coloration (Flint and Kishchinski 1982). The gloss of the throat area also tends toward greenish in *viridigularis* and purple in *pacifica*. Green-throated birds in the size range of *pacifica* have been found in the area of Barrow, Alaska (Bailey 1948), but generally nonoverlapping measurements appear to be typical of both bill length and bill height as well as tarsal length (over 75 mm for *viridigularis*, under for *pacifica*) for these two forms.

Ecology and Habitats

BREEDING AND NONBREEDING HABITATS. Breeding habitats have been analyzed by Davis (1972), who identified several characteristics typical of nesting ponds. On his major study site at McConnell River there were 250 pond areas ranging in size from 0.07 to 21.37 hectares, and the average pond size used during each of three years ranged from 2.13 to 2.73 hectares, with a maximum of five nests found one year on the largest pond. The smallest pond used was 0.225 hectare, but generally the birds used the largest pond available to

them. The birds selected ponds that offered islands or wet grassy areas for nests, especially islands. Although there was a high level of territorial fidelity from year to year, nest sites varied greatly. Pond depth of those ponds used by loons ranged from 30 to 90 centimeters, averaging 52 centimeters, and these ponds were rich in bottom sediments that could be stirred up to facilitate escape by chicks. Adults fed their chicks on foods obtained on the nesting territories, which consisted of fish and various invertebrates. Finally, areas that offered excellent visibility and protected the nest from wave action were apparently favored. In Sweden the area of the lake (minimum 56.5 hectares, average 212 hectares) is apparently important, as is transparency of the water for fishing opportunities (Lindberg 1968). Studies by Dunker (1974) in Scandinavia suggest that breeding territories there are associated with lakes at least 50 hectares in area, in conjunction with an estimated minimum territorial requirement there of at least 20 hectares and the need for an extensive feeding area because of generally low densities of available food. Marginal vegetation for hiding nests was also regarded as important. Dunker suggested that breeding habitats of all loons are generally associated with shallow lakes of glacial origin throughout the Holarctic region.

Nonbreeding habitats consist primarily of coastal areas that are usually fairly close to land, but there is little use of inland lakes, reservoirs, or large rivers. In the Gulf of Alaska and eastern Bering Sea, bay and shelf habitats are used almost exclusively, with very few birds occurring on shelfbreak and oceanic areas (table 5).

SOCIALITY AND DENSITIES. Davis (1972) noted that about 35 pairs of birds nested on his study area of 16.5 square kilometers at McConnell River, giving an overall breeding density of one pair per 47 hectares. Bergman and Derksen (1977) found an average five-year breeding density of one pair per 125 hectares and a three-year nest density of a nest per 160 hectares on their study plots at Storkersen Point in northern Alaska. Derksen, Rothe, and Eldridge (1981) found similar densities of up to one pair per square kilometer, with preferential use of deep *Arctophila*-dominated wetlands, especially for brood rearing. King (1979) reported that arctic loons composed an estimated 72 percent of the total summer population of loons on the National Petroleum Reserve. The average loon density of 0.6–1.7 loons per square kilometer over 42,000 square kilometers of the coastal plains habitat suggests that in such regions the arctic loon may have an overall breeding density of about one pair per 200 hectares. Some additional estimates of breeding densities are shown in table 30.

Like all loons, arctic loons are highly territorial and

intolerant of conspecifics during the breeding season. Bergman and Derksen (1977) found an average internest distance of 640 meters for 41 nests and a minimum observed distance of 250 meters. Davis (1972) found that arctic loon territories coincided with shoreline boundaries and scarcely varied at all from year to year. Thirty of 35 territories were used in all of three years, and 3 of the remainder had nests in two of the three years. The 2 remaining territories had nests one year and birds present during a second year. Actual territorial size was estimated as 3.66 hectares, ranging from 1.125 to 9.54 hectares. Davis believed that this represented a nearly saturated density, though territorial defense was not very time consuming and was apparently not a major drain on the time budget for these birds. Furthermore, arctic loons only very rarely entered the territories of red-throated loons, though they dominated them socially when equally matched numerically. Evidently the two species maintain mutually exclusive territorial behavior, which is probably facilitated in large measure by their marked tendency to select different-sized ponds.

COMPETITORS AND PREDATORS. As just noted, red-throated loons are strong competitors with arctic loons, but differences in territorial requirements tend to reduce the incidence of interspecific competition on breeding areas. When interspecific territorial intrusion occurs arctic loons eliminate red-throated loons faster than they do conspecifics, but at about the same rate as red-throated loons evict arctic loons when the latter trespass on their territories. Generally, red-throated loons are less likely to intrude on arctic loon territories than the reverse, and arctic loons repel such intruders faster than they evict intruders of their own species, indicating a highly directed type of aggression (Davis 1972).

Predators of arctic loons eat mainly eggs and small young. Bergman and Derksen (1977) judged that the arctic fox (*Alopex lagopus*) might have been a major nest predator in their study area and believed that jaegers (*Stercorarius* spp.) and glaucous gulls (*Larus hyperboreus*) were also responsible for some nest losses. Similarly, Peterson (1976, 1979) found that the red fox (*Vulpes vulpes*), jaegers, and glaucous gulls were the major cause of egg losses in her Alaskan study areas. She judged that foxes caused most destruction of shoreline nests and that jaegers and gulls were major predators on island sites. In one year an estimated 95 percent of the nests were destroyed, apparently because of the relative absence of alternative prey species. Furthermore, loon eggs tended to be taken late in the incubation period, especially after the goose hatching period, until which there was an abundant supply of relatively conspicuous

eggs for the nest predators. In Scandinavia the hooded crow (*Corvus corone*) is an important nest predator of this species (Lehtonen 1970). As with other loons, there is probably little danger from predation after fledging, but the birds are susceptible to heavy losses from fishing nets (Parmelee, Stephens, and Schmidt 1967).

General Biology

FOOD AND FORAGING BEHAVIOR. The best summary of foods taken during the fall and winter season is that of Madsen (1957), who examined 145 birds obtained between October and February and found food remains in 123. All of these contained fish remains, and over half had been feeding on gobies (Gobiidae, especially *Gobius* spp.). The second most important food was cod (*Gadus callarias*), including fish to 25 centimeters long, and the third-ranked food source was sticklebacks (*Gasterosteus* spp.). Collectively these three food sources accounted for about 90 percent of the total diet, and 80 percent of the samples contained nothing but these three types of food. About half the birds had eaten only a single kind of fish, another third two kinds, and about a tenth three kinds. Several other types of fish were present as minor food types, and there were also small quantities of invertebrates (crustaceans, polychaetes, mollusks, insects), at least some of which probably were in the stomachs of fish prey. Such crustaceans as crayfish, crabs, and prawns may be taken locally from the sea or in fresh water, judging from other studies. Based on all available sources, a summary of major foods of this species is provided in table 13. Generally small shoal fish seem to be the preferred foods at nearly all times of year, based on available information. Cramp and Simmons (1977) have summarized the European data effectively.

Foods taken on the breeding grounds are less well known, but Davis (1972) reported that the adults fed their chicks on sticklebacks (*Pungitius pungitius*), grayling (*Thymallus thymallus*), fairy shrimp (Anostraca), tadpole shrimp (Notostraca) and other unidentified invertebrates. Likewise, Lehtonen (1970) found that for the first 10 days after hatching the chicks are fed on aquatic invertebrates but later shift to fish eating.

Prey are captured by sustained dives, in waters of varied depths, but with most dives averaging about 45 seconds and occurring in water 3–6 meters deep (Cramp and Simmons 1977). A maximum dive duration of 5.04 minutes and a maximum lateral movement of more than 500 meters have been observed (Lehtonen 1970). However, most fishing dives last less than a minute, and only rarely are they over 2 minutes in duration. Apparently the maximum reported depth of dive for this species is 46 meters according to Lehtonen.

MOVEMENTS AND MIGRATIONS. Takeoff is achieved only with difficulty in this species and requires a running start of from 40–60 meters (against a headwind) to 120–200 meters (in windless conditions). In still weather a flight speed of about 65–80 kilometers per hour can be attained. Flight is seen especially often in later summer, between August 15 and October 1, primarily during midmorning hours (6–10 A.M.) and again in the evening (6–10 P.M.) (Lehtonen 1970).

As with the other loons, little is known of migration routes and relative numbers of migrating birds. Unlike the red-throated loon, virtually all the wintering of this species occurs along the Pacific coast, fairly commonly extending north to Vancouver Island. There it is mainly a migrant, with the spring migration primarily from late April onward, peaking in May or early June. The fall migration begins in September and is largely offshore. Average flock size from October through March was 4.3 birds (range 1–25) (Hatler, Campbell, and Dorst 1978). Farther south, spring migration along the California coast begins in the latter half April and extends through early May, with most migration apparently during daylight and with some tendency for the birds to migrate in pairs. By early May the birds begin arriving in southeastern Alaska, and they move east along the arctic coast of Alaska by the first week of June. Nesting areas of northern Alaska are reached by the second or third week of June. There is some limited evidence of an overland migratory route from the arctic coast to the southwestern shore of Hudson Bay. The fall migration begins in late August, with the first arrivals appearing in California in September and large numbers passing the coastline of northern California as late as mid-November (Palmer 1962).

Social Behavior

MATING SYSTEM AND TERRITORIALITY. All the evidence suggests that this species exhibits monogamous mating, sustained or renewed annually to produce essentially lifelong pair bonding. Sjolander (1978) found that banded birds remained paired for at least several years. Davis (1972) observed a very high incidence of territorial stability from year to year, suggesting such permanent pair bonding as well as a high level of territorial site tenacity. The age of initial pair bonding is still unknown, although Lehtonen believed that in Finland these birds take up territories when 5 or 6 years old but do not actually lay eggs until the following year, when they are 6 or 7 years old. Thereafter most birds breed

every year (Lehtonen 1970). Sjolander (1978) doubted that such a long period before breeding is typical.

The birds are strongly territorial, and Davis (1972) has found that these territories are used for copulation, nesting, brooding, and feeding of young as well as for other general adult activities. The territories often encompassed more than one pond, with the nonnesting pond used for escape, resting, and feeding of young. Territory sizes averaged (for 35 territories) 3.66 hectares in the three years of Davis's study, with a range of 1.125 to 9.54 hectares. Smaller territories appeared to be less productive of young than larger ones, although statistical significance was lacking. These territories are far smaller than those that have been estimated in Scandinavia, where large and relatively deep lakes are typically used and where breeding on lakes of less than 10 hectares is extremely rare (Dunker and Elgmark 1973). Lindberg (1968) estimated that 16 territories in Sweden averaged 212 hectares, and Lehtonen (1970) judged that territories in Finland ranged from 100 to 150 hectares. Dunker (1974) estimated that 12 territories in southern Norway ranged from 43 to 96 hectares, and the average of 9 was 64 hectares. Within these 9 territories, most of the birds' surface and foraging activities were in areas no more than 15 meters deep, and such areas appear to be a critical foraging component of territories in these relatively deep lakes.

VOICE AND DISPLAY. Various persons have described displays and calls of this species, but the observations of Sjolander (1968, 1978, summarized by Cramp and Simmons 1977) are probably most useful and comparable to those of other species. Sjolander recognized seven call types of adults plus two intensities of chirping contact calls of chicks. Adults of both sexes utter a croaking call that is long, pulsed, and hoarse, lasting several seconds, similar to a repeated raven (*Corvus corax*) call. It is sometimes used in territorial encounters, during antiphonal calling by the pair, and also during disturbance. A wolflike wailing call consisting of two or three notes of increasing volume and pitch is uttered during territorial encounters and also between neighboring territorial pairs. In 6 ascertained cases it was uttered by a male. A yodel or long call, lasting a second or more and consisting of a short, weaker note and a longer, higher-pitched note, is the major territorial call. It is often repeated, with yodellike breaks between calls. This is the loudest of the calls, and in 18 ascertained cases it was performed by the male of a pair. A similar short yodeling is sometimes also uttered. A short call or "kuik call," with a rising inflection, precedes a splash dive and is uttered by both sexes. Both sexes also utter a low contact call that sounds like human humming. Finally, a somewhat louder moaning call is used as a contact note toward the mate or young and is characterized by a sudden rise in pitch in the middle.

Displays of the arctic loon have been thoroughly described by Sjolander (1978) and also by Dunker (1975); Sjolander's terminology is generally adopted here. During normal swimming (fig. 31A) the neck is only moderately raised, but in the raised neck display it is greatly extended, with the neck feathers sleeked (fig. 31B). This posture is adopted by both sexes and by young from 2 weeks old. A high-front variant is similar (fig. 31B, inset) but the neck feathers and crown feathers are more ruffled. During "flattening" (fig. 31D) the bird stretches out on the water with much of the body submerged. When the bill is held very near the water this appears to be an escape or fear-associated response, while a similar posture characterized by a distinctly uptilted bill (fig. 31C) seems to serve as an aggressive display (*"niedrige Angriffsstellung"* of Lehtonen 1970). Bill dipping is often performed from a high front or raised neck posture (fig. 31E), and in nonsocial situations it grades into normal peering behavior. A combination of bill dipping,

neck stretching, and head nodding characterizes "jerk swimming" (fig. 31F), which is usually performed by two birds approaching one another. In the short neck posture (fig. 31G) the bill is lowered toward the breast and the neck appears very thick. It is used by females before copulation and by males when showing a nest site. Probable primary threat displays include splash diving (by both sexes) and rushing (fig. 31H) over the water, often in rapid pursuit of another bird. However, rushing sometimes has also been observed in connection with apparent courtship. Other intensely aggressive displays are fencing (fig. 31I) in an upright posture while treading water and bow jumping (fig. 31J), a rapid alternation of bill dipping and an upward jump while fencing, with or without associated wing spreading. At times a "penguin dance," which is apparently a low-intensity form of bow jumping without the associated bill dipping, has been observed. Finally, in circle dancing (fig. 31K) two or more birds approach one another in a raised neck or high front posture and begin circling a common center point. As many as 32 birds have been seen participating in such circle dances, and up to four revolutions have been observed. This display appears to be a ritualized threat behavior. Other behaviors, such as "mock sleeping" and "rolling preens," are only questionably ritualized into signals, though they sometimes occur during social display.

Copulation is preceded by "search swimming" by the female; she swims close to shore in a pronounced short neck posture, typically following this with "going ashore," and lying down while remaining in her short neck posture. These behaviors are the only ones that consistently precede copulation, although jerk swimming is also fairly frequently performed (Sjolander 1978). The mechanisms of actual pair formation are obscure in this and other loon species, though Dunker (1975) believed that social display in flocks may serve as a kind of premating courtship and may help to synchronize as well as induce pair bonding. He noticed that the highest incidence of spring-to-fall flocking occurs in August, when territorial calling is at a low incidence and when a good deal of circle dancing may be seen. However, Sjolander (1978) questioned the possible sexual significance of such social interactions during flocking and regarded the major purpose of late-summer flocking as foraging. He believed that pair bonding might occur in spring, when mate-seeking females visit territories.

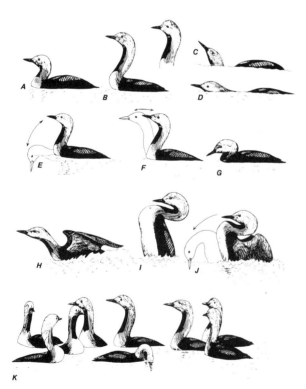

31. Social behavior of the arctic loon (mainly after Sjolander 1978): A, normal swimming; B, raised neck and raised neck with crown raising; C, bill tilting in flattened posture; D, flattened posture; E, bill dipping; F, jerk swimming; G, short neck posture; H, rushing; I, fencing; J, bow jumping; K, circle dancing.

Reproductive Biology

BREEDING SEASON AND NESTING SUBSTRATE. Peterson (1979) reported that nesting in the Yukon-Kuskokwim delta area of Alaska began about a week after the peak arrival of pairs on ponds, with 50 percent of the nests begun by June 9, 11 days after peak arrival. All clutches were completed during the period June 4–22. Davis (1972) stated that arctic and red-throated loons arrived on the Hudson Bay coast during the first week of June and moved to inland breeding ponds as these thawed. Nest building began as shoreline nesting areas emerged from spring floodwaters, with nest building and egg laying beginning a day or two later. In one year this occurred in mid-June, and in another in late June. Seven egg dates for Ontario range from June 23 to July 4 (Peck and James 1983). On Victoria Island the birds appear by mid-June, and eggs may be laid as early as June 19 (Parmelee, Stephens, and Schmidt 1967). Lehtonen (1970) found that in Finland nesting began 5–20 days after initial occupation of the territory.

Davis (1972) has analyzed the characteristics of nesting ponds of this species, which were summarized earlier. He found that island sites were preferred to shoreline locations, and this characteristic also has been found typical in Alaska (Bergman and Derksen 1977), Sweden (Lindberg 1968; Sjolander 1978), and Finland (Lehtonen 1970). Island sites appear to be more successful than shoreline sites, perhaps because they are less accessible to foxes. Bergman and Derksen (1977) found that islands used by arctic loons for nesting averaged larger than those used by red-throated loons (106.9 vs. 4.1 square meters) and were usually of land substrate rather than islandlike platforms of vegetation. However, Peterson (1976, 1979) found that among 79 nest sites the use of shoreline and island locations did not differ significantly. She reported average distances of 297 and 386 meters between neighboring nests in two different years.

NEST BUILDING AND EGG LAYING. The nest site is apparently selected by the male in this species (Sjolander 1978; Lehtonen 1970). Davis (1972) found that arctic loons that nest on wet substrates construct a nest platform of mud and aquatic vegetation, while nests on dry substrates are simply scrapes on flat ground, with the weight of the bird's body forming a shallow depression. Eggs are normally laid at two-day intervals, and the normal clutch is two eggs. However, Lehtonen (1970) noted that among 85 clutches there were 12 single-egg clutches and 2 three-egg clutches, or an average clutch size of 1.88 eggs. He also noted a range in laying intervals of from 2 to 7 days. Seven of 85 clutches were replacements in his study. Davis (1972) found an average clutch size of 1.74 eggs for 91 clutches exclusive of renesting attempts, the number of which were not specifically noted. Peterson (1976, 1979) found an average clutch size of 1.93 eggs for 43 nests and did not mention any possible renesting efforts.

INCUBATION AND BROODING. Both sexes incubate, though most is done by the female (Lehtonen 1970). Davis (1972) concluded that two-egg clutches were less likely to be lost than one-egg clutches, and even when birds with two-egg clutches lost one egg the remaining one had a higher probability of survival than did eggs from single-egg clutches. In his study area the major predators were birds, especially parasitic jaegers (*Stercorarius parasiticus*). He concluded that two-egg clutches are advantageous for arctic and red-throated loons because of the occasional raising of the second chick in broods of two and the better chance of hatching at least one. He regarded the immediate onset of incubation and the resulting staggered hatching period as an adaptation ensuring that one of the two chicks would have an advantage in feeding competition with its sibling and increasing the probability that at least one chick would survive to fledging. Sjolander (1978) determined a 27–29 day period between the laying of the second egg and the hatching of the first. He observed that the empty shells were carried out on the water and dropped away from the nest after hatching, an adaptive behavior since the nest is then used for brooding for several days.

GROWTH AND SURVIVAL OF YOUNG. Davis (1972) found that young arctic loons had a higher survival rate than young red-throated loons at his McConnell River study site, owing to better survival of the first-hatched chick in each nest as well as to increased survival of the second-hatched in nests hatching both eggs. His overall estimated rate of chick survival was 56 of 90 hatched chicks, or 62 percent. Lehtonen (1970) reported that 34 out of 72 hatched young survived to fledging, or 47 percent. The fledging period of chicks in the wild was estimated at 59 to 64 days by Lehtonen (1970) and at 46 to 53 days by Parmelee, Stephens, and Schmidt (1967). Sjolander (1978) observed that a 62-day-old chick was unable to take off in calm weather, though a 72-day-old bird flew without apparent difficulty. He also observed that the chicks remained on the nest the first day after hatching, and they were fed in decreasing amounts on the nest during the first 4 days. Brooding on the water was first seen at 2 days, with each parent typically taking a single chick under its wing. The young were fed not only until they fledged but also considerably beyond, with the latest observed feeding being of a 91-day-old juvenile. The earliest age at which successful food catching was observed was 36 days. The chirping calls of the young gradually developed into a garbled version of wailing; croaking was initially heard at 55 days and the low call at 64 days. The juveniles remained in their parents' territory until the autumn migration, or 124 days of age. However, after 90–100 days parental contact became limited, and by about 80 days of age most young began sleeping away from their parents.

BREEDING SUCCESS AND RECRUITMENT RATES. Lehtonen (1970) reported that 34 young were fledged from a total of 159 eggs laid in his study area, representing a breeding success of 22 percent. These were produced from 85 nests, a production rate of 0.40 young raised per nesting pair. Davis (1972) reported a hatching success of 94 eggs out of 138 (68.1 percent) and a rearing success of 62 percent, for an overall breeding success of 42.2 percent. With an average clutch size of 1.74 eggs, the resulting production rate should be approximately 0.73 reared young per nesting pair, exclusive of any re-nesting efforts. Nilsson (1977) estimated that arctic loons have an annual adult survival rate of about 89 percent and that a breeding success rate of 0.4–0.5 fledged young per pair would be needed to maintain the population. Nordstrom (1962, 1963) also estimated from banding recoveries a high survival rate and an average life expectancy of more than 8 years for birds 2 years old or older. There are still no good data on recruitment rates for this or any other species of loons, or any direct evidence as to when sexual maturity is attained and breeding begins.

Evolutionary History and Relationships

See the account of the red-throated loon for a discussion of the evolutionary history of this species.

Population Status and Conservation

There are no population estimates available for this species, either for North America or for the larger world range. Like the other loons, it is relatively vulnerable to oil pollution on migration and wintering areas; a summary of thirteen oil spills tabulated by Clapp et al. (1982) included 197 arctic loons among a total of 41,431 oiled birds, or 0.47 percent of the total. King and Sanger (1979) assigned an oil vulnerability index of 58 to the species. It should be possible to obtain good estimates of the North American arctic loon populations by winter surveys along the Pacific coast from southern Alaska to Baja California, but such figures do not seem to be available, perhaps because of the difficulty of identifying loons in winter. The summer population of arctic loons on the 95,000 square kilometer National Petroleum Reserve was in the vicinity of 28,000 birds in the late 1970s (King 1979), suggesting a breeding season density of about one bird per 4 square kilometers. If this is a typical density the arctic loon is almost certainly the most abundant of all the loon species, with a probable world population in the millions.

Common Loon (Great Northern Diver)

Gavia immer (Brünnich)

OTHER VERNACULAR NAMES: Islom (Danish); plongeon imbrin (French); Eistaucher (German); himbrimi (Icelandic); shiroeri ohamu (Japanese); gagara (Russian); somorgujo comun (Spanish); islom (Swedish).

Distribution of Species (See Map 3)

BREEDS from western and central Alaska (Seward Peninsula, Aleutian Islands, and the Brooks Range), northern Yukon, northwestern and southern Mackenzie, central Keewatin, northern Manitoba, northern Ontario, southern Baffin Island, Labrador, and Newfoundland south to northern California (at least formerly), northwestern Montana, western Wyoming, northern North Dakota, southern Minnesota, Wisconsin, Michigan, southern Ontario, New York, Vermont, New Hampshire, Maine, and Nova Scotia; also along both coasts of Greenland, Iceland, Scotland (in 1970), and (probably) Bear Island. Formerly bred to Iowa, Indiana, Ohio, and Pennsylvania.

WINTERS in North America primarily along the Pacific coast from the Aleutians south to Baja California and Sonora, and along the Atlantic and Gulf coasts from Newfoundland south to southern Florida and west to southern Texas; and in the western Palearctic along the Atlantic coast south to northwestern Africa.

Description (modified from Witherby et al. 1941)

ADULTS IN BREEDING PLUMAGE (sexes alike). Whole head and all of neck velvety black, slightly glossed green on back of head, neck, and throat and with mauvish tinge on sides of head, lower part of throat and entire neck much more strongly glossed greenish purple; in middle of throat a prominent line of short white parallel vertical streaks, on each side of lower neck a similar line of longer streaks not quite meeting in center of throat or back of neck; mantle black glossed greenish and thickly spotted with white, each feather having a pair of subterminal spots of more or less square shape, those on upper mantle very small, becoming large on lower mantle and still larger on scapulars, back, and rump with very small spots but upper tail coverts unspotted; breast and belly white, sides of upper breast glossy black streaked white, feathers being much like those forming white streaks on neck; flanks as upperparts, feathers with small white spots and those nearest breast streaked with black and white, with

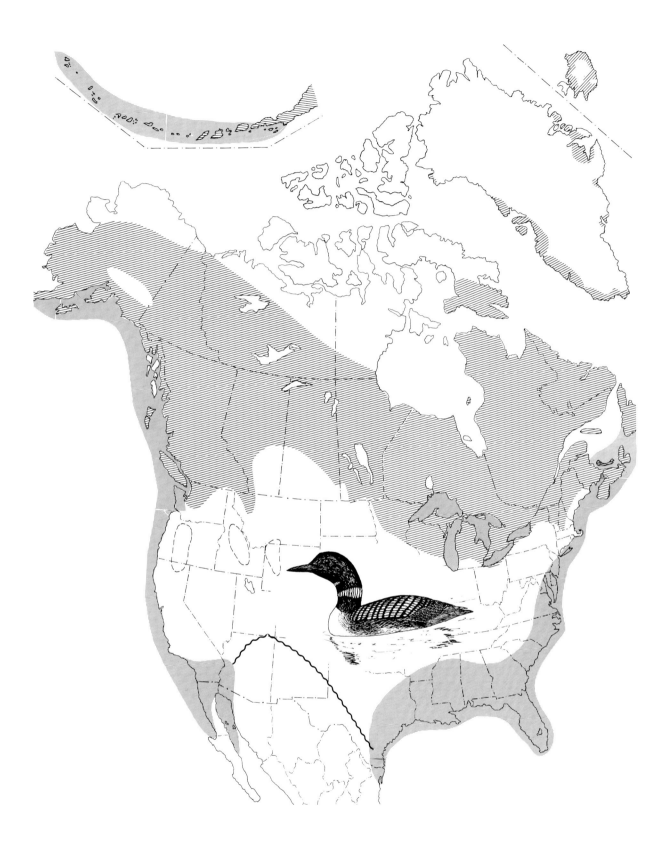

3. Current North American distribution of the common loon; symbols as in map 1. Peripheral or casual breeding areas are shown by a broken line.

white subterminal spots, or edged with white; under tail coverts brownish black tipped white; axillaries and under wing coverts as winter but central streaks darker; tail feathers (rectrices) glossy black with no white tips; primaries and secondaries as winter but outer webs and tips black glossed purplish; innermost secondaries with white spot on outer webs; primary coverts as primaries; wing coverts as rest of upperparts but not quite so glossy, greater coverts sometimes with a second pair of duller white spots.

WINTER PLUMAGE. Forehead, crown, and back of neck dark brown, feathers of neck somewhat downlike and whitish bases often ill concealed; whole of rest of upperparts dark brown, feathers margined grayish brown to gray; outermost underscapulars with dull white edges or white subterminal bands; between nostrils and eye feathers brown; tipped white; under eye, chin, throat, and rest of underparts white, but feathers along sides of neck tipped brown, giving somewhat mottled appearance, feathers of sides of lower throat and upper breast with brown centers and white edges, giving somewhat streaked appearance; sides of breast and flanks brown, some feathers with white outer webs and edges; across vent narrow band of brown; under tail coverts brown, broadly tipped white; axillaries white with dark brown median streaks; under wing coverts white, lower ones with pale brown median streaks; tail feathers dark brown, tipped white; primaries and secondaries with blackish brown tips and outer webs, dark brown shafts, paler brown inner webs; wing coverts as rest of upperparts but many lesser with small ill-defined whitish subterminal spots and innermost sometimes with white subterminal bars.

JUVENILES. Much like adult winter but edges of feathers of upperparts paler gray and more prominent (feathers more rounded, less square tipped than in adult and gray margins more even); under eye, ear coverts, and sides of neck more finely streaked brown; throat more or less finely streaked or freckled brown; flanks paler brown; axillaries with paler brown and broader median streaks; under tail coverts brown, tipped grayish brown instead of white as adult; tail feathers tipped pale brown instead of white.

DOWNY YOUNG. First down rather short, thickly covering whole body and especially thick on underparts. Upperparts dark mouse brown, whole chin, throat, and foreneck grayish brown, all around breast and belly brownish gray shading to white that covers whole of center of underparts. Second down like first but decidedly paler on upperparts. Iris reddish brown, legs and feet dark gray, bill dark gray to dusky (Fjeldså 1977).

Measurements and Weights

MEASUREMENTS. Wing (unflattened): males 339–81 mm (average of 10, 360.9); females 315.8–360.0 mm (average of 10, 337.9). Culmen: males 72.5–90.1 mm (average 81.5); females 73–86 mm (average 79.9). Eggs: average of 20, 90.9 x 57 mm (Palmer 1962).

WEIGHTS. Adult males (all seasons) 2,200–6,100 g (average of 12, 3,523); females (all seasons) 1,468–5,662 g (average of 13, 2,677) (various sources). Estimated egg weight 167 g (Schönwetter 1967). Newly hatched young weigh 52–95 g (Fjeldså 1977).

Identification

IN THE FIELD. In breeding plumage this species is likely to be confused only with the yellow-billed loon, and distinguishing features between the species are outlined in that species' account, although the simple combination of black bill and white neck collar should suffice for birds in breeding plumage. Birds that are not in breeding plumage are more difficult. Its large size and relatively heavy, tapered bill help to separate the common loon from the arctic and red-throated loons, though distinction from the arctic loon is more difficult. A pale "eyebrow" area is usually present in the common loon in this plumage, but not in the arctic loon. Compared with the common loon, the yellow-billed loon in nonbreeding plumage has much paler throat flecking and foreneck coloration, the sides of the neck and posterior auricular (ear covert) area are much paler, there is a more distinct "eyebrow," and the mantle area is more strongly cross-banded and generally paler throughout. When swimming, common loons tend to hold the bill more level, rather than uptilted, and the bill itself retains a dusky culmen ridge for most of its length.

IN THE HAND. Wing length and culmen length measurements readily separate this species from all others except the yellow-billed loon. The common loon always has a variably dark bill, which ranges from entirely blackish in breeding plumage to one that is variably paler in younger and nonbreeding birds, but the culmen invariably has a dusky culmen ridge along its entire length. Nearly all bill measurements of the common loon and *adamsii* exhibit considerable overlap, but as noted in the account of the yellow-billed loon the forward extension of forehead feathers relative to the tubercle in the middle of the nostril (typically beyond in *adamsii* and barely reaching the tubercle in the common loon) and the forward extension of feathering in the chin (reaching a point directly below the nostrils in *adamsii* but not in the common loon) provide simple and useful distinctions. Furthermore, the color of the primary shafts in *adamsii* is invariably paler, varying from white to yellowish brown, while in the common loon the shaft color is dark brown. Another useful distinction is that in *adamsii* the two rami of the lower mandible are completely fused, often forming a bulge at the point of fusion, while in the common loon the two rami show a distinct groove along their path of fusion (Burn and Mather 1974). Finally, though the feature may be of limited use, the eye of the common loon reportedly is slightly larger than that of *adamsii*, especially in height, with the proportionate difference greater than in bill measurements (Binford and Remsen 1974).

Ecology and Habitats

BREEDING AND NONBREEDING HABITATS. McIntyre (1975) analyzed the use of lakes by breeding loons in Minnesota to try to determine essential or optimal habitat features, primarily including lake size, the presence of nesting islands, and human disturbance factors. She found that very few lakes smaller than 10 acres (4 hectares) supported territorial pairs, but that nearly all lakes larger than 50 acres (20.2 hectares) had territories. Few pairs on lakes smaller than 25 acres (10.1 hectares) successfully raised young, and none of these raised two-chick broods, but over half of the pairs on lakes larger than 50 acres raised young. She also found that adding artificial nesting islands to lakes without them did not significantly enhance the value of those lakes to loons, though the presence of islands did influence nest site choice. Recreational use of lakes was found to have no effect on the choice of a lake for territorial use by loons, and no direct evidence was found that recreational use had an adverse effect on loon productivity. Only on small lakes having intensive recreational use was a decline in two-chick broods evident. Water clarity and associated prey visibility were not directly tested, since all the lakes studied were eutrophic, and McIntyre concluded that the ultimate factor in breeding habitat selection by common loons may be lakes with an abundant supply of small fish as well as water clear enough to allow efficient foraging. Small, eutrophic lakes are used provided these conditions exist and shoreline development and recreational use are kept at a minimum. Trivelpiece et al. (1979) found that loons in Adirondack Park used public lakes less than private ones, with nearly 50 percent of the breeding pairs occurring on public lakes accessible only by trail or on private estates with restricted usage. Vermeer (1973) similarly reported an inverse correlation between the amount of human disturbance and the numbers of breeding loons, as well as a positive correlation between

numbers of breeding loons and the presence of islands for nesting in Alberta lakes. Sawyer (1979) noted a reduction in breeding loons in wildlife management units of Maine with a high degree of recreational development, and similar trends in other areas (New Hampshire, Wisconsin, etc.) have been attributed to increasing human disturbance.

Nonbreeding habitats consist of coastal areas, especially bays, coves, channels, inlets, and other shallow sites, and also lakes or reservoirs, less commonly rivers. Information in table 5 suggests that the highest coastal densities occur in bays and that deeper waters are little used at any season. In studies along the Virginia coast, McIntyre (1975) found that wintering loons feeding during flood tide were in deeper waters and those feeding during ebb tide were in shallower water, apparently feeding on prey species that lagged behind the receding tide. Thus loons stayed in place or even moved closer to shore during ebb tides. Most feeding depths were from 6 to 15 feet, comparable to depths observed during summer. During migration in spring almost any open waters seem to be used, especially ice-free rivers or lakes with river inlets. Fall stopover sites appear to be traditional lakes along the migratory routes.

SOCIALITY AND DENSITIES. Breeding densities of common loons are quite variable (table 37), apparently depending on general habitat suitability for breeding, but in all cases typical densities are appreciably lower than those reported for red-throated loons and arctic loons. This certainly is in part a measure of the larger body size and increased food requirements for breeding common loons. Barr (1973) estimated that a loon pair and two chicks eat about 1,050 kilograms of fish during a 15 week breeding season, and thus food supplies are likely to place an upper limit on breeding density in most areas. The actual territory size is difficult to measure in most cases, but Minnesota studies suggest that territories often center on islands, with a minimum territorial requirement of 100–200 acres (40–80 hectares) per pair. Many pairs that occupied small lakes used an adjacent small lake for supplementary foraging, and although they were not observed defending the second lake it may also have been part of the breeding territory (McIntyre 1975). Olson and Marshall (1952) judged that 52 territories they studied ranged from 15 to more than 100 acres, and they invariably included a suitable nesting site and an area of open water for foraging, escape, display, and loafing.

COMPETITORS AND PREDATORS. Except for a possible small degree of breeding contact between the common loon and the yellow-billed loon, it seems unlikely that any other loon species provides any significant degree of

Table 37: Average Breeding Densities and Production Rates for Common Loons

	New Hampshire,[a] 1976–77	Minnesota,[b] 1972, 1976	Minnesota,[c] 1975–76	Saskatchewan,[d] 1973–74	New York, 1977–78
Territorial pairs	89	25.5	73	99	80.5
Water area (ha)	44,739	1,098	5,304	3,847	28,229
Hectares per territorial pair	502.9	43.7	72.7	39.0	350.6
Total young fledged	43.5	7.5	35.5	53	67.5
Young fledged per territorial pair	0.48	0.29	0.49	0.53	0.84
Hectares per fledged young	1,048	151	148	74	417

NOTE: Adapted from tabular material in Trivelpiece et al. 1979.
[a]Hammond and Wood 1977.   [b]McIntyre 1978.   [c]Titus 1979.   [d]Yonge (cited in Titus 1979).

competition with the common loon. Reimchen and Douglas (1980) reported only rare interactions between common and red-throated loons during summer on a lake in the Queen Charlotte Islands, and when interactions occurred the common loons were dominant except when the red-throated loons were close to shore.

Few if any predators attack healthy adult common loons, but eggs and young are taken by a large number of predators. McIntyre (1975) saw American crows (*Corvus brachyrhynchus*) eating loon eggs and judged from track evidence that raccoons (*Procyon lotor*) and skunks (*Mephitis mephitis*) might be significant mammalian predators of nests. Olson and Marshall (1952) were usually unable to determine the causes of nest predation in their studies, but they judged that at least crows, minks (*Mustela vison*), and muskrats (*Ondatra zibethicus*) might have been in part responsible. Fox, Yonge, and Sealy (1980) determined that 82 percent of the egg predation they observed resulted from herring gulls (*Larus argentatus*) and common ravens (*Corvus corax*), and minks were the only mammalian nest predators they identified. Chick losses are sometimes fairly high, but such factors as starvation, freezing, and other nonpredator dangers are probably more often direct causes of chick mortality than is predation on healthy chicks. Olson and Marshall (1952) reported one case of a chick taken by a large northern pike (*Esox lucius*). In general the loss of eggs to predators seems a rather secondary cause of breeding failure in common loons (table 32).

General Biology

FOOD AND FORAGING BEHAVIOR. There are few comprehensive studies of foods and foraging in this species. Madsen (1957) was able to study only 3 specimens from coastal waters off Denmark in winter, and he found only fish remains. He summarized the available information on the species' diet and listed the previously reported prey species, of which codlike forms (*Gadus* spp.), sculpins (*Cottus* spp.), herrings (*Clupea* spp.), gobies (Gobiidae), and flatfish (Pleuronectidae) seem to figure prominently. Cramp and Simmons (1977) have summarized more recent European literature, and Palmer (1962) has done the same for North American sources. The North American literature includes more freshwater fish than does the European information, and most of these are nongame varieties. Barr (1973) studied the feeding biology of common loons in the oligotrophic lakes of Algonquin Park, Ontario, and concluded that common loons are facultative predators of fish and crustaceans and select the most readily available prey species, mainly in the size range of 5–20 centimeters. In that area the yellow perch (*Perca flavescens*) was most commonly taken, with coregonids of secondary importance. Stomach contents of 27 loons from Ontario and the Great Lakes states included yellow perch and representatives of at least six other fish genera as well as unidentified cyprinids and catastomids (Olson and Marshall 1952).

Foraging is performed by extended dives and underwater chases, their length probably depending upon such variables as water depth, density, size, and elusiveness of prey, and possibly other variables. Mean diving times from various areas range from 34.0 to 64.4 seconds (Reimchen and Douglas 1980), which certainly is well below the species' reported capabilities for prolonged underwater submersion (tables 11 and 12). Reimchen and Douglas (1980) reported a success rate of 50.8 percent on 187 foraging dives, with sticklebacks (*Gasterosteus aculeatus*) the sole prey species they saw taken.

MOVEMENTS AND MIGRATIONS. There are probably few if any movements by adults out of their large territories during the breeding season. However, young birds begin to leave their natal lakes when they are 11–13 weeks old and apparently begin to move to larger lakes. McIntyre (1975) found that all but 2 of 23 young birds moved out of their parents' territories when 11–14 weeks old, and the 2 remaining ones remained until they were more than 5 months old. This movement probably normally occurs shortly after fledging and sometimes involves displacements of several miles.

Migration in common loons occurs over a broad geographic front and is much less confined to coastal areas than in the other loons. Fall migration occurs over several months, with departures from interior breeding areas beginning in late August or early September and continuing until freeze-up. The Great Lakes, especially Lake Huron, are a major fall concentration area. There are three major spring migration routes, including a Pacific coastal route, an Atlantic coastal route, and a third route funneling birds from the Gulf and Atlantic coasts through the Great Lakes to northwestern Canada. Early coastal departures begin in mid-March, and peak numbers pass along New England in late May. Many birds have reached their Minnesota breeding grounds by late May, with arrival times partly dependent on thawing periods of the lakes. Early arrivals were birds that occupied territories and consisted either of already paired birds or of males whose mates arrived from a day or two to as much as two weeks later. It thus seems likely that most birds do not retain pair bonds over winter but instead renew them each spring through common territorial fidelity. However, observations in Iceland led Sjolander and Agren (1972) to believe that there the birds are permanently paired, since they arrived on breeding areas as pairs and began egg laying only 5 days later.

Social Behavior

MATING SYSTEM AND TERRITORIALITY. Most authorities agree that the pair bond in this species either is permanent or is renewed yearly as long as both members of the pair remain available, so that probably little actual pair-bonding behavior is necessary in adults after they initially acquire mates. However, birds of either sex that lose their mates may acquire new ones fairly rapidly, and territories that are vacated are often filled immediately (McIntyre 1975), suggesting that a "floater" population of available single birds or even pairs is probably available in at least some areas. McIntyre believed that initial breeding probably does not occur until the birds are over 4 years old, based in part on the incidence of nonbreeding birds in nuptial plumage on the breeding range. Like other loons, a high degree of territorial fidelity seems typical of this species, though it has not been so well documented as in arctic and red-throated loons. Such territorial site tenacity is indicated by a high use of the same nesting sites from year to year (Olson and Marshall 1952; Fox, Yonge, and Sealy 1980). Territories are established in early spring immediately after males or pairs return to their breeding areas, and the incidence of territorial defense gradually wanes over the summer. Sjolander and Agren (1972) judged that common loon territories are up to 25 hectares in area and, as noted earlier, Marshall and Olson (1952) made similar estimates.

VOICE AND DISPLAY. Vocalizations of this species have been described by Sjolander and Agren (1972) and summarized by Cramp and Simmons (1977). Yodeling (or the long call) is the most familiar and variable call, consisting of a prolonged series of notes with sharp breaks, usually with the pitch rising on the third note and undulating thereafter. It is uttered only by territorial birds, especially early in the reproductive cycle. A probable low-intensity version of this is the wail, which sometimes resembles a wolf's howling and is typically uttered during less intense territorial encounters. A tremolo call is produced during agitated disturbance, especially where there may be a tendency to flee; it often functions as a distraction display when uttered as a duet by a pair defending a nest or young (Barklow 1979). It exhibits a high degree of modulation in frequency as well as amplitude and thus has potentially high information content. A moaning call resembling human moaning may be uttered when one adult is searching for its mate or young. It is often repeated at irregular intervals and carries well in spite of its apparent low amplitude. Chicks utter chirping sounds and other high-pitched notes that are still only poorly described.

Displays in common loons have been described by several authors (e.g., Olson and Marshall 1952; McIntyre 1975), but those of Sjolander and Agren (1972) seem most comparable to descriptions of other species, and they have carefully separated territorial displays from courtship. The only behavior other than copulation that these authors interpreted as courtship was a formalized bill dipping followed by mutual splash diving; all the rest were referred to territorial signaling. These displays include neck stretching ("raised neck" posture) (fig. 32D), often with raising of the forehead feathers and sometimes with associated wailing or long calling (fig. 32B), bill dipping (fig. 32C), and splash diving, all of which are very frequently performed. Circle dancing, with two or more birds circling a point while perform-

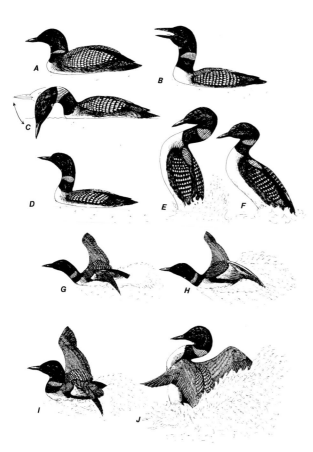

32. Social behavior of the common loon (after various sources): A, normal swimming; B, calling; C, bill dipping; D, neck stretching and crown raising; E, F, early and late stages of upright or fencing posture; G–J, aggressive rush followed by upright posture with wing raising.

ing neck stretching, bill dipping, and diving, is also a common ritualized form of aggressive behavior during territorial confrontations. More intense territorial displays include the "upright" or fencing posture (fig. 32E,F), in which a penguinlike posture is maintained by treading water, sometimes with wing spreading (fig. 32J) and occasionally with jumps out of the water. This posture is often preceded or followed by long rushes over the water (fig. 32G–I), with the wings opened and flapping. McIntyre (1975) described some additional possible displays, such as jerk diving, "hunched" posture (similar to flattening in arctic and red-throated loons), and others, but these in general do not seem clearly defined as to occurrence or possible functions.

As in other species, copulation is preceded by the female's search swimming along the shoreline, finally going ashore and crouching down. The male typically follows the female closely and immediately attempts to copulate. Although special copulation platforms have been mentioned by some authors, they do not appear to be characteristic of the common loon or other loon species (Sjolander and Agren 1972).

Reproductive Biology

BREEDING SEASON AND NESTING SUBSTRATE. In Ontario the records of eggs from 113 nests extend from May 11 to August 25, with 57 records concentrated from June 2 to 23. In a Minnesota study, nest establishment began in early June and peaked June 15–28. Hatching extended over a 16 day period, July 11–26, with the late-hatched nests the result of renesting efforts (Olson and Marshall 1952). Broods have been observed in the Aleutian Islands as early as June 6 (F. Zeillemaker, pers. comm.). McIntyre (1975) found the egg-laying period in Minnesota to extend from May 7 to June 28, with almost half the first nests initiated in the third week of May. Renesting efforts typically are begun 5–14 days after the loss of a nest, whether the first or second effort, and probably average about 10 days. A maximum of three laying cycles was reported by McIntyre (1975). Of 15 replacement clutches, 11 pairs chose new nest sites and 4 used the old site (Palmer 1962). Of 15 sites that had previously been successful, 10 pairs used the same site the following year, while of 32 sites that were unsuccessful, 28 pairs changed their site for the following nesting effort, in McIntyre's study.

Nest site characteristics have been studied by Olson and Marshall (1952), who noted that for 54 nest locations the major criteria seem to be sheltered locations, availability of small islands, and proximity to open water. Cover density was only of incidental significance, although some kind of vegetative cover was present in nearly all cases. McIntyre (1975) found that proximity to food did not seem to influence nest site choice, but when islands were available 88 percent of all nests were built on them. Lakes with islands also significantly increased the probability of nesting success, though increased distance of the island from shore did not. Vermeer (1973) found that 20 of 26 nests studied were in sites sheltered from wave action, and all but one afforded easy underwater exit by the sitting bird. Nineteen were on islands less than 2 acres in area, and only one was on the mainland. There was a positive relation between the distribution of breeding pairs and of islands and a negative correlation between abundance of breeding pairs and relative human disturbance.

NEST BUILDING AND EGG LAYING. Evidently either sex might select the nest site (McIntyre 1975), although

Sjolander and Agren (1972) suggested that the male does this. Certainly both sexes participate in nest building, and McIntyre (1975) observed that one pair spent 4 days between the onset of building and the laying of the first egg of an initial nest; replacement nests may be built more rapidly. Eggs are normally laid at 2 day intervals, and in most cases the clutch is of 2 eggs. Peck and James (1983) reported two-egg clutches for all of 91 Ontario nests, and McIntyre (1975) indicated that 51 Minnesota clutches had an average clutch of 1.67 eggs. There are a few records of two females' laying in the same nest, raising the clutch to 3 or 4 eggs.

INCUBATION AND BROODING. Incubation begins with the laying of the first egg and is performed by both sexes, with participation fairly evenly divided, though the male possibly does much of the daytime incubation. There is a very high level of nest attentiveness, the nest usually being left only for territorial defense. McIntyre ascertained four incubation periods to be 26 to 31 days, averaging 27.5. Peck and James (1983) mentioned two nests with incubation periods of 29 days, and likewise Olson and Marshall (1952) reported two nests as having 29 day incubation periods. The eggs typically hatch a day apart, and the chicks usually remain in the nest for a day after hatching. Brooding on the back of the parent begins while the chicks are still in the nest, and they often eat nest vegetation before leaving the nest (McIntyre 1975).

GROWTH AND SURVIVAL OF YOUNG. At the time the chicks leave the nest they can make short dives of a few seconds, and they soon begin to beg for food by pecking at the base of the parents' bills. When they are 1 or 2 weeks old they begin to capture some of their own fish, although their efficiency rate is extremely low (Barr 1973). By the time the chicks are 6 to 7 weeks old the adults begin to leave their territories, but they continue to feed their chicks until fledging or until they leave their natal lakes. Fledging probably occurs at about 11 weeks of age, although by 2 months of age the flight feathers are nearly as long as in adults and flight might be possible then (McIntyre 1975). Most brood mortality probably occurs within the week following hatching, and very little occurs after the chicks are about 2 weeks old. Causes of early brood losses are still very uncertain,

Table 38: Reported Reproductive Success Rates in Common Loons

	Minnesota,[a] 1978	Vermont,[b] 1978	New Hampshire,[c] Eight-Year Average	Saskatchewan,[d] Three-Year Average
Lakes reported	59	33	—	1
Lakes with territorial pairs	55	20	54.8	1
Total territorial pairs	68	23	94.9	98.6
Total nesting pairs	55	16	69.1	76
Pairs with known hatching results	50	—	—	76
Total broods hatched (% of nests)	44 (80%)	8 (50%)	49.5 (72%)	39.3 (52%)
Total chicks hatched	65	12	66.8	—
Average brood size at hatching	1.48	1.5	1.5	—
Hatched chicks per nesting pair	1.18	0.75	0.96	—
Hatched broods per nesting pair	0.8	0.5	0.64	0.52
Pairs with known fledging results	46	—	—	—
Total fledged broods	40	—	—	35
Total juveniles fledged	54	12	55.4	53
Brood size of fledged young	1.35	1.5	1.24	1.48
Ratio of fledged/hatched young	0.92	1.0	0.83	—
Juveniles fledged per nesting pair	0.98	0.75	0.79	0.73
Juveniles fledged per territorial pair	0.79	0.52	0.58	0.53

[a]McIntyre 1979.   [b]Metcalf 1979.   [c]Unpublished data for 1976–83, provided by J. Fair (in litt.).
[d]Fox, Yonge, and Sealy 1980, 1973–75 data.

though predation by fish has been implicated as one possible factor (Olson and Marshall 1952).

BREEDING SUCCESS AND RECRUITMENT RATES. Breeding success rates for this species have been summarized in tables 37 and 38, which suggest that in various studies covering a variety of areas the number of young reared per territorial pair has averaged close to 0.5 (range 0.20–0.84). Common loons seem to have a higher potential for renesting than the other loon species, but they also require considerably larger breeding areas than do arctic or red-throated loons. Estimates of annual recruitment rates are still lacking, although Zimmer (1982) estimated that Wisconsin's loon population consisted of 40 percent successful breeders, 32 percent unsuccessful breeders, and 28 percent nonbreeders. If this is a reliable statistic, a production of 0.5 per territorial pair would represent a recruitment rate of 12.8 percent. Munro (1945) noted that the late summer population of loons in his study area (possibly including some nonbreeders) was 150 adults to 21 young, or a maximum recruitment rate of 12.2 percent. Such recruitment rates might indeed be typical of the species, based on what is known of adult survival rates in the arctic loon, and some indirect estimates of adult survival (based on a variety of breeding success data) suggest an annual survival rate in the range of 80–85 percent.

Evolutionary History and Relationships

See the yellow-billed loon account for a discussion of this topic.

Population Status and Conservation

There are still no good continentwide estimates of common loon populations, although several states have now instituted annual breeding surveys. Minnesota probably has the largest breeding loon population south of Canada, possibly numbering as high as 5,000 pairs and representing a fairly stable population. Maine's population is probably next largest and also is apparently stable. It has numbered in the vicinity of 1,100 pairs since the late 1970s and possibly totals about 3,000 birds. In New Hampshire the population rose from 271 birds in 1976 to 363 in 1983, and the Vermont population declined from 56 in 1978 to 34 in 1983. New York supports about 200 nesting pairs, mainly in Adirondack Park, where the possible effects of acid rain have not yet become apparent. Wisconsin's population is approximately 1,300 birds, and about 100 are left in the Upper Peninsula of Michigan. The Montana population is about 125 birds,

and a few pairs probably breed annually in northern North Dakota, northwestern Wyoming, Idaho, and northern Washington. Thus the current population south of Canada is probably no more than 15,000 birds. Increased human recreational use of lakes in prime breeding areas pose problems for loons, as do pesticides (Fox, Yonge, and Sealy 1980), the possible effects of acid rain on prey populations, and water-level fluctuations in reservoirs. However, considerable success has been attained with artificial nesting sites (moored rafts) in areas lacking good natural sites or having extreme water-level fluctuations. Like other loons, this species is susceptible to being caught in fishing nets and has suffered local mortality from botulism (on Lake Michigan). Clapp et al. (1982) reported a loss of 374 common loons among 33,255 birds (1.12 percent) involved in 12 different oil spills. An oil vulnerability index of 47 has been assigned the species by King and Sanger (1979). Recent observations by Alexander (1985) indicate loons may have a serious sensitivity to mercury poisoning as well; perhaps as many as 7,500 birds died in the Saint George Sound area of Florida during the winter and spring of 1983, and most of the birds autopsied showed lethal or near-lethal levels of mercury.

Yellow-billed Loon (White-billed Diver)

Gavia adamsii (Gray)

OTHER VERNACULAR NAMES: Hvidnaebbet lom (Danish): plongeon à bec blanc (French); Gelbschnabel-Eistaucher (German); svabrusi (Icelandic); hashijiro (Japanese); beloklyuvaya gagara (Russian); vitnäbbad islom (Swedish).

Distribution of Species (See Map 4)*

BREEDS in North America from northern and western Alaska (south to Saint Lawrence Island and the southern Seward Peninsula) east to Banks, Victoria, and Prince of Wales islands and northern Keewatin, and south to east-central Mackenzie and east-central Keewatin; and in Eurasia from extreme northwestern Russia east to Siberia (including Novaya Zemlya). Summers

*This form is considered by some (e.g., Dementiev and Gladkov 1968) to be a subspecies of *G. immer*, and at least in Eurasia the two forms apparently exhibit nonoverlapping breeding ranges. Proof of breeding sympatry in North America is apparently lacking, though there appears to be a broad zone of contact and possible overlap in northern Canada (Godfrey 1966).

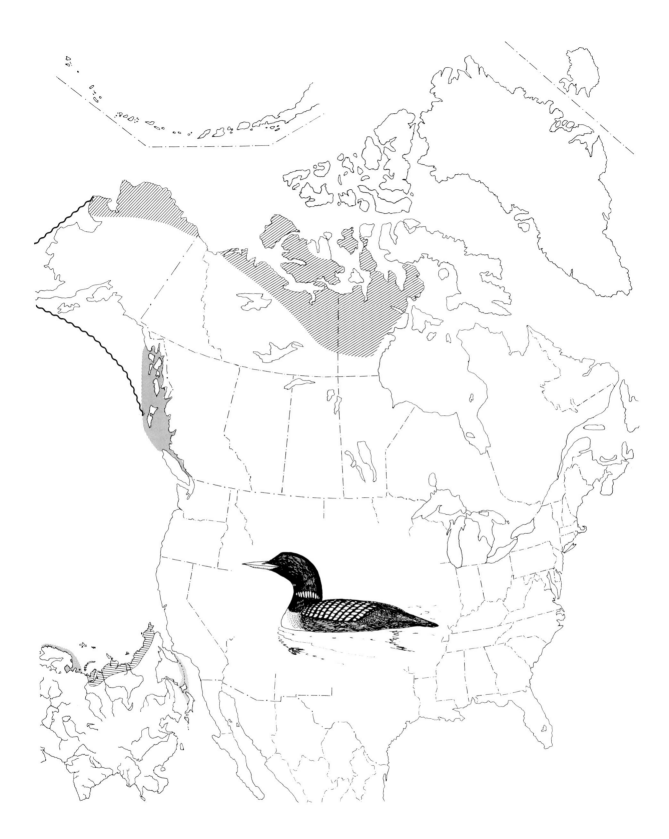

4. Current North American distribution of the yellow-billed loon; symbols as in map 1.

outside the breeding range east to northeastern Keewatin (Melville Peninsula) and northern Baffin Island, and south to southern Mackenzie (Great Slave Lake) and southern Keewatin.

WINTERS in North America along the Pacific coast of Alaska, casually south in coastal areas to California and extreme northern Baja California, and inland to Alberta; and in Eurasia in the breeding range, casually west to Greenland and south to southern Europe, China, Korea, and Japan.

Description (after Various Sources)

ADULTS IN BREEDING PLUMAGE (sexes alike). Similar to that of the common loon, but with a purplish rather than greenish gloss on the throat and neck, fewer and broader white stripes on the throat (6 vs. 12) and half-collars (ca. 10 vs. 18), and larger white markings on the upperparts, especially the scapulars. The white spots on the sides of the rump average smaller, however. The shaft color of the primary feathers is generally whitish rather than brownish. The bill is ivory to pale straw, the legs and feet are dark gray, and the iris is reddish brown. Rectrices usually 18.

WINTER PLUMAGE. Similar to that of the common loon but with the face and sides of the head paler, with whitish feathers extending from the cheeks to above the eye, large pale spots on the side of the nape whitish and more contrasting with the darker areas behind, less contrast between the darker upperparts and the underparts, and a more complete pale brown neck collar. The bill is pale cream, with a variable dusky cast over the bases and the sides of both mandibles near the head but the culmen pale (Cramp and Simmons 1977).

JUVENILES. Similar to the common loon but with a less definite head and neck pattern, more grayish brown upperparts, with noticeably wider and paler tips to the feathers of the mantle, scapulars, and wing coverts. The iris is browner than in adults and the bill is variable in color, but a pale culmen is typical.

DOWNY YOUNG. Similar to the common loon but relatively fuscous, more grayish brown on head, with a nearly white belly, but a drab gray underwing. The bill is also lighter (Fjeldså 1977).

Measurements and Weights

MEASUREMENTS. Wing (unflattened): males 366–88 mm (average of 10, 376.4); females 361–87 mm (average of 8, 368.7). Culmen: males 89–97 mm (average of 6, 91.3); females 87.5–96.0 mm (average of 8, 91.1). Eggs: average of 15, 90.45 x 56.4 mm (Palmer 1962).

WEIGHTS. Adult males (breeding season) 3,677–6,750 g (average of 15, 4,968); breeding females 4,025–6,350 g (average of 15, 4,759) (various sources). Estimated egg weight 160 g (Schönwetter 1967). Newly hatched young weigh about 85 g (Fjeldså 1977).

Identification

IN THE FIELD. The large size of this species makes separation from the red-throated and arctic loons relatively simple in any plumage. In breeding plumage it can be distinguished from the common loon by its entirely yellowish bill, which is somewhat dusky along its culmen ridge to the nostrils or somewhat beyond. The body appears somewhat bulkier and the neck somewhat thicker than in the common loon, and the bill is usually held uptilted. The white collar and patch on the upper foreneck have broader and fewer white lines, and the white spots on the mantle are fewer and larger, while they are reduced or absent on the upper rump and lacking from the longest upper tail coverts. The white streaks on the side of the chest average wider. In nonbreeding plumage the yellow-billed loon is substantially paler all over, especially on the head, where the only dark areas are the auricular patch and the crown, while the foreneck, sides of the neck, and hindneck are all distinctly paler than in the common loon. The pale "eyebrow" area is more extensive, tending to encircle the entire eye, and the dark auricular patch is likewise somewhat encircled by paler feathering. Even in immature birds the dusky ridge on the upper mandible extends no more than halfway to the tip of the bill (Binford and Remsen 1974). In juvenile or first-winter birds the bill-shape differences that characterize older birds are not developed, but in the yellow-billed loon the upperparts are a paler grayish brown, and the pale tips to these feathers are wider and paler. By the second winter pale areas in the scapular feathers begin to show up; these are larger and more distinct in the yellow-billed loon than in the common loon (Burn and Mather 1974).

IN THE HAND. In addition to the general characteristics mentioned above, some measurements are of value in separating common and yellow-billed loons. There is no actual difference in the angularity of the lower bill profile, but the yellow-billed loon is more likely to have a straight or slightly recurved culmen profile. The adult bill averages 9.7 mm (12 percent) longer in exposed culmen length in *adamsii* and is 3.0 mm (13.8 percent) higher at the anterior edge of the mandible, although overlap occurs. However, the bill of *adamsii* is usually more flat-sided and less convex, especially anterior to the nostrils, and the feathers at the base of the upper mandible in *adamsii* apparently always extend several millimeters beyond the nasal tubercle, while in the common loon they reach no farther than the tubercle (or a millimeter beyond). Further, in *adamsii* the chin feathers extend farther anteriorly, reaching a point below the posterior edge of the nostrils, while in the common loon they end about halfway to the nostrils. When measured relative to the anterior edge of the malar feathering, this distance is generally under 17 mm in the common loon and over 17 mm in *adamsii* (Binford and Remsen 1974; Burn and Mather 1974). According to Cramp and Simmons (1977) there are 20 rectrices in the common loon and only 18 in *adamsii*, although Palmer (1962) lists 18–20 rectrices for the common loon.

Ecology and Habitats

BREEDING AND NONBREEDING HABITATS. Breeding habitats of this species have been variously described as consisting of large tundra lagoons (Bailey 1948), low-rimmed freshwater tundra lakes and rivers (Snyder 1957), and larger and deeper lakes in tundra areas that provide a combination of nesting islands and food (Parmelee, Stephens, and Schmidt 1967). Similarly, Derksen, Rothe, and Eldridge (1981) noted that all their observations of this species in the National Petroleum Reserve of northern Alaska were on deep and open lakes or large, flowing bodies of water. However, Sage (1971) noted that the birds he studied in northern Alaska did not use rivers for breeding, nor did they nest closer to the coast than about 110 kilometers. He believed that neither size of the lake nor its food resources were critical, since long foraging trips seemed to be typical and one nesting occurred on a lake of only about 18 hectares. On a larger lake of 43 hectares nesting also occurred, and on neither of these sites were there any isolated hummocks or islands that might have provided optimum nesting sites. Sage judged that nesting habitat might be influenced locally by contact with common loons at the southern edge of the breeding range of *adamsii*, and it is possible that habitat use is slightly different in Asia, where common loons do not occur and both rivers and coastal areas seem to be favored for nesting. Thus, Portenko (1981) reported that during summer on Wrangel Island the species inhabits large lakes and rivers with deep pools below rapids where feeding opportunities exist. However, the latter sites may simply be used for foraging rather than nesting, judging from his accounts. Shallow to deep tundra lakes or ponds, with low islands or suitable shoreline vegetation, and with a fish supply within easy flying distance, are probably generally typical habitats.

Outside the breeding season the birds are primarily pelagic, usually seen on ice-free waters close to the breeding grounds, and are rather rarely found on freshwater areas far from the coast.

SOCIALITY AND DENSITIES. This species seems to be the most widely dispersed of all the loons, at least on its breeding grounds. Derksen, Rothe, and Eldridge (1981) estimated a summer density of only one bird per 10 square kilometers (or 200 hectares per pair) in one study area of the National Petroleum Reserve. In aerial and ground surveys of 42,000 square kilometers of coastal habitats of the National Petroleum Reserve, average loon densities during 1977 and 1978 ranged from 0.6 to 1.7 loons per square kilometer, with *adamsii* composing about 15 percent of the total population, or 0.09–0.13 bird per square kilometer (King 1979).

COMPETITORS AND PREDATORS. It is certain that this species has widespread breeding contact with the arctic loon and red-throated loon, and it is probable that local contacts also occur with the common loon in North America. However, it is easily the largest of the loons and should be able to dominate these other species socially. Too little is known of its foods to judge possible foraging overlaps among these species.

Nest predation, perhaps by foxes (*Alopex?*), has been reported by Sage (1971), and it seems likely that predation is most apt to be a problem during the nesting phase of the life cycle. Parmelee, Stephens, and Schmidt (1967) judged that adult loons have few natural enemies, though Eskimos kill a few. They judged that entanglement in fishing nets is a significant source of mortality in fledged birds.

General Biology

FOOD AND FORAGING BEHAVIOR. Relatively little is known about the food of this species. Cottam and Knappen (1939) reported on a small sample of 4 birds (summarized in table 13), for which various genera of Cottidae constituted most of the identified food. Bailey (1922) noted that a specimen from Alaska had been feeding on rock cod (*Sebastes*). Otherwise there is little definite information. Sage (1971) observed that the only apparent food fish in one Alaskan nesting lake was the stickleback (*Pungitius*), although nearby lakes held grayling (*Thymallus*). Sage observed that a pair with young exhibited intensive foraging behavior at about 9:00 A.M. and again in late afternoon and early evening. Members of this pair flew up to 5 kilometers from their nesting lake to forage, and he judged that on one occasion a bird might have flown as far as 30 kilometers to feed. Foraging dives on various nesting lakes that he observed averaged about 62–65 seconds, but sometimes the birds remained submerged as long as 90 seconds.

MOVEMENTS AND MIGRATIONS. Very little is known of the seasonal migrations of this species, which spends most of the year at sea, well away from most human contacts. Certainly during spring the birds move northward along the Bering Sea coastline, past Point Barrow, and then eastward along the arctic coasts of Alaska and Canada. They remain offshore on open leads until the breeding waters become ice-free (Sage 1971). However, surveys of the eastern Bering Sea and Gulf of Alaska have so far indicated only very low densities of the species, mainly in bays and areas within the continental shelf (see table 5). On Victoria Island the birds arrive by early to mid-June, and some probably remain until well into September, depending upon ice conditions (Parmelee, Stephens, and Schmidt 1967). Presumably most of the North American breeding population winters in waters off the coast of British Columbia and southern Alaska, but there is no good documentation of this. Likewise, there is little evidence of a major overland migration of birds from Pacific coast wintering areas across Alaska or the Yukon, though there are some spring records for birds in the vicinity of Fort Simpson on the Mackenzie River.

Social Behavior

MATING SYSTEM AND TERRITORIALITY. There is no reason to believe this species differs from the other loons in its mating system, namely monogamy with pair bonds presumably reestablished yearly. Sjolander and Agren (1976) suggested that a lifelong pair bond may be present in this and all other species of loons. Sage (1971) noted that on one lake where a pair with a brood had been found in 1969 he observed nests in exactly the same location during the next two summers. This suggests a strong site tenacity in the species, as is also typical of other loons. Sage believed that the same individuals return to each lake to breed in subsequent years, supporting the idea of a potentially permanent monogamous breeding system. Sizes of territories have not been carefully estimated for this species, but a breeding dispersion at least as great as that of the common loon is evidently typical, and pairs are separated by at least 600 meters (Cramp and Simmons 1977). Parmelee, Stephens, and Schmidt (1967) found two pairs of loons nesting on a single lake that was about 1.5 miles long, with the birds nesting on islands at opposite ends of the lake. Sjolander and Agren (1976) reported that the smallest lake on which they found nesting birds was about 20 hectares, and the areas of 10 of 14 measured nesting lakes ranged from 30 to 50 hectares. The greatest distance from a nest where the bow jumping display was seen used against an intruder was over 400 meters.

VOICE AND DISPLAY. According to Sjolander and Agren (1976) the calls of this species are very similar to those of the common loon but are uttered more slowly in at least some cases and are pitched about half an octave lower. They identified eight calls, including three variations of yodeling. Typical yodeling consists of a prolonged series of calls lasting up to about 2 seconds, with sudden breaks between tones. This call may be heard for more than 8 kilometers. In 5 ascertained cases the call was uttered by a resident male. A variation of yodeling is "short yodeling," in which the call stops after the first break, and another variation is "choked yodeling," in which the call is uttered by a bird in the "fencing posture"; apparently because of the posture the sound quality is also altered. The tremolo call, equivalent to the "laughing" note of the common loon, is similar to the call of that species but is lower in pitch and slower. In most cases it is uttered antiphonally between pairs, with the male's voice distinctly lower in pitch than the female's. Wailing is similar to yodeling but is a wolflike howl that lacks sharp breaks and remains on almost the same pitch, with a slight rise in the middle of the call. Moaning is similar to human moaning, but when performed at higher intensities a break to a higher note occurs in the middle. Moaning is performed by both sexes. The minor calls include a weak and low-pitched call that can be heard only fairly close by and a chirping note uttered by young birds.

Sjolander and Agren (1976) regarded the low call and moaning as contact calls between pairs, wailing and the various forms of yodeling as territorial calls, the tremolo as an alarm call, and chirping as a juvenile contact call that may be the ontogenetic precursor to yodeling and wailing.

Displays of the species have also been described and illustrated by Sjolander and Agren (1976); the postures appear to be virtually identical to those of common loons and thus are not illustrated here. Neck stretching ("raised neck") is performed by both sexes, with the bill pointed obliquely upward. In a slightly less neck stretched ("high front") position the forehead feathers are sometimes raised in a distinct bulge. The tremolo vocalization often accompanies this posture. In a "short neck" posture the bill is held downward, hiding the throat. This posture is performed by females, though males sometimes were seen in a similar posture when searching for nest sites. Bill dipping often occurs during neck stretching or the high front posture and is performed by both sexes in a manner more rapid than normal peering behavior. Both sexes may also perform "mock sleeping" with the bill under the wing feathers but the eyes open. Conspicuous, high-intensity aggressive displays include rushing, fencing, and bow jumping. During rushing the bird half flies, half runs over the water for as much as 450 meters. It is performed by both sexes, most often when an individual is within a meter of another bird. During fencing the bird rises almost vertically and treads water, with its bill held close to the breast and the wings either folded or extended. The bird may remain almost stationary in this posture or may even leap out of the water occasionally. While leaping, the head is moved forward and backward, and the movements may become elaborated into bow jumping, which is an alternation of fencing and bill dipping postures. Fencing or bow jumping are apparently typically performed by resident males and by males with young. Splash diving, done by both sexes, is a loud splashing dive, sometimes performed after rushing or fencing. Both sexes also perform jerk swimming, a repeated forward and backward movement of the head, especially as two birds approach each other. Finally, females perform search swimming while swimming along the shoreline in a short neck posture, or they may go ashore and lie down in the short neck posture while searching for suitable nest sites.

Sjolander and Agren (1976) regarded neck stretching as a general arousal signal, such as when the bird is frightened, the high front posture as similar but more aggressive, and both bill dipping and splash diving as indicating excitement or alarm. Fencing, bow jumping, and rushing are all essentially associated with territorial defense and are basically aggressive in motivation rather than related to pair formation or copulation.

The only behaviors consistently shown in connection with copulation are mutual jerk swimming and, immediately before copulation, search swimming and going ashore. The female typically goes ashore and may make nest building movements with her bill. The male typically follows the female, mounts her, and copulates for 9 to 12 seconds (2 observed cases). He then dismounts and returns to the water, while the female may remain ashore for several minutes (Sjolander and Agren 1976).

Reproductive Biology

BREEDING SEASON AND NESTING SUBSTRATE. In arctic Canada and Alaska the breeding season is fairly late and compressed. Sage (1971) judged that the eggs he found in northern Alaska had been laid during the first half of June, and Parmelee, Stephens, and Schmidt (1967) found one nest on Victoria Island in 1960 that they believed had been incubated since June 25 and probably begun by June 22. In 1962 another clutch was found that had probably been laid about June 20. Incubation in one apparent renesting effort had probably begun about July 3.

The eggs of 6 pairs of loons studied by Sjolander and Agren (1976) in northern Alaska hatched between July 18 and 21, suggesting that they must have been laid shortly after the middle of June. Bailey (1948) reported on 22 clutches obtained in the vicinity of Barrow, Alaska, between June 15 and July 8. All of these dates suggest that egg laying is probably confined to the period between mid-June and early July.

As in other loons, the nest is at the water's edge, either along the perimeter of a water area or on an island. The nests are typically in turf from which the grass cover has often been removed; they may be up to about 6 feet from water but usually are within a foot or so of the water's edge. All of 10 nests found by Sjolander and Agren (1976) were within 2 meters of shore and within a meter above the water, and most were at the very edge of water. Three were on small islets, 2 were on small peninsulas, and the other 5 were on the shorelines of lakes lacking islands or peninsulas.

NEST BUILDING AND EGG LAYING. No specific information is available on nest-building behavior or the length of time required to complete a nest. The eggs are probably laid 2 days apart, based on the staggered period of hatching. Virtually all reported clutches of this species are of 2 eggs.

INCUBATION AND BROODING. Both sexes incubate the eggs, for periods that range from as little as 220 seconds to at least 14 hours, based on the observations of Sjolander and Agren (1976). They also noted that the eggs were turned at irregular intervals of from 12 seconds to more than 6 hours. The incubating bird sits with its head toward the water and may sleep for short periods without tucking the bill under the scapulars. The birds never called while on the nest and never allowed a human within 40 meters of the nest before leaving it. The incubation period has not been established, but it is likely to be similar to the 29–30 day period known for the common loon. Sjolander and Agren (1976) judged that it might be 27 to 29 days.

As noted earlier, one case of renesting has been reported (Parmelee, Stephens, and Schmidt 1967), involving a pair whose clutch was collected on June 30 and had been incubated an estimated 5 days. Successful renesting occurred, and the date of hatching of this second nest (by August 1) indicated that incubation of the second clutch must have begun no later than July 4, only 4 days after the first clutch was collected, unless the incubation period of this species is shorter than that of the common loon. On the other hand, Sage (1971) observed that no replacement clutch was laid by a pair after nest losses to predators in early July during two separate years. The eggs of 6 pairs of loons observed by

Sjolander and Agren (1976) all hatched between July 18 and 21, suggesting a highly synchronized breeding period in that area.

GROWTH AND SURVIVAL OF YOUNG. The chicks hatch asynchronously, about 2 days apart. The eggshells are evidently removed from the nest, and during the first 3 days after hatching the young are evidently brooded ashore, either at the nest or close to it. After 9 days the young are brooded under the parents' wings or on their backs as the adults are swimming. For the first 12 days the young remain within about 2 meters of the nearest parent, but gradually this distance increases to at least 30 meters. Both parents feed their chicks, with fish 4–6 centimeters long and also at least some plant materials. Feeding of the young continues until the young are at least 35 days old and probably longer, judging from the account of Sjolander and Agren (1976). Although these authors stated that all food was captured at the nesting lake, Sage (1971) saw breeding adults flying to other lakes to forage. By the time the young are 30 days old they are able to swim at least 40 meters underwater. As in the other species of loons, one of the two chicks often is lost during the first few days of life, though the reasons for this are still uncertain. The fledging period is still unreported.

BREEDING SUCCESS AND RECRUITMENT RATES. No data on these topics are yet available.

Evolutionary History and Relationships

Certainly the yellow-billed loon and common loon are extremely close relatives, based on both morphology and behavior. Sjolander and Agren (1976) found no major differences in reproductive or territorial behavior between these two forms and supported the idea that they should be considered subspecies. This position has been frequently taken by Soviet ornithologists (e.g., Portenko 1981; Dementiev and Gladkov 1968). However, there is an absence of intermediate connecting forms in both Asia and North America, which seems to justify continued species-level separation. The yellow-billed loon may well be a Pleistocene disjunct population that originated in the vicinity of the Bering Sea and has become adapted to high arctic breeding, keeping it ecologically isolated from the more subarctic-adapted common loon.

Population Status and Conservation

So little is know of this species' population size that not much can be said of conservation needs. It both breeds and winters well away from most human contacts, and except for the unlikely possibility of major oil spills or other habitat destruction on major wintering or migratory routes, little current concern seems justified. It has been assigned an oil vulnerability index of 65 by King and Sanger (1979).

Grebes (*Podicipedidae*)

Least Grebe

Tachybaptus dominicus (Linnaeus)

OTHER VERNACULAR NAMES: Mexican grebe; grèbe de Saint Dominique (French), Schwartzkopftaucher (German); zambullidor chico, zaramagullon chico (Spanish).

Distribution of North American Subspecies (See Map 5)

Tachybaptus dominicus brachypterus (Chapman)
RESIDENT from southern Texas through Mexico (except Sonora and Baja California) and central America to Panama.

Tachybaptus dominicus bangsi (Van Rossen and Hachisuka)
RESIDENT in the southern half of Baja California and southern Sonora (casually); one breeding record for southeastern California.

Tachybaptus dominicus dominicus (L.)
RESIDENT in the Bahamas and Greater Antilles.

Description (Primarily after Palmer 1962)

ADULTS IN BREEDING PLUMAGE (sexes alike). Crown greenish black, throat black (some feathers tipped whitish), rest of head and most of neck lead colored (lower foreneck brownish), bill largely black, iris yellowish orange, upperparts slaty brown with greenish sheen, underparts white, except for buffy brown or fuscous on breast, flanks, and along sides. Feet brownish olive.

WINTER PLUMAGE. Crown blackish brown, chin and upper throat white (with more or less black intermingled), white of underparts extending well up sides. Wing mostly fuscous, except basal portion of outer primary white; extent of white increasing on each succeeding feather until it wholly includes 7th–12th feathers, beyond which it decreases.

JUVENILES. Generally gray, with striped head and neck, changing by winter to a lead gray head with a darker crown and whitish throat, a brownish to dusky neck and breast, and dusky sides.

DOWNY YOUNG. Crown mostly black with V-shaped patch in center, at first almost scarlet and naked, but by about end of first week covered with cinnamon down; lores orange; throat with dark mustachial and medial stripe; white stripe above reaches lores; iris dark brownish. White and blackish stripes on sides of head and neck continue on blackish brown upperparts as three narrow broken stripes that become more obscure as the chick grows. Underparts white. Feet medium gray; bill pale with black tip and culmen and broken black band near base (Harrison 1978). The iris color changes from brown through whitish and yellowish to the final orangy yellow during growth.

Measurements and Weights

MEASUREMENTS (of *brachypterus*). Wing (unflattened): 6 males 88–93 mm; 3 females 90–92 mm. Culmen: 6 males 22.7–24.0 mm; 3 females 22.5–23.6 mm. Eggs: average of 20, 33.4 x 23.3 mm (Palmer 1962).

WEIGHTS (of *brachypterus*). All seasons, males 122–150 g (average of 18, 130.4); females 110–34 g (average of 16, 116.6) (various sources). Estimated egg weight, 10 g (Schönwetter 1967).

Identification

IN THE FIELD. This tiny grebe is unlikely to be mistaken for any other; it is smaller than the pied-billed

5. Current North American distribution of the least grebe, showing residential distribution (*crosshatched*), peripheral or casual range (*broken line*), and an extralimital breeding record (*dot*).

and has yellow eyes and a blackish, sharply pointed bill. Young pied-billed grebes have strongly striped faces and dark brown eyes, as well as pale brownish or buffy bills. The birds are found in shallow and weedy ponds, often in association with pied-billed grebes. In nuptial plumage the chin and throat are black, but in winter these areas are white, as are the underparts. The birds evidently fly more often than do other grebes and exhibit large white wing patches while in flight. Probably the loudest call is a high-pitched *beep* or *gamp* that serves as an advertising note, while a loud, rapid trill is often produced that may last several seconds.

IN THE HAND. The very small size (wing under 100 mm) and yellowish or orange eyes immediately identify this grebe in the hand.

Ecology and Habitats

BREEDING AND NONBREEDING HABITATS. Apparently rather flexible in its breeding sites, this species occurs in freshwater habitats ranging in size from lakes and quiet sections of rivers or streams down to shallow intermittent ponds or roadside ditches. Some of these areas are surrounded by tall trees or other dense vegetation, while others provide little or no vegetational cover (Palmer 1962). Reservoirs and similar man-made impoundments are quickly colonized by this species, probably as a result of localized dispersion tendencies. Nonbreeding habitats are evidently much the same as breeding habitats, inasmuch as the birds are essentially nonmigratory except at the northernmost limits of the range.

SOCIALITY AND DENSITIES. There is no specific information on these topics, but the species is probably fairly tolerant of crowding. Gross (1949) noted that least grebes nested on three of the artificial ponds of the Atkins Garden and Research Laboratory in Cuba, with the defended areas apparently extending radially for only about 40 feet from the nest. Not only were other least grebes expelled from this area (except for the youngest of the pair's successive broods), but there were hostile interactions between the resident least grebes and pied-billed grebes that also nested on the pond.

COMPETITORS AND PREDATORS. As just noted, pied-billed grebes competed for breeding areas on the pond Gross (1949) studied in Cuba. However, least grebes are much more insectivorous than are pied-bills, and thus the two probably do not seriously overlap in food requirements. Because of their tiny size, these birds may well be preyed upon by a variety of predators; Gross suggested that turtles might be egg predators, and nests

submerged by turtles have been observed in Texas (Palmer 1962). Obviously the tiny chicks might fall prey to a wide diversity of bird, mammal, turtle, or fish predators.

General Biology

FOOD AND FORAGING BEHAVIOR. A very limited sample of stomachs from Texas birds indicate that the foods of this species are almost entirely insects, especially aquatic beetles (Dytiscidae, Hydrophilidae, Haliplidae) and true bugs (Belostomatidae, Corixidae, Notonectidae), with smaller numbers of other insect groups (Cottam and Knappen 1939). Similarly, birds from El Salvador contained aquatic insect remains, and some from the West Indies contained insects, algae (*Chara*), crustaceans including crayfish, and small fish (apparently *Limia*) (Palmer 1962). Eight birds from Guatemala had eaten beetles (Hydrophilidae, Dermestidae, Gyrinidae, Scarabidae, Curculionidae), ants, spiders, bugs, dragonflies, shrimps, crabs, and fish bones (Storer 1976a). Some feathers have also been found in the stomachs of this species; they are generally believed to be associated with fish eating but may also help form pellets for chitin removal, according to Storer.

Foraging is done by diving or by surface feeding, which involves immersing the head and neck, snapping at passing insects, or picking objects off emergent vegetation. Probably this last method of foraging accounts for the fairly high frequency of ants being eaten (7 of 8 samples) that Storer reported. Storer suggested that the opportunistic feeding methods of least grebes probably help them utilize small and temporary water areas that might support only limited invertebrate faunas.

MOVEMENTS AND MIGRATIONS. No specific information on this is available, but limited dispersal does occur, perhaps through nocturnal movements stimulated by rainfall patterns. In Texas the birds tend to visit the upper coast during winter months, but there is no clear evidence of a regular migration there. The birds take wing more readily than do other North American grebes and thus are probably able to make considerable movements when necessary.

Social Behavior

MATING SYSTEM AND TERRITORIALITY. Least grebes exhibit strong seasonal pair bonding, and it is possible that in some areas the bond is maintained almost permanently. Gross (1949) reported that a pair in Cuba produced a series of eight clutches in less than a year, with little if any interruption of their efforts. Although this may not be typical, it suggests a frequent succession of nesting efforts by the same pair, with each clutch laid shortly after the hatching of the previous one. He estimated a limited area of territorial defense (approximately 5,000 square feet) in this particular pair. Storer (1976a) noted that two pairs of least grebes were resident on a pond of about 300–400 by 75–100 feet, suggesting a maximum territorial size of no more than about 13,000 square feet (1,400 square meters) per pair.

VOICE AND DISPLAY. Calls of this species have been described by Storer (1976a), who recognized four distinct vocalizations. The species' advertising call is a loud, high-pitched note sounding like *gamp*, uttered with the head held high, the neck thin, the body feathers slicked down, and the white of the underparts showing at the anterior end of the body. Both sexes utter the call, and it is regularly used as a signal when the pair is separated or when the calling bird is somewhat alarmed. A second call is a loud, rapid trilling, uttered most often when the pair members are together. Frequently the birds trill simultaneously but with alternating notes. The trill may serve as a triumph note and may help in reinforcing pair bonds or in territorial announcement. During aggressive encounters a high-pitched nasal note is sometimes uttered, and a final note Storer recognized was a soft contact note given by birds when swimming together or when approaching the nest or one another.

Displays have also been described by Storer (1976a), although Zimmerman (1957) has also provided a few observations. Probably the advertising call, with its associated neck stretching, contributes to pair formation. This posture was not illustrated by Storer, but he stated that it is similar to the corresponding posture of horned grebes (fig. 34C). He did not observe the "rushing" display described and illustrated by Zimmerman (1957), who observed 6 instances when two birds simultaneously rose in the water and glided parallel for about 3 feet with their throat feathers expanded, bills tilted slightly upward, and necks extended (fig. 33A). They then stopped and began preening. During the gliding performance one of the birds repeatedly uttered a high-pitched call, possibly the trill described by Storer. Storer did observe several copulations. As with other grebes, copulation occurs at the nest or prospective nest site. Storer judged that the soliciting posture of the species is unique, with no rearing or wing quivering and with the head held drawn back and the bill pointed downward (fig. 33D). It also strongly differs from the aggressive Z-neck posture of the species (fig. 33B). Mounting is not preceded by any specific displays, and during copulation the submissive bird raises its head and strokes the breast of the dominant bird with its nape (fig. 33C). Af-

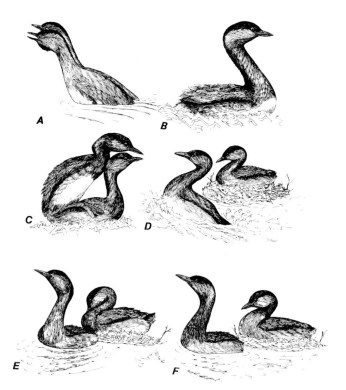

33. Social behavior of the least grebe (mainly after Storer 1976a): A, rushing; B, Z-neck posture; C, copulation; D–F, postcopulatory sequence.

ter dismounting, the male treads water rapidly while the female performs head flicking movements, probably calls, and remains on the nest (fig. 33F). The male also performs head flicking movements while his neck is greatly extended and his bill is tilted upward. Reversed mounting was observed by Storer more often than normal copulation, and it differed from normal copulatory behavior only in minor details.

Reproductive Biology

BREEDING SEASON AND NESTING SUBSTRATE. In Texas the breeding season is extended, with nesting almost throughout the year and eggs reported as early as January and as late as December 17. The peak of the Texas breeding season is probably from mid-April to the end of August (Palmer 1962). In most areas breeding seems to be associated with the tropical wet (summer) season, but in Cuba nesting by a single pair was almost continuous from at least July to the following May (Gross 1949). As for other grebes, the nesting substrate consists of floating or emergent masses of vegetation in the water.

NEST BUILDING AND EGG-LAYING. Nests are constructed among living or dead vegetation over water that may be 50–80 centimeters deep (Marchant 1960) or at times up to 5 feet deep (Palmer 1962). Invariably some kind of anchoring vegetation is present, and material is often almost constantly added to adjust for rising water levels. The same nest may be used for successive broods, or a new nest may be built on the same site or close by. Observations in Cuba suggest that the eggs are normally laid daily, a clutch of 4 eggs being completed in as little as 5 days. For 8 Cuban clutches, the average was 4.4 eggs (Gross 1949), and Marchant (1960) reported 5 Ecuadorean clutches of 4–6 eggs, averaging 5. Seventeen clutches in Texas ranged from 2 to 5 eggs, averaging 4.2 eggs (Palmer 1962).

INCUBATION AND BROODING. The incubation period is 21 days, beginning with the first-laid egg, so that the young hatch over several days. Both sexes participate in incubation, with the male apparently guarding the territory during much of the time he is not actually engaged in incubation or in tending the young of a previously hatched brood. Gross (1949) reported that a male occasionally delivered food to the young of two older broods, and both tended some 3-week-old chicks and a clutch of 4 eggs. Five successive broods, involving 23 hatched chicks, were produced by this pair in the 6 months from July to January, while a sixth clutch was laid in late February, a seventh brood was hatched in late April, and a final (unsuccessful) nest was constructed in May.

GROWTH AND SURVIVAL OF YOUNG. Growth rates and fledging periods in this species are still unreported; in the closely related little grebe the fledging period has been estimated at 44–48 days, with the young becoming independent at 30–40 days. Gross (1949) believed that pied-billed grebes may have killed some of the least grebe young in his study area.

BREEDING SUCCESS AND RECRUITMENT RATES. The only available information on this comes from Gross's (1949) observation. He noted that of a total of 35 eggs, 27 were hatched and 24 young were successfully raised, representing a remarkable 68.6 percent breeding success. In Texas the hatching success of 12 out of 17 clutches was over 90 percent; the remaining clutches were destroyed or infertile (Palmer 1962).

Evolutionary History and Relationships

Storer (1976a) considered this species closely related to a subgroup of grebes that also includes *ruficollis* and its Old World relatives. These he collectively placed in the genus *Tachybaptus*, with the comment that *dominicus* possibly warrants subgeneric recognition, in part because of its peculiar soliciting posture and postcopulatory displays.

Population Status and Conservation

No specific information is available on these topics. The species is locally common in southern Texas (such as at Santa Ana and Laguna Atascosa National Wildlife Refuges), mainly in the Rio Grande valley and along the central and lower coastal areas, and probably is sufficiently adaptable to adjust rapidly to local changes in surface water condition. It also has an extremely wide range through Central and South America to northern Argentina and in the Antilles. It is probably rarely, if ever, hunted for food or sport. Thus it seems to pose no special conservation problems.

Pied-billed Grebe

Podilymbus podiceps (Linnaeus)

OTHER VERNACULAR NAMES: None in general English use; grèbe à bec bigarré (French); Bindentaucher (German); svartvitnäbbad dopping (Swedish); zambullidor pico pinto, zaramagullon grande (Spanish).

Distribution of North American Subspecies (See Map 6)

Podilymbus podiceps podiceps (Linnaeus)
BREEDS from southeastern Alaska, central British Columbia, southern Mackenzie, north-central Alberta, central Saskatchewan, northern Manitoba, northwestern Ontario, southern Quebec, central New Brunswick, and Nova Scotia south, locally, to southern Baja California, Jalisco, the state of Mexico, Texas, Louisiana, and southern Florida. Another race breeds in the West Indies and in Central America.

WINTERS from Vancouver Island, southern British Columbia, rarely south to central Arizona, Utah, and central Texas eastward, north to the line of winter ice in the Mississippi and Ohio valleys and the Atlantic coast.

Description (Mostly after Cramp and Simmons 1977)

ADULTS IN BREEDING PLUMAGE (both sexes). Forehead brownish black shading to dark brown on crown and greenish brown on nape; sides of head and neck grayish brown, contrasting with clearly defined black throat patch bordered with whitish lines at sides (this pattern usually more clear-cut in males). Upperparts dark brown, often with hoary look in fresh plumage; breast and flanks barred and blotched with dark brown on

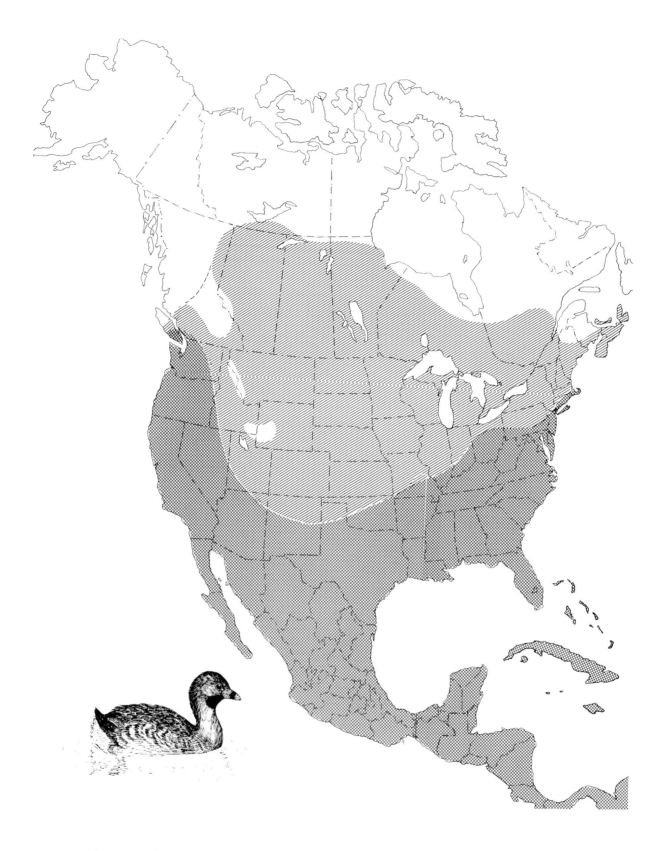

6. Current North American distribution of the pied-billed grebe, showing summer distribution (*hatched*), wintering and residential range (*crosshatched*), and peripheral or casual range (*broken line*).

whitish ground, contrasting with usually conspicuous under tail coverts. The bill is multicolored, with a broad black band in the middle, and otherwise mostly bluish white. The iris is brownish red, and the feet are dark lead color.

WINTER PLUMAGE. Forehead and crown dark grayish brown; throat whitish, sometimes with traces of black; rest of head grayish buff with slightly darker line through eye. Neck, breast, and flanks grayish buff tinged reddish to rich reddish tawny. Otherwise as in breeding plumage, though white rear often more conspicuous. The bill is yellowish to dusky gray with a dull greenish tinge but no black band.

JUVENILES. Bill yellowish brown; no orbital ring. Sides of head and neck variably dark to grayish brown, blotched and streaked irregularly with white; throat white; rest of neck brownish buff. Upperparts blackish brown; breast and flanks rather dark brown mixed with buff; belly and under tail coverts white, though latter not as conspicuous as in adult.

DOWNY YOUNG. Black above, with four long, narrow white stripes on the back and neck; on either side of the occiput there is a reddish buff or pale chestnut patch, with another patch or traces of this color on the forecrown; forehead white, extending back as two conspicuous white stripes over each eye, separated by black; the side of the head white, striped and spotted with dull black; throat white, spotted with sooty black, the bill and bare lores dull yellow; the culmen, a narrow line near the base of the lower mandible, and tip black, with a white egg tooth. Legs and feet greenish black; iris dark brown (Roberts 1949).

Measurements and Weights

MEASUREMENTS (of *podiceps*). Wing: males 124–35 mm (average of 13, 130); females 115–26 mm (average of 18, 120). Culmen: males 20–24 mm (average of 13, 22.7); females 17–21 mm (average of 18, 19.3) (Cramp and Simmons 1977). Eggs: average of 20, 43.9 x 30.1 mm (Palmer 1962).

WEIGHTS (of *podiceps*). In breeding season, males 438–85 g (average of 6, 458); females 274–379 g (average of 6, 339) (various sources). Hartman (1961) reported slightly lower averages for a more general sample. Estimated egg weight, 22 g.

Identification

IN THE FIELD. This small, dumpy-looking grebe is the only member of the group having a short, chickenlike

bill, dark brown eyes, and a whitish eye ring. In most plumages the birds appear to be a rather drab brown except for a white rump area and a throat that is either black (in breeding plumage) or pale buffy (immatures and winter-plumage birds). There is otherwise little patterning, though young birds retain a striped head pattern for some time, perhaps through their first fall. In flight (which is extremely rare) the wings appear mostly brown, but the secondaries exhibit highly variable amounts of white on their tips and inner webs. The birds are more often heard than seen, and the usual vocalizations are a complex and prolonged series of notes that ends with *pow* or *wup* sounds reminiscent of cuckoos. Generally the notes increase in loudness and speed as well as pitch and then fall away toward the end. Heavily overgrown ponds are favored habitats.

IN THE HAND. This is the only grebe with brown eyes, secondary feathers that are mostly brownish, at least on their outer webs, and in which the bill is less than 25 mm long but more than 10 mm in basal depth. Criteria for determining age in this and other North American grebes have yet to be developed, but based on Kop's (1971) studies of the great crested grebe it is possible that first-year birds can be externally identified by their smooth rather than serrate posterior tarsal surface, their bare skin areas on the crown and upper lores (hidden by longer contour feathers), evidence of worn remains of down-tipped contour feathers, and variable remains of the striped head pattern of juveniles.

Ecology and Habitats

BREEDING AND NONBREEDING HABITATS. Breeding habitats of pied-billed grebes are invariably associated with fairly dense emergent vegetation, whether in freshwater or brackish habitats. Estuaries provide suitable breeding areas where tidal fluctuations are slight, though the birds have been found to tolerate tidal fluctuations of as much as 2 feet. Sizes of ponds on which breeding occurs range from as small as half an acre (1.3 hectares) to more than 100 acres, and in depth from little as a foot to 5 feet. Apparently depths of less than a foot are avoided, although in some breeding sites the water level may drop as much as 2 feet in a 6 week breeding period. The type of aquatic vegetation seems to be unimportant provided it is adequate to support and variably conceal a nest (Miller 1942). Yocom, Harris, and Hansen (1958) noted that 80 percent of the broods they observed were on water areas of 1–5 acres, and 17 percent were on areas over 5 acres (2.5 hectares). Faaborg (1976) found breeding birds on ponds of 0.6–7.0 hectares, especially those that were rather heavily vegetated, with 20–40 percent open water. Stewart (1975) reported that the birds used fresh to moderately brackish seasonal to semipermanent ponds and lakes of an acre or more.

Nonbreeding habitats are similar to those used during breeding, but more exposed waters are more frequently used. In coastal areas the birds move increasingly into brackish waters with colder weather, where the dangers of freezing into the ice are reduced, although use of open salt water seems to be very small (Miller 1942). During spring migration the birds favor marshes with open-water areas that are mostly 15–25 inches deep (Glover 1953).

SOCIALITY AND DENSITIES. Although certainly not gregarious in any sense, pied-billed grebes are surprisingly tolerant of crowding, and nesting densities occasionally are quite high in favored habitats having sufficient emergent vegetation to eliminate most visual contacts. Miller (1942) mentioned finding two active nests no more than 50 feet apart on a 1.5 acre marsh and another pair of nests within 100 feet of each other on a much larger marsh. Single pairs sometimes occupy ponds as small as half an acre, and generally those up to 10 acres support only single pairs (Palmer 1962). However, at times the nesting densities are certainly considerably greater than this. Chabreck (1963) found 107 active nests on a brackish impoundment of 200 acres (81 hectares), a density of 1.3 nests per hectare. Outside their rather small territories the birds feed commonly without conflict, especially in deep, open-water areas of the marsh. Glover (1953) judged the home range of a pair to be only about twice as large as its nesting territory (which had an estimated radius of 150 feet).

COMPETITORS AND PREDATORS. At least in North America there are no close relatives of this species that are likely to be significant competitors. It evidently is one of the most omnivorous of the North American grebes, and as such is likely to be able to exist in close proximity to other grebes. Its nesting requirements are similar to those of the least grebe, and sometimes considerable aggressive interaction occurs between these species, with the larger pied-billed grebe usually favored (Gross 1949). Apparently it readily tolerates both coots and gallinules in its breeding areas (Miller 1942), although Glover (1953) reported territorial defense against a coot. Probably few predators can capture adults on the water, though the birds are poor fliers and, like all grebes, helpless on land. Nest predators probably include the usual array of egg-eating birds and mammals, but only raccoons (*Procyon lotor*) have been specifically identified as major nest predators (Glover 1953).

General Biology

FOOD AND FORAGING BEHAVIOR. Wetmore's (1924) analysis of 180 stomachs (174 with food remains) is the best source of information. He found that 24.2 percent of the food was fish, including a wide variety of species, but especially catfish (*Ictalurus* and *Ameiurus*), eel (*Anguilla*), perch (Percidae), and sunfish (Centrarchidae). Crayfishes contributed 27 percent of the total food and probably are a major food item of the pied-billed grebe, judging from its heavy bill that is certainly well adapted for crushing crustaceans. Other crustacean remains included shrimps (*Crago*), prawns (*Palaemonetes*), and fiddler crabs (*Uca*). Insects composed 46.3 percent of the total food and were predominantly represented by aquatic bugs (Heteroptera) and beetles (Coleoptera), especially predatory forms, and the nymphs of damselflies and dragonflies (Odonata). Other observers have mentioned leeches, frogs, and various other prey species, and Miller (1942) saw the birds taking roach (*Abramis crysoleucas*) and carp (*Cyprinus carpio*) up to 5 inches in length. Most foraging dives are in quite shallow water and are rather short in duration. Bleich (1975) reported an average dive duration of only 7.58 seconds for foraging dives and an average lateral distance moved from point of submergence of 3.69 meters (in water 2–3 meters deep). Likewise, the dive durations of a bird observed in England were from 3–24 seconds, with a maximum lateral movement of 20 yards and with small fish the apparent prey. These observations suggest that little prolonged underwater chasing normally occurs. Certainly most of the prey species reported above are slow moving and thus should be readily captured without extended pursuit.

MOVEMENTS AND MIGRATIONS. During the breeding season nesting birds appear to be highly sedentary, with very small home ranges. Migrant birds move as singles, as pairs, or in flocks of various sizes, but such flocks are never very cohesive. Spring migrants are among the earliest of the grebes, typically following the thaw line closely, moving through the southern states in March and the central and northern states during late March and April. Fall migration is protracted and tends to coincide with major movements of ducks, probably peaking in October and November in central and southern states; nearly all birds have left breeding areas by mid-November. Almost nothing is known about rate and distance of migration in individual birds, since there have been relatively few banding efforts on this species. The birds migrate nocturnally and generally in inconspicuous numbers, with maximum recorded concentrations of about 20,000 birds at Salton Sea, California, in November (Palmer 1962).

Social Behavior

MATING SYSTEM AND TERRITORIALITY. At least seasonal monogamy is typical of this species, and territorial behavior is well developed although very restricted in area. Glover's (1953) estimate of a territory as having a radius of about 150 feet around the nest seems to be the only estimate of its size. Chabreck (1963) stated that the average distance between nests he studied was 180 feet (55 meters) and never less than 75 feet (23 meters). The territory is usually defended by the male, but occasionally the female will join her mate in such defensive efforts.

VOICE AND DISPLAY. Miller (1942) described the variations in the calls of this species, including a territorial "love song," an evening note the male used to bring his mate, a note the male uttered as he pursued the female, a general alarm note, a call to warn the young of danger, and some peeping calls of the chicks. Perhaps the best description of the complex territorial "song" of this species is by Ladhams (1969). He described it as consisting of three characteristic sections, in the same order each time. There was an opening trilled phrase lasting about 2.5 seconds, followed after a short pause by a series of clear, bell-like notes that were highly variable in number but averaged nine and that passed into a final series of wails and hiccups; these last two types of notes were repeated several times. Generally, when the middle section of bell-like notes was protracted the final section was compressed, and vice versa, so that the average length of the song was about 15 seconds and the longest 25. Typically the song was repeated at intervals of from 2 to 10 minutes, and its duration did not seem to be related to either the weather or the time of day. When singing, the bird would raise its neck, tilt its bill upward, and lower its breast below the water while simultaneously elevating its rear. The head feathers are somewhat fluffed, especially those of the throat, while the neck is compressed, producing a distinctive large-headed appearance. The bill is opened only slightly, but the song can readily be heard for half a mile. The female's version of this call is very similar but lacks the final series of hiccuplike *cow* notes, and duets without the *cow* ending are the usual means of greeting by a pair as they meet, turn, and swim side by side (Nuechterlein and Storer 1982).

Certainly the major "display" of this species is its territorial song, which is uttered in the posture just described. Mather (1967) reported that other displays consist of a quick dive and resurfacing, over a distance of about 20 yards, terminated by vigorous wing flapping and associated splashing of water, and also a prolonged wing flapping of as long as 12 seconds while "standing" almost vertically in the water. Kilham (1954) described several possible courtship displays, including wing flapping in place, head turning, bill touching, and a swimming chase. Displays or possible displays mentioned by Palmer (1962) include mutual splashing and diving, bill touching, simultaneous diving, emerging, and racing over the water surface, picking up and carrying weeds, a dive toward the female with a subsequent short chase by the male, and a "circle display." In this display the members of the pair approach each other, and when about a foot apart they stop and make a half-turn, while both perform head turning movements and the presumed male raises his back, neck, and wing feathers. After a brief pause they swing back to face one another and continue this pivoting behavior for some minutes. Comparable "pivoting" behavior occurs in eared grebes during food presentation and to a limited degree in horned grebes during triumph ceremonies (Fjeldså 1973c, 1982a). There seem to be none of the elaborate display sequences such as the discovery ceremony that are typical of the *Podiceps* grebes, and the postures appear to be more clearly agonistic in origin. Fjeldså (1973c) stated that the pied-billed grebe and little grebe share several simple courtship displays, including diving and swimming together, turning in an upright posture (rarely resembling penguin postures), duetting while in a forward threat or flared out posture, racing and chasing, mutual diving, swimming, and flying, the weed trick (fetching and presenting weeds), slow head shaking, springing dives, bill touching, and rarely courtship feeding. He judged that duetting in a hunched posture might be comparable to the hunched display in the eared grebe. It seems unlikely that the mutual racing or chasing behavior of pied-billed and least grebes is highly ritualized and comparable to the elaborate races of western grebes, but presumably all these forms of behavior, from slow "barging" to rapid rushes or races, have their origins in antagonistic chases (Fjeldså 1973c).

Copulation in pied-billed grebes has been described in detail by McAllister and Storer (1963), and it takes a form very much like those of other species, contrary to earlier descriptions that were apparently based on misinterpreted aggressive behaviors. It occurs on the nest or a nestlike platform and is preceded by the two usual grebe precopulatory displays, rearing (with or without associated wing shaking) and a low inviting posture, with the head and neck stretched forward in line with the body. During treading the female was seen to retract her head and touch the male's breast with her nape feathers while extending her neck and tilting her bill upward. During treading the male called, and afterward he flopped back into the water without any associated postcopulatory display other than to bring back a few loads of weeds to the nest and to bathe in the usual way. The female shuffled her wings and then dived into the water and swam away.

Reproductive Biology

BREEDING SEASON AND NESTING SUBSTRATE. Chabreck (1963) reported active nests in Louisiana between May 3 and September 10, with a hatching peak in early June. In an Iowa study the period of eggs in 138 nests ranged from May 2 to August 8 (Glover 1953), and 131 Ontario egg records are from May 3 to August 22 (Peck and James 1983). All of these suggest a protracted breeding season, although the only proven case of double brooding seems to be that reported by Miller (1942). He noted that a second clutch was hatched approximately 5 weeks after the hatching of an earlier brood, indicating that a second laying began only about 4 days after the first clutch hatched. As in all grebes, the nest is built over water, in vegetation of varying densities. Sealy (1978) found a minimum water depth of 12.7 centimeters for 31 nests and an average distance from open water of 1.3 meters. Peck and James (1983) stated that records of 387 nests indicated nests were placed in sites from as little as a few centimeters deep to more than a meter, and usually in vegetation 2.5–15.0 meters from open water but rarely also in open water and attached to logs or dead trees. Glover's (1953) analysis of nest sites indicated that the birds favored relatively high vegetation density but used many different emergent vegetation types, especially *Eleocharis*. The average distance to open water in his sample of 138 nests was 25.8 feet (7.8 meters), but there was no significant correlation between nesting success and vegetation density at the nest site. In Chabreck's (1963) study the birds more often nested in open water than in stands of emergent vegetation, with nests constructed of submerged widgeongrass (*Ruppia*) in areas of unobstructed visibility. Sites of 82 North Dakota nests had an average water depth of 25 inches and a range of 11–37 inches (Stewart 1975). Otto (1983) concluded that the two prime determinants of nest placement in his study area were water depth (at least 25 centimeters) and vegetation density (minimum of 10 square centimeters of stem basal depth per square meter of surface).

NEST BUILDING AND EGG LAYING. In Glover's (1953) study, nest building was determined to require 3–7 days, depending on the availability of nesting materials and perhaps also the physiological state of the female, who did most of the construction. He found that most birds built two nests; the first was an unproductive effort and eggs were placed only in the second. Sealy (1978) has summarized data on clutch size in this species from many parts of its range, which suggest that at

least in Canada and the United States the average clutch size ranges from 4.3 to 7.0, with most means between 6.1 and 6.8 eggs. Typically an egg is laid every day, though a day may be skipped toward the end of the egg-laying cycle. Chabreck (1963) observed that clutch sizes in Louisiana tended to decrease through the breeding season and that most clutches took about 2 weeks to complete as a result of apparently irregular egg-laying rhythms. Otto (1983) judged that the seasonal decline in clutch size in his area occurred because renesting birds produced lower average clutch sizes in replacement nests.

INCUBATION AND BROODING. Both sexes incubate, with incubation beginning with the first-laid egg. Thus hatching is staggered over a period ranging from as little as 2 days to one comparable to the time spent in egg laying, or as much as 2 weeks. The incubation period in Otto's (1983) study averaged 22.7 days (120 eggs, range 20.5–27.0 days), with some variation that was probably associated with the seemingly variable intensity of incubation behavior. Chabreck (1963) reported a nesting success of 89.6 percent (96 of 107 nests hatching one or more young) and a hatching success of 91.3 percent (327 of 358 eggs in 49 successful nests). Wind, rain, and tides apparently caused most nest losses. Since the average brood size was only 4.4, there was evidently considerable mortality of eggs or newly hatched young. Glover (1953) reported a 70.4 percent nesting success (97 of 138 nests) and a hatching success of about 82 percent in successful nests. Wind effects and fluctuating water levels were major causes of nest losses in his study, with raccoons being the only predator of apparent significance. Weather effects and egg predators were the major factors affecting nest success in Otto's (1983) study of 150 Wisconsin nests. He estimated a nesting success of 76.7 percent, with most losses from predation and wave effects, and a hatching success of 90.7 percent. As a result of renesting efforts (a minimum incidence of 66 percent was estimated) there was a high production of hatched chicks, estimated by Otto to average 5.53 young per breeding pair.

GROWTH AND SURVIVAL OF YOUNG. The young attained initial flight abilities at about 35 days, and the adults stayed with their young for 25 days in one study (Kirby 1976). Brood sizes reported by Yocom, Harris, and Hansen (1958) averaged 2.9 for 95 broods of varying ages, with the oldest age-class averaging 2.6 young. This suggests a relatively low fledging success in spite of the high nesting success rates noted earlier for various studies.

BREEDING SUCCESS AND RECRUITMENT RATES. There are no estimates available of recruitment rates or annual survival rates for this species. However, if one assumes (based on sources mentioned above) an average clutch size of 6.5 eggs and a hatching success of 85 percent, the average brood size at hatching should be 4.1 young. If older broods average 2.6 young, the brood mortality would be approximately 37 percent and the recruitment rate about 56 percent, assuming that all pairs nest and none raise more than a single brood. This implies an annual adult survival rate in the vicinity of 40–50 percent, which is in line with what little is known of grebe life-history characteristics.

Evolutionary History and Relationships

Certainly the nearest relative of this species is the giant pied-billed grebe of Lake Atitlán, Guatemala (Bowes 1969), and fossil pied-billed grebes are also known from the Pleistocene (Storer 1976b). The two extant pied-billed grebes differ from their nearest relatives in a variety of anatomical ways (Storer 1976a), and these seem to represent specializations. Behaviorally, *Podilymbus* is distinctive in its sexually dimorphic "song" and its well-developed "pivoting" display, according to Storer. Fjeldså (1973c) listed a number of display similarities existing between the pied-billed grebe and the little grebe, and various additional anatomical similarities between *Podilymbus* and *Tachybaptus*. Downy plumages of these forms are also similar in lacking longitudinal stripes on the forehead and in having a rufous crown patch and a black area on the rear of the crown, according to Fjeldså.

Population Status and Conservation

Nothing can be said of the pied-billed grebe's population other than that it is the most ubiquitous of the North American grebes, and probably it or the eared grebe is the most common. It is highly adaptable to various small breeding habitats and is certainly not a concern for conservationists. Its avoidance of open ocean makes it quite safe from oil pollution, and it rarely is purposely killed by hunters.

Horned Grebe

Podiceps auritus (Linnaeus)

OTHER VERNACULAR NAMES: Slavonian grebe (British); nordisklappedykker (Danish); grèbe esclavon (French); Ohrentaucher (German); flórgodi (Icelandic); mimi kaitsuburi (Japanese); krasnosheynaya poganka (Russian); svarthakedopping (Swedish); zambullidor cornudo (Spanish).

Distribution of North American Subspecies (See Map 7)

Podiceps auritus cornutus Gmelin

BREEDS in North America from central Alaska, northern Yukon, northwestern and southern Mackenzie, southern Keewatin, and northern Manitoba south to eastern Washington, northeastern Idaho, southwestern and northern Montana, northern South Dakota, northwestern Minnesota, central Wisconsin, and extreme western Ontario (formerly from northern Ontario, southern Quebec and New Brunswick south to northern Utah, northwestern Nebraska, northeastern Iowa, northern Illinois, northern Indiana, and southern New England).

WINTERS in North America on the Pacific coast from the Aleutians and south coastal Alaska south to southern California and on the Atlantic and Gulf coasts from Nova Scotia south to southern Florida and west to southern Texas, rarely on inland waters from southern Canada and the Great Lakes southward.

Description (after Witherby et al. 1941)

ADULTS IN BREEDING PLUMAGE (sexes alike). Forehead, crown, and neck black glossed blue; mantle and scapulars glossy black, feathers narrowly edged gray or grayish; back brownish black; rump and tail as mantle but lower tail feathers white with black tips, sides of rump chestnut, feathers tipped black (sides of head and underparts); stripe from upper mandible above lores and eye through ear coverts to nape chestnut, portion from behind eye more golden (feathers very narrow, long, and silky); sides of head, chin, and upper throat glossy bluish black; rest of throat and upper breast chestnut, some feathers adjoining breast tipped white; breast and belly silky white; sides of breast and flanks chestnut, feathers mostly tipped grayish black and many adjoining breast and belly white with subterminal blackish gray spots or marks; vent grayish brown, feathers downlike and tipped white; axillaries and under wing coverts white; primaries brown with black shafts, pale brown with black shafts, pale brown inner webs and white bases, two innermost occasionally with small white tips; secondaries white, outer ones with varying amount of brown, two outermost often mostly brown with small white tips, varying number of middle ones pure white, inner ones white with black tips, and innermost black with white bases; wing coverts grayish brown, innermost white, tipped blackish brown.

WINTER PLUMAGE. Crown glossy black but not as bluish black as summer; neck browner black, on each side of lower nape a white patch not quite meeting in center of back of neck, feathers of which have long

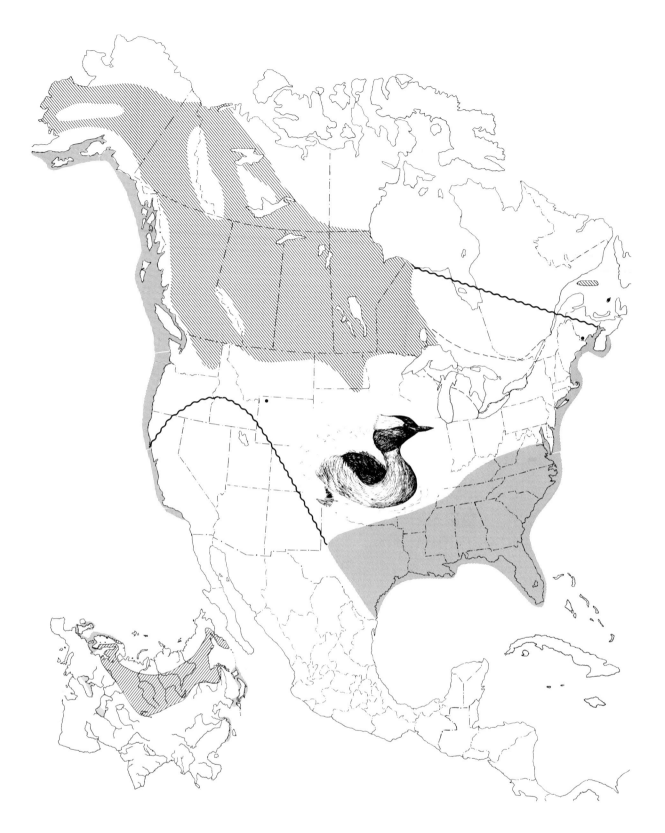

7. Current North American distribution of the horned grebe, showing summer distribution (*hatched*), wintering range (*shaded*), peripheral or casual range (*broken line*), and limits of major migratory corridors (*wavy line*). The Eurasian range is shown on the inset map.

brownish black tips and white bases; mantle and scapulars not as glossy as summer and grayish edges rather more obscure; no chestnut on sides of rump, feathers being white, tipped black; above lores a grayish spot; some feathers behind eye often more or less rufous; chin, cheeks up to eye, ear coverts, and upper throat white, often with a few ill-defined blackish spots; lower throat pale brown, feathers tipped white, occasionally whole throat white; sides of breast and flanks white, feathers tipped black (not chestnut); rest of plumage as summer.

JUVENILES. Much like adult winter but upperparts browner, sides of nape and sometimes cheeks slightly mottled brown, flanks more lightly marked brownish black.

DOWNY YOUNG. Chin and light stripes of head tinged buff, light stripes of neck and upperparts narrow and grayish white, general appearance of upperparts with clear, narrow stripes, not obscured as in eared grebe, upper mandible with two transverse black bands, proximal one not reaching cutting edge and neither one to lower mandible. Resembles the red-necked grebe in having well-defined black-and-white bands on head and neck and a median white spot on rear crown, but the banding across the bill does not continue to the lower mandible, and there is a larger bare pink spot on the anterior crown (Fjeldså 1977).

Measurements and Weights

MEASUREMENTS (of *cornutus*). Wing: males 138–51 mm (average of 23, 147.5); females 132.5–150.0 mm (average of 22, 141.5). Culmen: males 22–26 mm (average of 23, 24.1); females 21.0–24.5 mm (average of 22, 23.0) (Palmer 1962). Eggs: average of 45, 44 x 30 mm (Bent 1919).

WEIGHTS (of *cornutus*). All seasons, males 345–574 g (average of 13, 437), females 213–481 g (average of 15, 362 g) (various sources). Summer weights of nominate *auritus* average about 15 g heavier (Cramp and Simmons 1977). Estimated egg weight, 22.5 g (Schönwetter 1967). Newly hatched chicks weigh 14–20 g (Fjeldså 1977).

Identification

IN THE FIELD. In breeding plumage, this species is likely to be confused only with the eared grebe, which has a black rather than a brown neck and golden ear tufts that are more fanlike than hornlike. In winter the two species are more similar, but the horned grebe has a strongly bicolored head pattern (similar to that of the red-necked grebe), with no gray smudging extending into the ear region. The bill of the horned grebe appears slightly longer and more tapered, while the eared grebe has a shorter bill that is nearly straight dorsally and sharply tapered ventrally, producing an uptilted appearance. The calls of the two species are similar, but those of the horned grebe include a loud, accelerating trill, a double squeaking sound, *ko-wee, kowee* that serves as an alarm note, and several others.

IN THE HAND. This medium-sized grebe has a wing length that is similar to that of the eared grebe, as is its culmen length, but in the horned grebe the bill is slightly deeper than its basal width, and both the upper and lower mandibles are tapered rather symmetrically. White feathers are present on the upper wing coverts near the leading edge of the wing, and the nape area tends to be paler in immature and winter-plumage birds than in the eared grebe.

Ecology and Habitats

BREEDING AND NONBREEDING HABITATS. Ecological studies in Manitoba indicate that breeding horned grebes favor open, deep-water marshes ranging in size from 0.1 to 8.4 hectares over emergent deep marshes or shallow marshes (Ferguson and Sealy 1983). In Saskatchewan the birds were found to favor ponds larger than 0.2 hectares and those with areas of open water. In North Dakota one study indicated a use of ponds ranging from 0.1 to more than 128 hectares in size but with the heaviest use of ponds of less than 2 hectares and those with a predominance of open water (Faaborg 1976). In Iceland the birds prefer small, shallow water areas surrounded by rich vegetation, whereas in Finland the species extends into a broader habitat range, including open, oligotrophic lakes, but mainly those 1–10 hectares in area and having about 16 percent of the water area with emergent vegetation (Fjeldså 1973a). Stewart (1975) characterized the species' breeding habitats in North Dakota as ranging from fresh to slightly brackish waters and seasonal to permanent ponds and lakes having an expanse of open water and ranging from 0.33 acre to at least several hundred acres.

Nonbreeding habitats are diverse and include not only freshwater lakes and ponds but also coastal areas, where the birds use both estuaries and marine waters, sometimes in bays well out from shore. Probably most birds winter on brackish or marine waters rather than freshwater habitats such as larger lakes (Palmer 1962).

SOCIALITY AND DENSITIES. This is a relatively territorial grebe, with small ponds usually supporting only a single pair. If two or more are present, the nests are widely separated by open water or by a barrier of vegetation or land (Faaborg 1976). Faaborg reported that one 19.6 hectare pond supported 4 horned grebe pairs (5 hectares per pair), and Stewart (1975) noted that a 43 acre pond (17.4 hectares) had 5 nesting pairs (3.5 hectares per pair). Ferguson and Sealy (1983) reported that two ponds of 1.1 hectares each supported 2 pairs while one of 2.6 hectares had 4 pairs (0.55–0.65 hectare per pair). These are certainly unusually high densities and probably reflect very rich local food supplies. Fjeldså (1973b) found that aggressive levels of these birds in Norway and Iceland were related to relative food abundance, with minimum breeding areas of eutrophic waters of 0.12 hectare per pair and about 0.8 hectare per pair for ponds of low fertility. Distances of less than 10 meters between neighboring pairs were found typical only of waters larger than 5.7 hectares, and distances of less than 5 meters were typical only in lakes of more than 46 hectares. These cases of close nest placement reflected the distribution of very protected nest sites or localized patterns of emergent vegetation rather than a lack of territorial tendencies.

PREDATORS AND COMPETITORS. Known nest predators in North America and Europe include raccoons (*Procyon lotor*), mink (*Mustela vison*), gulls (*Larus* spp.), and hooded crow (*Corvus corone*), while suspected predators include American crow (*Corvus brachyrhynchus*), magpie (*Pica pica*), and American coot (*Fulica americana*) (Fjeldså 1973c; Ferguson and Sealy 1983). Newly hatched chick mortality typically is also quite high, but frequently such early death comes from fatigue and loss of heat or some other cause unrelated to predation (Fjeldså 1973c). There is good reason to believe that the pied-billed grebe is a serious competitor with the horned grebe in North America, and the more aggressive pied-billed grebe is favored in situations of local interaction (Faaborg 1976; Ferguson and Sealy 1983). There is also indirect evidence of competitive interactions between the horned grebe and several European grebes (Fjeldså 1983a).

General Biology

FOOD AND FORAGING BEHAVIOR. Wetmore's (1924) analysis of foods of this species was based on 122 stomach samples (114 with food remains) from nearly all months of the year. Fish remains occurred in 49 stomachs, and composed 34.6 percent of the volume. These were of a wide variety of marine and freshwater forms. Crustaceans were found in 29 stomachs and amounted to 17.9 percent of the volume. They were of various

larger forms such as crayfish, shrimps, and prawns, as well as amphipods and isopods. Insects made up 46 percent of the food volume, with beetles composing over half the total. Most of these were aquatic forms, but there were some terrestrial insects. Fjeldså's (1973a) study of foods and foraging behavior is more valuable and was based on 49 stomachs from breeding areas in Iceland. An analysis by weight indicated a predominance of fish in the diet, though by numbers the cladocerans and insects greatly outnumbered fish. Fjeldså concluded that the species is unspecialized and very flexible in its diet, normally foraging on nektonic prey of fish and arthropods but able to shift to whatever is readily available.

Foraging is preferably done in rather shallow areas, 0.3–1.5 meters deep, often along the edge of beds of emergent vegetation, and where some submerged vegetation is present and nektonic prey might be hiding. Besides diving for food, the birds "skim" prey from the water's surface and snatch insects from the air or from overhanging vegetation. When diving, the birds proceed at about 1 meter per second, and most foraging dives are in the range of 7–25 seconds duration, with extremes as long as 73 seconds (Fjeldså 1973a).

MOVEMENTS AND MIGRATIONS. Although migratory routes are still poorly known in North America, they seem to follow both coastlines and, to a much more limited degree, the Mississippi valley. When migrating overland, the birds move nocturnally, but coastal migration is often diurnal. Northward movement from wintering areas begins in March, most birds are gone from southern coastal wintering sites by the end of April, and by mid-May they have left northern waters. Arrival in inland breeding sites begins in March and may peak in April, though some northern areas may not be reached until late May or even early June. Like the other grebes, movement during the fall is fairly prolonged and related to the time of freeze-up, with birds arriving on most coastal wintering areas at various times from October to late November or even December. Some of the birds wintering on the Gulf Coast may arrive there via the Atlantic (Palmer 1962).

Social Behavior

MATING SYSTEM AND TERRITORIALITY. Horned grebes are seasonally monogamous, with possible longer retention of pair bonds. Fjeldså (1973c) reported that about 75 percent of the birds arrive on breeding lakes already paired and that those arriving already mated exhibit little elaborate display. He believed that pairs may be formed either through gradual development of individual associations in flocks or through complex displays performed by unpaired birds on the breeding grounds. Mate retention and nest site tenacity have both been observed among Swedish birds. In spring there is a rapid shift from the flocking tendency of migrating birds to dispersal and hostility. Territoriality in grebes is not a prerequisite to pair bonding but instead follows and is a consequence of it, according to Fjeldså. Fjeldså (1973b) found that only 16 of 560 birds remained unmated through the summer and that only 9.2 percent of the paired birds apparently laid no eggs, leading him to conclude that breeding is regular in first-year birds. However, such first-year birds were socially submissive and tended to be expelled from favored breeding areas to inferior habitats under the pressures of crowding. Territories are established around nests or mating platforms, and in Fjeldså's study area type A territories (which included exclusive foraging areas) were formed by pairs on small ponds or bays, and these averaged about 0.5 hectare. On larger lakes type B territories (with communal foraging areas) were more typical, and these averaged about 1 hectare for solitary pairs. On some large lakes loose colonies occurred, and there a small activity radius was used by each pair. Mutually exclusive use was typically limited to the area immediately around the nest (type C territoriality), with foraging done well away from the nesting colony (Fjeldså 1973b). Territory size was thus quite variable and largely related to the location and richness of the food supply.

VOICE AND DISPLAY. Vocalizations have been summarized by Cramp and Simmons (1977) and are described as part of general display behavior by Storer (1969) and Fjeldså (1973c). The advertising call, which is uttered by unmated birds during the breeding season and also used as a contact call to reunite the pair or family, is a loud, nasal call that tends to descend in pitch and ends in a trill or rattle. An alarm note is similar but is usually shriller. Duet trilling is a loud, accelerating trill uttered by a pair during the triumph ceremony, and during copulation a very similar trilled call is uttered by the active (treading) partner. Threat chittering is also similar, and it intergrades with duet trilling during hostile encounters between pairs. A contact chittering call is used by paired birds as a "conversational" call, and a repeated two- or three-syllable note is uttered by one or both of the birds during platform courtship. Various other notes have also been described for the species but are of uncertain importance or function.

Displays have been described in detail by Fjeldså (1973c), whose drawings are the basis for those included here (figs. 34 and 35). Unpaired birds utter the advertis-

ing call in a stretched neck posture with spread nuptial plumage (fig. 35C,J), a posture somewhat similar to the aggressive upright (fig. 35B) and also the appeasement posture (fig. 35A). Two grebes alternate advertising calls at a distance of about 10–15 meters, then one (of either sex) may dive. Simultaneously the other adopts the cat posture (fig. 34B). As the underwater bird approaches it briefly exposes itself in a "bouncy" posture (fig. 34A) as many as five times. When it has reached a distance of about a meter from the waiting partner, the submerged bird suddenly appears and looms out of the water in a ghostly penguin posture, with the neck initially bent forward and finally stretching out vertically as the bird turns toward its partner, which by now has adopted a raised neck attitude (fig. 34C) and is about to rise in the water to perform a mutual penguin dance with head shaking by both birds (fig. 34D,E). This dance may last a few seconds, after which the pair sinks back into the water and begins mutual habit preening (fig. 34G). Later, after pair formation, alternations of mutual head shaking and habit preening (fig. 34G,H) form a common type of ceremony. Alternations of habit preening and the penguin posture may last up to 40 seconds, and the

34. Social behavior of the horned grebe (after Fjeldså 1973c): A, bouncy posture; B, cat display; C, transition from cat to penguin posture; D, E, penguin dance; F, turning away; G, mutual habit preening; H, mutual head shaking; I, J, triumph ceremony; K, hunched phase and parallel swimming.

ceremony ends with a gradual sleeking of the plumage and a slow, ritualized turning away (fig. 34F). After this one of the partners may fly away and begin the entire discovery ceremony again with a cat display, or both might dive. Sometimes after such a dive the two birds will appear with weeds in their bills and swim toward each other (fig. 35D). Upon meeting, they rise in the water (fig. 35E), as in the penguin dance, but then turn and perform a parallel rush over the water for 5–10 meters (fig. 35F) while still carrying their weeds. They then subside and usually move slightly apart (fig. 35G), only to begin a new rush a few seconds later in a different direction. Although other North American grebes have a weed dance, and some also have slower barging ceremonies, only this species exhibits a weed rush display. A male has also been seen performing the penguin posture toward humans during nest defense (fig. 35H), after which the bird turned and moved about a meter in the same posture (barging) before finally lowering its head plumes and diving (fig. 35I). In the triumph ceremony,

performed only by well-established pairs, the two birds meet in a hunched posture (fig. 34I). They then turn parallel and swim together for a distance or make a series of turnings, ending with a parallel swim (fig. 34J) or with swimming toward the nest one after the other. Food presentation by a male to its mate has been rarely observed but does not seem to be ritualized into a display.

Platform behavior begins when one bird initiates soliciting by adopting an inviting posture, either on a platform (fig. 35K) or on the water. Water soliciting often precedes platform solicitation and actual copulation. During inviting the crest is lowered and the neck is slightly bent, producing a different appearance from the advertising posture (fig. 35J), which is sometimes performed on the nest by an incubating bird whose mate is nearby. A soliciting bird may also utter a platform call, and it usually also performs rearing, a standing posture with the neck bent and the bill pointing downward. Mounting is done from the rear by a quick jump onto the back of the soliciting bird, and during copulation the mounted bird may lift its head, but usually not to the point that it touches the breast of the treading bird. After copulation the treading bird walks over the head of the partner and lands in the water with a splash. It sometimes supplements this with water treading in a nearly erect posture and may terminate the sequence with head turning or simply turning to face the other bird. The bird on the nest may perform head turning or may rise and briefly adopt a posture similar to the red-necked grebe's ecstatic posture (Fjeldså 1973c).

Reproductive Biology

BREEDING SEASON AND NESTING SUBSTRATE. The egg-laying period in this species is relatively protracted. Egg records in North Dakota extend from April 6 to July 21, with a peak in June (Stewart 1975; Bent 1919). Egg records from Manitoba to British Columbia are from late May to early July, with a peak in early June, and from Ontario and Quebec there is a similar range from late May to early July (Bent 1919; Peck and James 1983). An even longer egg-laying period, from mid-May to September, was noted in Iceland and Norway by Fjeldså (1973b). In a Manitoba study egg laying extended for 53–57 days in two different years, with egg laying in initial nests completed by mid-June but renesting and second nestings lasting into July. This long nesting season reflected the fact that most pairs renested at least once following egg failure; 6 pairs renested three times, and 2 pairs (out of an apparent total of 70) produced second clutches following hatching of first broods (Ferguson and Sealy 1983). In this study all nests were

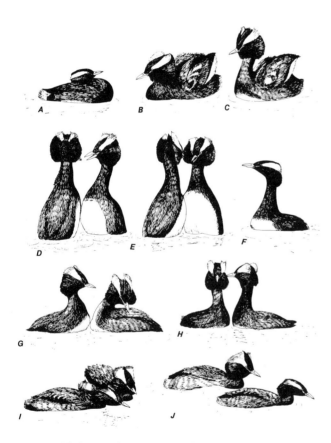

35. Social behavior of the horned grebe (after Fjeldså 1973c): A, appeasement; B, aggressive upright posture; C, advertising in stretched neck posture; D–G, weed ceremony; H, penguin posture; followed by I, barging and diving; J, advertising on nest; K, copulation invitation.

anchored to emergent vegetation, in water averaging 39.2 centimeters deep and rarely less than 20 centimeters deep, the apparent minimum depth needed for underwater access to the nest. Stewart noted a similar average depth of 40.6 centimeters for 11 sites, Fjeldså (1973b) a 32 centimeter depth for 741 nests, and Sugden (1977) a considerably deeper (86 centimeter) average water depth at 50 nest sites. Although nests are usually placed in stands of emergent vegetation, a variety of nest substrates are used, including such supports as submerged stones and even submerged vegetation without any associated emergent support for the nest (Fjeldså 1973b).

NEST BUILDING AND EGG LAYING. Nest building is done gradually by both sexes; Fjeldså (1973b) reported peaks of nest building beginning as early as 33 days before the initial egg. Several pairing platforms may also be built by the pair, especially by probable first-year birds, and even unpaired birds may build such platforms. The final nest is built during the last week before laying, sometimes in a matter of hours. The egg-laying period is highly protracted, with young birds tending to nest somewhat later than old birds. Birds that nested on sites that were unsuccessful the previous season nested somewhat later than those where breeding was successful. Relative crowding does not appear to affect the time of laying; thus pair bonding rather than any flocking effect seems to influence time of breeding in Fjeldså's view. Eggs are generally laid at 40-hour intervals, with a maximum clutch of about 7 or 8 eggs. The average clutch size in Norway was found by Fjeldså (1973b) to be 3.79 eggs (233 nests), compared with 3.75 for Iceland (537 nests), with seasonal declines in clutch size typical and with possible year-to-year variations as well. In North America the observed average clutch sizes include 4.5 eggs (13 nests) in North Dakota (Stewart 1975) and 5.9 eggs (79 nests) in Manitoba, the latter study indicating a decline in clutch size through the season similar to that observed by Fjeldså. Both these averages are substantially larger than those reported by Fjeldså for Old World birds, although a sample of 45 Finland nests averaged 4.5 eggs (Bauer and Glutz 1966). This variation supports the idea that local and perhaps yearly variations in clutch size do exist, probably as a reflection of local food supplies and associated condition of the laying females (Fjeldså 1973b).

INCUBATION AND BROODING. Incubation, or at least nest attentiveness, typically begins as soon as the first egg is laid. Ferguson and Sealy (1983) observed that incubation was almost equal between the sexes during the egg-laying period, but that after the clutch was complete the females spent appreciably more time on the nest than did males. They observed an incubation period of 23–24 days for 17 eggs but noted two extreme cases when hatching did not occur until 28 and 34 days. Fjeldså (1973b) reported an average incubation period of 22.8 days. If the clutch is lost most pairs will begin a new one by variously using the same nest, building a new one nearby, or even moving to a new territory some distance away. A few pairs may lay as many as three or four clutches; Fjeldså (1973b) reported an average renesting rate of 88 percent, with older pairs having a somewhat higher renesting tendency and an average 13 day renesting interval, compared with a 20 day interval among probable first-year pairs. Fjeldså also saw 9 second clutches laid after successful breeding but judged that such second breeding efforts are negligible in their influence on recruitment. Ferguson and Sealy (1983) observed a shorter (up to 9 days) renesting interval but a similar renesting incidence, with 6 pairs each making two and three renesting efforts. Most renestings were within the pairs' original territories but usually were on new sites. They observed only 2 cases of second nestings, which occurred 10 and 16 days after hatching of the initial clutches. Overall nesting success reported by Fjeldså (1973b) was 75.5 percent (721 nests), and he estimated an overall hatching success of 63.2 percent of all eggs laid. Nesting success was higher in older females and in nest sites protected from flooding or waves, even under crowded conditions, although under extreme crowding nest losses from desertion became significant.

GROWTH AND SURVIVAL OF YOUNG. Fjeldså (1973b) reported an average brood size at hatching of 3.58 young (274 broods), compared with an average brood of 1.92 young at 70 days, indicating a survival rate of 54 percent. He noted that the broods become fairly mobile after the first week, and by about the 10th day the young begin making foraging dives. At 3 weeks of age they may be avoided or even pecked by their parents, and in Fjeldså's study area they fledged by 55–60 days. Similarly Ferguson and Sealy (1983) reported that 75 percent of the known-aged chicks they observed were fledged by 45–50 days of age, with the earliest flight in chicks 41–42 days old. They observed a survival rate of 66 percent (of 155 chicks) between hatching and fledging, with most losses in the first 10 days after hatching. Chick survival was unrelated to initial brood size, indicating that late-hatched chicks in large broods did not specifically suffer from food competition with older siblings. Most of the chick losses in Fjeldså's study also occurred during the first few weeks; by 20 days the average brood size was already only 1.9 young (81 broods). Survival of chicks was better on food-rich lakes than food-poor ones and better for the broods of experienced females than those of first-year birds. There was also a drop in fecundity through the season, though this was related directly to the seasonal reduction in clutch size rather than to any higher rates of late-summer mortality.

BREEDING SUCCESS AND RECRUITMENT RATES. A six-year average of juvenile/adult ratios during August in Iceland was 0.53 young per adult for over 2,600 birds, indicating an overall recruitment rate of 34.6 percent, but with substantial year-to-year variation (Fjeldså 1973b). Ferguson and Sealy (1983) reported that 102 juveniles fledged from 155 hatched chicks (66 percent) and that hatching success was only 30.3 percent. Thus the overall breeding success was no more than about 20 percent, and with no nonbreeders and an average clutch of 5.9 eggs, no more than 1.2 young per pair (37.5 percent recruitment rate) would have been raised without renesting. Certainly renesting efforts increased this potential productivity somewhat, although clutch sizes of late nestings were lower and relatively few young were hatched after early July, when most renestings would have been hatching. It thus seems likely that recruitment rates for horned grebes probably range from about 40 to 50 percent, or fairly close to Fjeldså's (1973b) estimate of an annual 50 percent adult mortality rate, based on the rate of disappearance of adults between spring and fall. There are no published estimates of adult mortality based on banding.

Evolutionary History and Relationships

Fjeldså (1973c) reviewed the behavioral evidence on this species' relationships and concluded that it may be a possible link between the typical *Podiceps* forms and the *Podilymbus-Tachybaptus* group, partly because of its retention of such presumably ancestral traits as a triumph ceremony with duet trilling (as in least and pied-billed grebes), courtship chasing (also in little, least, and pied-billed grebes), courtship feeding (as in little and western grebes), splash diving during human disturbance at the nest (as in pied-billed and little grebes), the slight raising of the head by the passive partner during copulation (as in little, pied-billed, and sometimes western grebes). Fjeldså (1982a) later placed it at one end of *Podiceps*, nearest the red-necked and great crested grebes, in a published phyletic dendrogram of grebes. It shares with these species and the four typical *Podiceps* forms the complex discovery ceremony, though in the horned grebe and red-necked grebe the "cat display" component has apparently become rather simplified. In horned, great crested, and red-necked grebes the downy young have single lateral neck stripes and a white spot on the rear crown, although the horned

young is variable in its neck striping, according to Fjeldså (1973c).

Population Status and Conservation

No estimates of population size are available for either the Eurasian or the North American region. Clapp et al. (1982) considered the species one of the most vulnerable to oiling mortality in the southern United States. They summarized data for 8 oil spills in which horned grebes constituted 12.3 percent of oiled birds among a total mortality of 34,717 individuals, including one sample from Virginia in which over half the birds killed were of this species. However, the species is adapted to a rather wide array of foods, breeding habitats, and both wintering and breeding locations, so it is unlikely to suffer catastrophic losses from any single oiling event or from losses of local breeding habitats.

Red-necked Grebe

Podiceps grisegena (Boddaert)

OTHER VERNACULAR NAMES: Holboell's grebe; grastrubert lappedykker (Danish); grèbe jougris (French); Rothalstaucher (German); sefgodi (Icelandic); akaeri kaitsuburi (Japanese); seroshchyokaya poganka (Russian); grahakedopping (Swedish).

Distribution of North American Subspecies (See Map 8)

Podiceps grisegena holboellii Reinhardt
BREEDS in North America from western and central Alaska, central Yukon, northwestern and southern Mackenzie, northwestern Saskatchewan, central Manitoba, and western and south-central Ontario south to Saint Lawrence Island (at least formerly), the Alaska Peninsula, central Washington, northern Montana, northeastern South Dakota and south-central Minnesota, rarely to southwestern Oregon, northern Michigan, southern Quebec, and New Hampshire. Also breeds from Transbaikalia and Manchuria through northeastern Siberia (except Kamchatka, where another race is found) and on the Commander Islands.

WINTERS in Asia south to southern China and Japan; in North America from the Pribilof and Aleutian islands, Kodiak, and southeastern Alaska to central (rarely southern) California; and from Newfoundland (occasionally), Nova Scotia, and New Brunswick to Georgia and northern Florida.

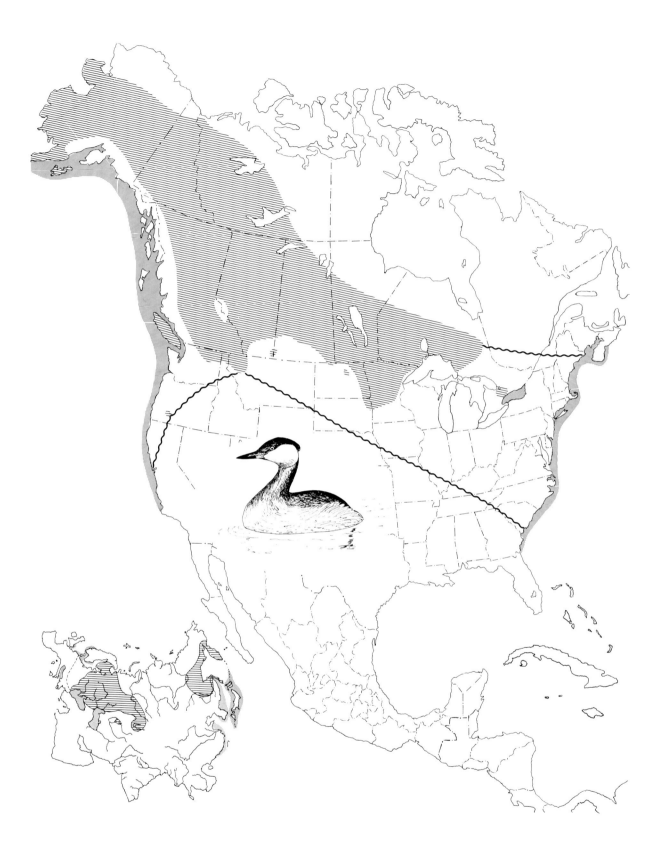

8. Current North American distribution of the red-necked grebe; symbols as in map 7.

Description (after Witherby et al. 1941)

ADULTS IN BREEDING PLUMAGE (sexes alike). Forehead crown, two tufts of elongated feathers on each side of crown black, feathers with white bases; back of neck black, feathers with chestnut bases; whole upperparts brownish black, feathers narrowly edged pale brown; chin, upper throat, and sides of head uniform gray divided from black of top of head by narrow white line (these gray feathers somewhat longish but not forming a tippet); rest of throat, sides of neck, and upper breast rich chestnut with bronzy tinge on breast; rest of underparts silky white more or less spotted blackish; feathers of sides of breast and flanks mostly brownish black or with large tips of brownish black and usually with considerable amount of rufous basally; axillaries and under wing coverts white; primaries grayish brown with black shafts; secondaries mostly white with shafts black basally and outer feathers with varying amounts of brown, inner feathers partly brown and white (sometimes tinged rufous), innermost blackish brown; wing coverts grayish brown except innermost, which are blackish brown and upper series of lesser which are white and often tinged rufous on inner feathers.

WINTER PLUMAGE. Forehead, crown (no tufts) and back of neck dark grayish brown; rest of upperparts darker brown (but paler than summer) and feathers edged brownish gray; under and behind eye finely mottled grayish brown; rest of sides of head, chin, and upper throat white; lower throat and sides of neck brown, feathers lightly tipped whitish; rest of underparts silky white, some small blackish brown dots on sides and body and larger ones on flanks, but considerably less than in summer; wings as in summer but innermost secondaries and greater coverts not quite so black.

JUVENILES. Resembles adult summer but upperparts with slightly more brownish tinge; sides of head, chin, and upper throat dull buffish white with two rather broad and fairly well defined brownish black stripes behind eye, a patch at base of lower mandible and varying amounts of brownish black mottling on chin and upper throat; rest of throat to base and sides of neck buffish chestnut varying in depth but never so dark and rich as in adult; underparts not so silky white as adult and flanks paler brown, partially concealed spots on upper breast and sides of body paler and not so distinct as in adult; wings as adult but white secondaries often more mottled with brown at tips.

DOWNY YOUNG. Head and neck striped, but black on each side crown wide and white patch in center of back of crown reduced to a stripe; light stripes on neck dull buffish; those down center of back of neck narrowed and obscured by black and those on upper mantle very narrow and scarcely noticeable; rest of upperparts, base of throat, sides of body, flanks, and vent brownish black, unstriped; center of breast and belly white; black stripes down sides of neck broken about middle, and two stripes down sides of lower throat joining and forming V. This species closely resembles the horned grebe except that both mandibles have two black bands. The pink spot of the anterior crown is vestigial, and the white median mark on the rear crown is very narrow. The black band above the eye is sharply bent and interrupted behind the ear, and there are two prominent slanting black bands on each side of the head in older young (Fjeldså 1977).

Measurements and Weights

MEASUREMENTS (of *holboellii*). Wing: adult males 185–212 mm (average of 20, 195.6); females 182–98 mm (average of 14, 189.3). Culmen: males 48.5–56.0 mm (average of 20, 50.2); females 45–50 mm (average of 14, 46.7). Eggs: average of 20, 55.7 x 36.1 mm (Palmer 1962).

WEIGHTS (of *holboellii*). All seasons, males 1,002–1,256 g (average of 6, 1,166); females 945–1,165 g (average of 3, 1,067). Weights of nominate *grisegena* are substantially smaller (Cramp and Simmons 1977). Estimated egg weight, 34.9 g (Schönwetter 1967). Newly hatched chicks (of *grisegena*) weigh 17–23 g (Fjeldså 1977).

Identification

IN THE FIELD. This is the largest of the native *Podiceps* grebes and the one with the most massive and loonlike bill. In all plumages the bill is usually yellowish basally and blackish toward the tip and along the culmen, separating the species from all other grebes except the distinctive western grebe. In breeding plumage the chestnut breast and foreneck, and the white cheeks and throat, are unique. Adults in winter plumage are generally grayish, except for a white throat that extends backward and upward behind the cheeks to form a white crescent-shaped area. First-winter birds often lack this feature but have longer and more massive beaks than any of the other *Podiceps* grebes. The birds are often found in fairly large water areas, including deep lakes, where loons also occur. The calls include loonlike roaring, wailing, howling, or hooting calls, quacking notes, clucking calls, and other vocalizations.

IN THE HAND. The large size (wing more than 150 mm) and heavy bill (which is more than 12 mm deep at its base) provide for ready distinction of this species from nearly all other grebes. The western grebe has a similar wing length, but its bill is much less robust, and all the other grebes have much shorter bills that are no more than 10 mm deep at the base of the upper mandible.

Ecology and Habitats

BREEDING AND NONBREEDING HABITATS. Habitats of the European race have been much more completely described than those of the North American form, and possibly some differences exist as a result of major dietary divergences between them. In Europe the species exhibits a generally inland distribution, breeding on relatively small waters having a fairly high incidence of emergent vegetation and associated shallow depths, often with low populations of fish (Cramp and Simmons 1977). Breeding waters there have minimum sizes variously estimated as from 0.4 to 3.0 hectares (Wobus 1964), and their main requirement seems to be the presence of islands or margins of emergent vegetation beds for nesting. Deep lakes are apparently avoided in Europe, and eutrophic or boggy lakes or ponds, often rather small, are used instead (Sage 1973). In North America the species favors shallow lakes that support good fish populations and sometimes are nearly devoid of emergent vegetation; at least in Minnesota, it is often found on fairly large lakes that also support breeding common loons. In North Dakota it breeds on fresh or slightly brackish permanent ponds and lakes that usually are at least 10 acres (4 hectares) in area, especially those with beds of submerged aquatic plants (Stewart 1975). In both North America and Europe it evidently favors water areas surrounded by well-grown forest over those in prairie areas, but it also extends north to scrub tundra habitats in some regions.

Nonbreeding habitats of both races are fairly large, often estuarine or coastal waters having good fish populations, and occasionally wintering birds move several miles out into open ocean (Palmer 1962; Cramp and Simmons 1977).

SOCIALITY AND DENSITIES. This is not a highly social species, and its breeding densities are relatively low. Wobus (1964) and Bauer and Glutz (1966) have summarized density data for the European population, and in general these densities average less than 1 pair per hectare, with large water areas having somewhat lower densities. The highest density these authors reported was a case of 3 pairs breeding on a 1.1 hectare marsh. One of the few accounts available for estimating North American densities is Munro's (1941) report of 15–40 pairs (in various years) nesting on a small lake in British Columbia that Wobus judged was about 44.1 hectares, representing a density of less than 1 pair per hectare. The birds are most often solitary breeders, but loose

colonies do develop at times. Thus Pruess (1969) estimated a total of 350–400 breeding pairs in 82 sites in Denmark, roughly averaging about 4 pairs per site. Nesting in small colonies has also been reported in North America; Bent (1919) observed a group of 7 nests in an area of emergent vegetation extending a hundred yards or more out into Lake Winnipegosis, and he mentioned accounts of similar-sized colonies for Minnesota and Montana.

PREDATORS AND COMPETITORS. No major predators of adults have been identified, though Wobus (1964) mentioned various large raptors as probable enemies of adults. Certainly the eggs are subject to the expected array of bird and mammal predators, and it is possible that pike (*Esox* spp.) are important predators of chicks (Wobus 1964). Wobus also suggested that the carrion crow (*Corvus corone*) may be the most important of the avian egg predators. Fjeldså's (1982b, 1983a) studies on the bill and jaw musculature of red-necked grebes suggest that the European race of this species is an arthropod forager mainly adapted to foliage gleaning but that the long-billed and fish-eating North American race is an ecological counterpart to the great crested grebe. Fjeldså attributes this North American population's divergence from the European race in bill shape and diet to ecological character release, in response to its allopatry with respect to the great crested grebe. The red-necked grebe also overlaps with the horned grebe and tends to displace it from sites that are suitable to both species (Fjeldså 1973d). Competitive foraging interactions between western grebes and red-necked grebes, which might be expected, are seemingly still unstudied, but red-necked grebes are not often found on the alkaline prairie marshes so strongly favored by western grebes.

General Biology

FOOD AND FORAGING BEHAVIOR. The best analysis of the foods of the North American race of this species is still that of Wetmore (1924), who examined 46 stomachs from all months of the year and found food remains in 36. Fish remains made up 55.5 percent of the food materials and occurred in 23 of the samples. Many of these remains were too fully digested to be identified. Crustaceans were found in 9 stomachs and composed 20 percent of the food remains. These included mud lobsters (*Upogebia*), shrimps (*Crago*), prawns (*Palaemonetes*), and crayfish (*Cambarus* and *Potamobius*). Insects occurred in 13 of the samples and composed 21.5 percent of the food remains, although in some cases they may have come from the digestive tracts of the fish consumed, as may also have been true of the small amounts of vegetation found. By contrast, aquatic and terrestrial insects and their larvae form a major part of the summer foods of red-necked grebes in Europe (Wobus 1964; Madsen 1957). However, fish are major foods during winter, when the birds are on marine habitats. Fish were present in all of 25 stomachs from Danish coastal waters, and over half of the birds had eaten more than one kind of fish. Gobies (*Gobius* spp.) and cod (*Gadus* spp.) were encountered in the largest numbers. Other food types found included crustaceans (in 18 stomachs) polychaetes (in 16 stomachs), insects (in 12 stomachs), and mollusks (in 5 stomachs).

Foraging is done by fairly extended dives in rather shallow water. Mean foraging dive durations generally depend on depth of the water, but dives typically average less than 30 seconds. Dewar (1924) indicated an average duration of 18.4 seconds, Sage (1973) an average of 22.7 seconds (water 1.5 meters deep), and Simmons (1970) averages of 24.8 seconds (water 2 meters deep) and 29 seconds (water to 4 meters deep). Escape dives would certainly average longer than these and probably are the sources of various reports of dives up to about a minute. Little or nothing has been written on maximum diving depths or rates and distances of lateral movement, but a possible maximum observed diving depth is 40 meters (Wobus 1964).

MOVEMENTS AND MIGRATIONS. Throughout its range this species is migratory, probably migrating primarily at night while moving overland, but also flying along the coast by day. Migration typically occurs in small groups, and it has been suggested that during fall older birds may precede the young of the year. No such segregation has been observed in spring. Very little is known of the routes taken in North America or the breeding origins and limits of birds wintering on the Pacific versus the Atlantic coast. Large flocks are unusual during migration, though Lake Ontario is an important stopover point in April and early May for birds moving up from the Atlantic coast. A similar inland movement from the Pacific coast also occurs in April, and the birds are virtually gone from the coast by May. Most breeding areas in the northern states and Canada are occupied by May, with only a few remote nesting areas reached as late as early June. Return to coastal wintering areas generally is essentially completed by mid-November, although some birds may remain on large inland waters until late November or early December (Palmer 1962).

Social Behavior

MATING SYSTEM AND TERRITORIALITY. At least seasonal, if not more prolonged, monogamous mating is typical of this species. Pair bonding evidently begins on migration if it has not already been initiated in the winter quarters (Wobus 1964). The degree of territoriality is somewhat variable, with some observers suggesting that a type A territory (including exclusive foraging areas) is defended while other observations suggest a type B territory (with foraging outside the defended area) is typical (Wobus 1964). Thus Wobus found some pairs defending each of these types of territories, encompassing areas of from 1 to 6 hectares. Munro (1941) stated that pairs defended stretches of shoreline associated with nesting sites of from 70 to 110 meters. All these observations suggest that no single description of territoriality is likely to fit the species and that local ecological conditions and associated food supplies are almost certain to influence the relative development of territorial behavior.

VOICE AND DISPLAY. This is a highly vocal species, with perhaps the most penetrating voice of any North American grebe, which correlates well with its generally high level of dispersal during the breeding season and its occurrence on rather large bodies of water. The species' calls and postures have been described by Wobus (1964) and more recently have been summarized by J. Fjeldså and R. Storer (in Cramp and Simmons 1977). Storer (1969) has also made some comparative comments on the displays of this species. At least seven types of calls have been described, of which the most significant is probably the "song and display call." This is a very loud, loonlike call uttered by birds on territory, either singly or as a duet, and sometimes as part of singing encounters by birds on different territories. In the North American race the call is perhaps best described as a series of somewhat loonlike *ah-oo* notes, usually followed by a group of *ah*, *whaa*, or other chittering sounds. As many as sixty notes may be uttered in sequence during territorial interactions, though during aquatic courtship the series may be about four to ten notes, with even fewer being typical during summer. Notes uttered by the female are lower in pitch and shorter than those of the male and may be more braying or whinnying in character. The advertising call of birds of the North American race is a loud, nasal *arrrrr*, used before the discovery ceremony and in combination with duet calls. A clucking call, given by both members of a pair during pair formation and also during platform courtship, is a rapid series of *ga* or *keck* notes. Disturbed birds produce a hissing note, especially at the nest, and utter an alarm note upon disturbance by humans. Finally, a purring note is uttered by birds soliciting on the nest, and the male during copulation utters a rattling call.

Displays of this species are illustrated in figure 36,

36. Social behavior of the red-necked grebe (mainly after Cramp and Simmons 1977): A, calling during "Sitzbalz"; B, triumph ceremony; C, head turning following triumph ceremony; D, cat display; E, approach and copulation invitation postures; F, rearing; G, H, copulation; I, postcopulatory ecstatic posture; J, postcopulatory water treading; K, postcopulatory head turning.

based on drawings by Wobus (1964) and by J. Fjeldså in Cramp and Simmons (1977). In general this species seems to be more vocal and correspondingly to have fewer elaborate postural signals than the other North American *Podiceps* grebes. Males utter the advertising call or "song" during most ceremonial encounters and apparently lack habit preening, a retreat ceremony, or a casual food presentation ceremony. The advertising call may be uttered as the male swims about with a slightly raised neck and crest, or as part of more elaborate ceremonies. The most frequent courtship ceremony is the "Sitzbalz," a mutual approach the two birds perform with erected necks, raised crests, and repeated head shaking ("low head waggles") while uttering the advertising call and holding their bills at slightly declining angles. At higher intensity, one (fig. 36A) or both (fig. 36B) birds may rise out of the water during the calling phase. In a typical "discovery ceremony" sequence one

bird approaches the other in a shallow approach dive, emerging at times in a "bouncy posture" or with only its head showing, and then rises and calls while in a "ghostly penguin" (fig. 36A) posture when 3–4 meters away from the other, which is typically in a cat posture (fig. 36D) with partially raised and folded wings and a retracted head. After the ghostly penguin and cat display the two birds may simultaneously rise and mutually call in a penguin dance (fig. 36B) and then subside and perform slow head turning movements (fig. 36C). Parallel swimming ("barging") may also occur before or after the penguin dance (Wobus 1964), or the penguin dance may be replaced by a parallel rushing of the pair over the water. A "weed trick" is also present, with one or both members of the pair emerging slowly from a dive with weeds in the bill, rising almost upright in a penguin dance while calling, and then subsiding. The weed trick is invariably performed after successful discovery ceremonies in this species (Fjeldså 1983a). After aggressive encounters the pair may perform a triumph ceremony by duetting from a hunched posture similar to the cat posture. Habit preening is also present, as it is in all typical *Podiceps* forms (Fjeldså 1983a).

Copulation is normally performed on a platform, although there have been observations of inviting behavior and attempted copulation on the water. Soliciting behavior includes a crouched inviting posture (fig. 36E) and a standing or "rearing" posture (fig. 36F). During treading the male calls, and the female holds her head low, keeping the bill level or tilting it up only slightly (fig. 36G,H). After treading is completed the male performs a ritualized retreat by swimming away in a nearly erect posture while water treading (fig. 36J), as the female also rises in an "ecstatic" posture (fig. 36I). Finally the male subsides and turns in the water to face the female (fig. 36J), and both birds perform a series of slow head turns. Chamberlin (1977) observed a simpler postcopulatory pattern in 6 cases, which apparently lacked the ritualized retreat of the male and the ecstatic posture of the female.

Reproductive Biology

BREEDING SEASON AND NESTING SUBSTRATE. The nesting season is moderately long; Peck and James (1983) indicate a spread of 60 egg dates in Ontario as May 15 to September 17, with a peak in mid-June. Records summarized by Bent (1919) indicate a spread of egg records from Minnesota, North Dakota, and Washington from mid-May to mid-July, and for the Canadian provinces from Manitoba to British Columbia from late May to early August, peaking in June. This distribution of dates suggests that double brooding is rare in North

America, but there is one record of double brooding for North Dakota (Mink and Gibson 1976). Nests are built over water as floating structures that are anchored to emergent vegetation or, more rarely, to submerged supports. Floating wooden platforms have occasionally also been used as nest sites. Gotzman (1965) found the commonest water depth at nests to be 0.50–0.75 meter, and most nests were within 20 meters of the edge of the reedbed, in plants of medium density. In Ontario nests have been found at depths ranging from 25 centimeters to more than 150 centimeters, and from about 1 to 18 meters from the nearest shore (Peck and James 1983).

NEST BUILDING AND EGG LAYING. The nest is built by both members of the pair, and it is frequently preceded by the construction of several platforms that are not completed but are used for copulation. The platform for the final nest may be initiated 4–8 days before egg laying (Wobus 1964). The eggs are laid daily or at 2-day intervals, and the clutch size seems to be quite variable. Wobus (1964) reported an overall mean of 3.8 eggs for 36 clutches, with a strong seasonal decline in average clutch size (from 4.9 to 2.5 eggs), and Onno (1960) reported a similar overall mean of 3.55 eggs for 44 clutches from birds of the nominate race. There are no comparable averages for North America, but Peck and James (1983) indicated a range of 1–7 eggs for 52 nests, with 26 nests having 3–5 eggs. Seven North Dakota clutches averaged 3.1 eggs (Stewart 1975). Palmer (1962) and Bent (1919) indicated that the normal clutch in North America is 4 or 5 eggs, at least as large as or perhaps slightly larger than in Europe.

INCUBATION AND BROODING. The incubation period is normally 22–23 days, but in cold weather it is sometimes protracted to as long as 27 days (Onno 1960; Wobus 1964). Incubation is performed by both sexes and begins with the laying of the first egg or variably later, so that staggered hatching of the young is typical, though simultaneous hatching is also fairly common. Typically one parent continues to incubate any remaining eggs as the other begins to tend the newly hatched chicks, but occasionally some unhatched eggs may be left in the nest.

GROWTH AND SURVIVAL OF YOUNG. There is little good information on development or growth rates in this species, and indeed the periods to juvenile independence and fledging are still unknown. Chamberlin (1977) noted that birds up to 54 days old were still being fed by their parents, their foods consisting of minnows, crayfish, and probably insects. Apparently unfledged young have been seen with their parents when 72 days old (Palmer 1962), but it seems unlikely that fledging

would normally require this long. Wobus (1964) judged that the young become independent when they are 8–10 weeks old. Mink and Gibson (1976) saw a pair begin a second nesting when the three surviving young of their first brood were still quite young. A total of 5 eggs was laid in the second nest. Remarkably, one of the young of the first brood apparently attempted to incubate the eggs of the second nesting but was eventually chased away from the nest by the parents.

BREEDING SUCCESS AND RECRUITMENT RATES. Little information exists on mortality rates and brood sizes, but Munro (1941) reported that 35 Canadian broods had 65 young, or 1.85 young per brood, based on counts taken mainly in July. Munro also reported an August average of 0.65 young per adult, suggesting a recruitment rate of approximately 39 percent. Munro judged that during an average year only about 70 percent of the adults attempting to breed are successful. There are no estimates available of adult mortality rates in this species, either in North America or in Europe. The age of maturity has been estimated as 2 years (Palmer 1962; Cramp and Simmons 1977), but there seems to be no hard evidence confirming this except perhaps for the fairly high incidence of nonbreeding, presumed first-year birds in the breeding range and on larger lakes along the migratory route. The attainment of full adult nuptial plumage in the first year (Cramp and Simmons 1977) certainly suggests a physiological capability for breeding then. Perhaps, like the great crested grebe, which also attains adult plumage the first year, first-year birds often mate and establish territories, but successful breeding may not occur until the second year. In that species a total of 715 pairs raised 873 young (3 British samples cited by Cramp and Simmons 1977), representing a recruitment rate of 37.9 percent, very similar to the estimated 39 percent recruitment rate for the red-necked grebe and suggestive of rather similar demographic characteristics. The two species also have very similar average clutch sizes of about 3.5 eggs.

Evolutionary History and Relationships

This species is probably a fairly close relative of the great crested grebe, with which it shares a number of similarities in behavior, both social display characteristics and also foraging and nesting adaptations. Perhaps the ancestral red-necked grebe evolved in North America while the great crested grebe was differentiating in Eurasia. A later invasion of Eurasia by the red-necked grebe may have resulted in a foraging shift in the latter to arthropod eating as a means of reducing competition between these two very similar species.

Population Status and Conservation

No information is available on the population size of this species, either the North American or the Eurasian component. It is certainly one of the least common of the North American grebes, with seemingly low population densities even in the middle of its breeding range. On coastal wintering areas it is also well dispersed and thus is unlikely to suffer major local losses from oil spills. Clapp et al. (1982) listed 186 deaths of this species owing to oil among a sample of 37,750 birds involved in 12 oiling incidents, or 0.49 percent of the total losses.

Eared Grebe

Podiceps nigricollis Brehm

OTHER VERNACULAR NAMES: Black-necked grebe (British); sorthaleset lappedykker (Danish); grèbe à cou noir (French); Schwarzhalstaucher (German); stargódi (Icelandic); hajiro kaitsuburi (Japanese); chernosheynaya poganka (Russian); svarthalsad dopping (Swedish); zambullidor orejudo (Spanish).

Distribution of North American Subspecies (See Map 9)

Podiceps nigricollis californicus (Heermann)
BREEDS from central interior British Columbia, northwestern Alberta, central Saskatchewan, southern Manitoba, central Minnesota, northwestern Iowa, and northern Nebraska south to east-central California, southern Nevada, central Arizona, northern (rarely southern) New Mexico, and locally in southern Texas. Also breeds locally south to central Mexico.

WINTERS coastally from southern British Columbia to Chiapas; interiorly in California, Nevada, Utah, New Mexico, and southern Texas. Mono Lake, California, is a primary migratory staging area for coastally wintering birds.

Description (after Witherby et al. 1941)

ADULTS IN BREEDING PLUMAGE (sexes alike). Forehead, crown, and neck jet black (feathers of crown somewhat elongated and wedge shaped, broader at tip than at base); mantle and scapulars not quite so intensely black as crown; back dark brown; center of rump and upper tail feathers black, sides of rump and lower tail feathers chestnut with black tips and some

with white bases (sides of head and underparts); chin and throat jet black (feathers at sides somewhat elongated); below and above back of eye through ear coverts a broad patch of golden chestnut (feathers very narrow, long, and silky); upper breast with some mixture of white and a little chestnut giving a mottled appearance; sides of breast and flanks chestnut and black; vent pale brown, feathers downlike and tipped whitish or rufous buff; whole of rest of breast and belly silky white; primaries brown with black shafts, pale brown inner webs and white at extreme base, 7th (from the inside) and sometimes 5th or even 6th tipped white and with inner web mottled white, 3d and 4th same but sometimes with inner webs mostly white except at base, 2d mostly white with some black at base of outer web, first same but often with some black at tip (amount of white on inner primaries thus varies); secondaries white, inner ones with more or less black at tip and innermost entirely black; all wing coverts blackish brown.

WINTER PLUMAGE. Crown, mantle, and scapulars black with slightly more brownish tinge than in summer, on each side of lower nape a white patch not quite meeting in center of back of neck, feathers of which have long black tips and white bases; back blackish brown; sides of rump and tail white, with long black tips; ear coverts and patch below eye brown, chin, cheeks, and upper throat white; lower throat brown, feathers lightly tipped white; breast and belly silky white, feathers of sides of breast with subterminal blackish gray spots, flank feathers with long black tips, vent brownish, feathers tipped white.

JUVENILES. Much like adult winter but upperparts more brownish black, sides of nape tinged buff, throat paler, flanks not so much marked with black.

DOWNY YOUNG. Upperparts blackish with light stripes very ill defined, very narrow grayish stripes on each side of crown, over eye, and down back of neck, but those on mantle and back scarcely noticeable and much broken and hidden; stripes on sides of head and sides of neck better defined and more prominent but considerably broken and not regular; narrow inverted V-shaped blackish stripe on each side of chin and sometimes a narrow short one in center, but usually center of chin as throat, white without stripes; center of breast and belly white, sides black with flecks of white on flanks, vent blackish. Bill flesh colored with two transverse blackish bands across both mandibles. Bare spot on crown and lores bright pink. Unlike the horned and red-necked grebes, the head has a diffuse black cap reaching below the eyes, with only narrow white lines and no median white spot or line on the rear crown (Fjeldså 1977).

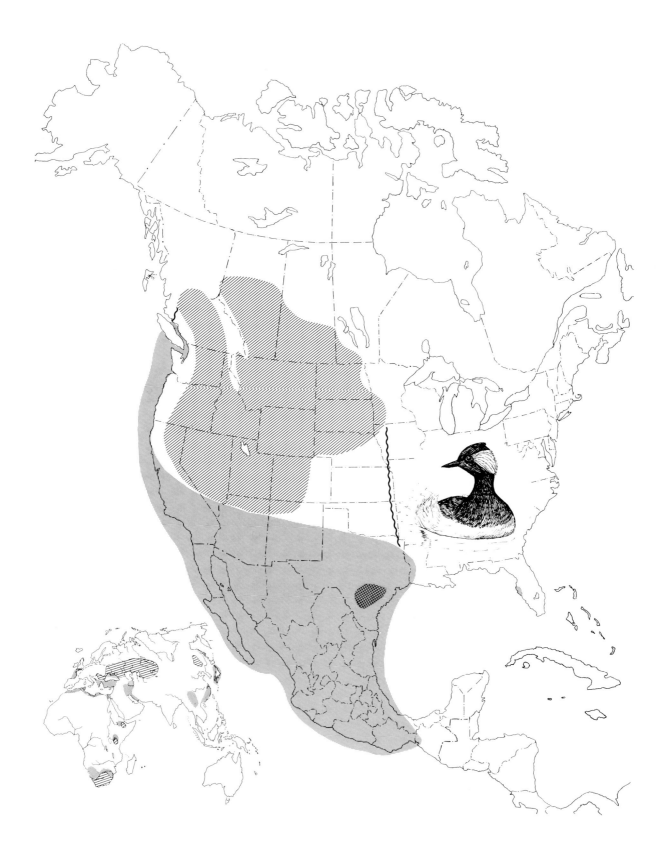

9. Current North American distribution of the eared grebe; symbols as in map 7.

Measurements and Weights

MEASUREMENTS (of *californicus*). Wing: males 130–36 mm (average of 16, 132.9); females 123–31 mm (average of 8, 127). Culmen: males 25.5–29.0 mm (average of 16, 26.5); females 22–24 mm (average of 8, 23.7). Eggs: average of 20, 44.2 x 29.7 mm (Palmer 1962).

WEIGHTS (of *californicus*). In breeding season, males 310–439 g (average of 5, 359 g); females 291–396 g (average of 5, 330.6). In nonbreeding (fall and winter) birds, males 187–358 g (average of 6, 276); females 160–476 g (average of 11, 250.1) (various sources). Estimated egg weight, 20.6 g (Schönwetter 1967). Newly hatched young weigh 12–15 g (Fjeldså 1977).

Identification

IN THE FIELD. This small grebe can be readily recognized in breeding plumage by its distinctive black neck and the golden tuft of feathers fanning out from behind the eye. Apart from this area, the general coloration of the above-water parts is black, compared with a generally brownish color in the horned grebe. In nonbreeding plumage the horned and eared grebes are extremely similar, and apart from the minor differences in bill shape there are few good field marks. Generally the eared grebe has a less sharply two-toned head color; instead the ear area tends to be smudged with gray, shading to black above and white below. The pale area behind the ears does not extend as far back on the nape as in the eared grebe, and there are no white feathers on the leading edge of the upper wing coverts should the bird be seen in flight. The calls are poorly studied but include a shrill chittering display trill, loud threat chittering notes, an ascending, plaintive, flutelike *poo-eee* note that serves as an advertising call, and various other calls.

IN THE HAND. In any plumage, the combination of a short beak (culmen 21–26 mm) that has a straight or even concave culmen shape and a distinctly angulated lower mandible should serve to identify this species. Unlike the similar horned grebe, no white is present on the upper wing coverts.

Ecology and Habitats

BREEDING AND NONBREEDING HABITATS. This species breeds in shallow and generally rather extensively overgrown areas of lakes or marshes, especially highly eutrophic waters with rich plant and invertebrate life, especially insects and mollusks. Small, often quite transitory ponds are frequently used, including sewage ponds, fishponds, and newly flooded areas as well as more permanent waters such as reservoirs and river backwaters (Cramp and Simmons 1977). In North Dakota the birds typically breed on slightly brackish to subsaline waters that range from seasonally present to permanent and usually are at least 10 acres (4 hectares) or larger, but they also breed on shallow river impoundments and occasionally on stock ponds or sewage lagoons. An expanse of open water and extensive beds of submerged aquatic plants are typically present (Stewart 1975). Although in North Dakota the birds breed on ponds as small as 0.4 hectare, these ponds support only single pairs, and colonies tend to develop on larger ponds offering a combination of rich feeding areas in open water and sheltered areas of emergent vegetation. Ponds having about 80 percent open water were found to be most heavily used in one study (Faaborg 1976).

Nonbreeding habitats include a variety of inland water types including freshwater and saline lakes as well as coastal waters such as estuaries, bays, channels, and arms of the sea (Palmer 1962; Cramp and Simmons 1977). In accord with their tendency to invade new breeding habitats, nonbreeding birds seem to be highly opportunistic in their ability to exploit newly available areas.

SOCIALITY AND DENSITIES. This is among the most social of North American grebes, and colonies often are quite large. Yocom, Harris, and Hansen (1958) reported several colonies of more than 100 birds in Washington, and Stewart (1975) mentioned 7 North Dakota colonies of at least 100 pairs. Even rather large ponds or marshes might have nearly all their nests in a small, compact area providing the most sheltered nesting locations, and there nests may be only a few feet apart (Stewart 1975). Faaborg (1976) tabulated 7 North Dakota colonies on ponds and lakes ranging in size from 19.6 to 128.9 hectares, with a collective total of 218 pairs on 334.6 hectares, or 1.5 hectares per pair. Faaborg also noted that the average area used per pair on 3 smaller ponds (under 7.5 hectares) was 1.5 hectares.

PREDATORS AND COMPETITORS. Surprisingly little definite information is available on nest predators, in spite of the conspicuous aspect of the large colonies characteristic of this species. Certainly predation is locally a problem, and Broekhuysen and Frost (1968) suggested that the Hartlaub's gull (*Larus hartlaubii*) might have been the primary egg predator in their study area. No other particular predators have been mentioned, but they are likely to be significant factors in the survival of both eggs and young chicks. Major competitors are likely to be other insectivorous grebes, especially the horned and pied-billed grebes, which have similar ecological adaptations. Faaborg (1976) observed ecological overlap with both horned and pied-billed grebes in North Dakota and suggested that the pied-billed grebe is dominant over horned and eared grebes on ponds it occupies. This may tend to limit eared grebes to the ponds that have too little vegetation to attract pied-billed grebes and are larger than those favored by horned grebes.

General Biology

FOOD AND FORAGING BEHAVIOR. Wetmore's (1924) analysis of 27 stomachs, still the best study of eared grebe foods in North America, included birds taken during most months and from many western localities. Fish remains were found in only 5 stomachs and composed only 9.8 percent of the total food. Insects composed almost 85 percent of the total, and nearly half the total was primarily aquatic bugs (Heteroptera). Damselfly and dragonfly nymphs, aquatic beetles and their larvae, and miscellaneous insects made up most of the remaining insects. Information on the European population as summarized by Cramp and Simmons (1977) indicates that there too insects are the predominant food source, together with mollusks, crustaceans, amphibians, and small fish, depending on time of year and locality. Even during winter fish are apparently a minor dietary item, and crustaceans may be as important a food source as fish in marine habitats (Madsen 1957).

Foraging is frequently done by picking objects from the surface, for which the dorsoventrally flattened bill is apparently specifically adapted or by immersing the head while surface swimming. At times the birds also capture flying insects (Cramp and Simmons 1977; Prinzinger 1974, 1979). However, much of their food is obtained by diving, usually in rather shallow waters up to about 5.5 feet (1.7 meters) deep. They may remain submerged for up to nearly a minute, but dives usually last less than 30 seconds (Madsen 1957). Prinzinger (1974) reported mean foraging dive durations of 28.8 seconds for adults and 21.7 seconds for juveniles, mainly in waters 0.5–1.0 meter deep.

MOVEMENTS AND MIGRATIONS. Probably all of the North American population is migratory, with much of the movement apparently nocturnal. The Salton Sea is the major wintering area for western birds, and Mono Lake is a primary moult/migration area, supporting as many as 750,000 birds at its peak fall population (Jehl 1983; Storer and Jehl 1985). Other shallow, rather alkaline lakes that attract large numbers of migrating birds are Malheur National Wildlife Refuge, Oregon, and Bear River National Wildlife Refuge, Utah (Palmer 1962). Some wintering occurs in coastal waters, but the birds tend to remain close to shore, in shallow waters, especially in the zone just beyond the breaking surf (Munro 1941). The southern limit of birds breeding in North America is probably northern Central America, with wintering common at least to the volcanic lakes of Guatemala. Northward movement from Central America begins in late February and March and tends to follow the Pacific coast, with first arrivals in the northern states and western provinces by April and peak numbers by May. The fall migration begins early, in August, and probably peaks in October and November, with some stragglers remaining until freeze-up (Palmer 1962). As noted earlier, the birds are unusually adept at finding and exploiting newly available habitats such as recently flooded areas.

Social Behavior

MATING SYSTEM AND TERRITORIALITY. Pair bonding is certainly monogamous and is probably of seasonal duration. There is no information on possible re-pairing by birds of the previous year, but given the extreme degree of coloniality, finding previous mates by territorial site tenacity seems unlikely. McAllister (1958) observed that there was apparently no territorial behavior in the colony she studied and that although some apparent pair bonds existed on arrival, these proved to be transitory and definitive pairing was not seen until a few days before nesting. Territorial defense, if present at all, is limited to the area immediately around the nest, and even this tends to break down before hatching. Nests have been found very close together; Stewart (1975) mentioned one large colony where several nests actually touched one another and another where the nests were only 16 inches apart. In one colony about 40 nests were within an area of approximately 1,090 square feet, averaging only 27 square feet per nest. All of this suggests that neither mate fidelity nor territorial fidelity occurs in the eared grebe and that it is thus the most flexible and opportunistic of all North American grebes in its reproductive strategies.

VOICE AND DISPLAY. Vocalizations have been described by Prinzinger (1974, 1979), McAllister (1958), and R. Storer (in Cramp and Simmons 1977). A major call is the advertising call, a three-noted *poo-ee-chk* with the final syllable audible only at close range. It is ascending and flutelike and is uttered both by unmated birds and by paired birds seeking their mates or young. A related call is the "display trill," which is similar to the first but given more rapidly and uttered before and perhaps during the penguin dance ceremony. Threat chittering is very similar to the display trill but is loud and prolonged, with each of two alternating notes distinctly accented. A copulation call is uttered before and during treading, and a similar inviting call is produced by a bird soliciting copulation. The soliciting bird's partner utters a trill at this time, resulting in a "copulation duet." One or more types of alarm calls also exist, and the young utter contact and food calls.

Displays of the eared grebe are varied and have been described in detail by McAllister (1958), by Fjeldså (1982a), and by others in less complete form. The drawings shown here (fig. 37) are based on sketches by Fjeldså (1983a, and in Cramp and Simmons 1977). Hostile interactions are characterized by aggressive forward displays (fig. 37A), with associated chittering calls. The most frequent display and call of unmated birds is probably the advertising posture (fig. 37B), in which the stationary or swimming bird calls with body feathers fluffed, especially those of the neck and head, to attract a potential mate or achieve contact with an actual mate or young. Habit preening occurs commonly; McAllister

37. Social behavior of the eared grebe (mainly after Cramp and Simmons 1977): A, aggressive posture; B, ghostly penguin posture; C, cat display; D, advertising; E, penguin dance; F, barging; G, mutual head shaking; H, hunched phase of triumph ceremony; I, barge diving.

reporting that it apparently has a courtship function, while Fjeldså stated that it is weakly developed and occurs when two birds meet, when strange birds pass a pair, or after a hostile interaction. When performed by two birds it is typically synchronized; habit preening also grades into normal preening behavior. Mutual head shaking is a common display (fig. 37G); it is typically performed as the birds face one another (described for European form) or may also occur as they swim in parallel (American observations). The head is turned from side to side in a horizontal plane, the two birds moving their heads alternately or synchronously and sometimes interspersing this behavior with a quick habit preen. The penguin dance is a major display in eared grebes. While facing each other, the partners tread water and rise up breast to breast, with crests raised, head shaking, and uttering a trilled call (fig. 37E). Dropping back to the water, they continue head shaking while facing each other, may begin habit preening, or may go directly from the penguin dance to a "barging" display (fig. 37F), moving in parallel for several meters. Sometimes barging is terminated by diving in unison, and the birds may reappear a few meters away to begin barging again. An alternative to the penguin dance is for the pair to turn parallel and splash across the water for up to 50 meters, then submerge and sometimes follow this with a penguin dance. The most complex ceremony of all is the discovery ceremony. It begins with one bird display calling (fig. 37D) while another approaches it and dives. The first bird, facing the point where the other dived, performs a cat display (fig. 37C). The approaching bird emerges briefly several times, usually in a retracted neck "bouncy" posture, and finally emerges in the "ghostly penguin" posture (fig. 37B), usually oriented away from the other bird and with the bill held nearly horizontal while the neck is stretched maximally. Its partner then lowers its wings and extends its neck, and the two birds barge together and perform a penguin dance. This complete ceremony may occur only very early in pair formation; later versions are variously simplified. The weed trick display is sometimes performed; a single bird will obtain a weed and present it to another, either by dropping it in front or possibly even throwing the weed across the other's breast or back. Established pairs will also perform a "hunched display" or triumph ceremony. The birds approach breast to breast, with body plumage expanded, necks retracted, and bills lowered (fig. 37H), then turning parallel, they swim together while calling and lowering their back feathers. Finally, paired birds sometimes touch bills then quickly pivot on the water so that they end tail to tail ("pivoting"). This may be a variation of food presenting behavior (Fjeldså 1982a).

Copulation occurs on the nesting site or other platform, which is built by both sexes. Inviting behavior may be performed by either sex and consists of leaning the head forward and lying motionless on the platform or on water while uttering low calls. A second soliciting display is rearing, during which the bird stands and leans forward with the bill held horizontal. This posture may end by the bird's lowering its body to an inviting posture. The other bird responds to these displays by eventually swimming up behind its mate, raising its head feathers, and sometimes pecking at its partner's rear. Mounting is done with a jump, and during treading both birds utter a "duet" (Franke 1969). Afterward the male slides forward into the water and either performs no postcopulatory display (McAllister) or glides away with neck outstretched while making head turning movements (Fjeldså, Franke).

Reproductive Biology

BREEDING SEASON AND NESTING SUBSTRATE. The breeding season is fairly extended, the egg records extending in North Dakota from May 21 to August 9 (Stewart 1975). Colorado and Utah records are from late April to late July, and those from California, Oregon, and Washington are from late April to early August (Bent 1919). Some replacement clutches are laid if the nest is lost early in the breeding season, but otherwise the nest may be abandoned (Palmer 1962). Double brooding has been reported elsewhere for the species (Cramp and Simmons 1977) but not yet for any North American populations. Nests are constructed in water with submerged or emergent support, in stands of vegetation ranging from semiopen to fairly dense. Gotzman (1965) found most nests in medium-density stands of vegetation, usually within 40 meters of the edge of a reedbed and often over 100 meters from the feeding area. Depths of water at nest sites ranged from 4 to 48 inches (0.1–1.2 meters) in seven North Dakota colonies, averaging about 22 inches (0.6 meter) (Stewart 1975). McAllister found nests in water 1.0–3.5 feet (0.3–1.1 meters) deep, in the densest portions of the reedbeds. At least in Europe, nesting among gull colonies is frequent (Gaukler and Krause 1968), and Franklin gulls (*Larus pipixcan*) are also common nesting associates in North America (Bent 1919).

NEST BUILDING AND EGG LAYING. Nest building is normally done by both sexes, although McAllister (1958) saw only females building. She noted that a nest can be built in 3 hours, but additional materials may be added later. Egg laying may begin early in the nest-establishment period; these eggs are laid on flimsy plat-

forms and subsequently abandoned. Typically the eggs are laid at daily intervals, although a day or two is often skipped in the sequence before the clutch is completed. The birds are apparently indeterminant layers and tend to replace lost eggs. McAllister reported a significant difference in clutch sizes during two years, and there may also be locational differences in clutch size as well. Her collective figures for 293 clutches range from 1 to 6 eggs and average 3.48. She also observed a high level of synchronization of nest initiation and egg laying in her colonies, with all initial eggs in 180 clutches begun in a 13 day period. A sample of 101 North Dakota nests ranged from 2 to 8 eggs and averaged 3.8.

INCUBATION AND BROODING. Incubation begins with the first egg and is done by both sexes. Hatching is thus asynchronous, with an average incubation period of 20.5–21.5 days (Palmer 1962), sometimes as long as 22 days. The young chicks are carefully tended by the parents for at least 12 days, and in some cases the brood may be divided between the two adults, which then tend to avoid each other.

GROWTH AND SURVIVAL OF YOUNG. There is little information on the development of the young, but according to one report they are relatively independent when only 21 days old. This seems rather unlikely, though the insect-eating tendencies of the young may allow for rather early independence. By 17 days of age they are already diving for their own food (Bauer and Glutz 1966). Information on brood sizes is difficult to obtain because of the colonial nature of the species and the resulting problems of identifying broods. Gaukler and Krause (1968) noted that 110 families had from 1 to 4 chicks, with an average brood size of 1.87. Yocom, Harris, and Hansen (1958) judged an average production of about one young per adult in late July. These observations suggest rather high mortality of eggs or young, although the causes likely are unknown.

BREEDING SUCCESS AND RECRUITMENT RATES. Very low nesting success was reported by Broekhuysen and Frost (1968), who reported that only 12 young hatched successfully from a total of 223 eggs. Losses in that study were mainly caused by waves and flooding during storms. Friley and Hendrickson (1937) reported an 87.5 percent hatching success in a sample of 15 nests, with a resulting average brood size at hatching of 2.8 chicks. McAllister (1958) noted no significant egg predation but did not analyze nesting success. She did, however, mention that many chicks died from exposure shortly after hatching. It seems likely that egg losses from waves or flooding, together with early mortality of chicks, are likely to be the major factors influencing breeding success in eared grebes. There are no published estimates of recruitment rates or adult mortality rates. The oldest known age attained by a banded bird in North America is only slightly over 5 years, suggesting rather high adult mortality.

Evolutionary History and Relationships

Fjeldså (1983a) has discussed the probable evolutionary history of the eared grebe and concluded that it is part of a cluster of four species (including *gallardoi*, *occipitalis*, and *taczanowskii*), and may be the "lowest side branch" of this cluster, based on its less specialized social displays. This group in turn is probably related to the horned, red-necked, and great crested grebes, and these form the typical *Podiceps* group, in Fjeldså's view.

Population Status and Conservation

From a worldwide view this is probably the most common of all grebes, since it has the most widespread distribution and at least locally forms much the largest concentrations. The size of the North American population is unknown, but Jehl (1983) reported peak numbers of about 750,000 in fall at Mono Lake, which may be the major part of the North American population or even virtually all of it. The birds seem somewhat vulnerable to oil spills because of their coastal migration and marine tendencies, but so far very few have been involved in oiling incidents (Clapp et al. 1982).

Western Grebe*

Aechmophorus occidentalis (Lawrence)

OTHER VERNACULAR NAMES: Swan grebe; grèbe de l'ouest (French); Renntaucher (German); achichilique (Spanish).

Distribution (Including *clarkii*; See Map 10)

BREEDS from southeastern Alaska, through northwestern Alberta, south-central Saskatchewan, and south-central Manitoba to central and south-central California, western Nevada, northern Utah, and southern Wyoming eastward locally to northeastern New Mexico, southern Colorado, western Nebraska, South

*Including "Clark's grebe," which was recognized as a distinct species *A. clarkii* after this manuscript went to press (*Auk* 102:680).

Dakota, and southwestern Minnesota. Also breeds locally in Mexico (Williams 1982).

WINTERS mainly near the coast from southeastern Alaska and British Columbia to Baja California and Jalisco; locally inland to western Nevada and south to Puebla. Birds breeding in Mexico are evidently residential.

Description (after Roberts 1949)

ADULTS IN BREEDING PLUMAGE (both sexes). Crown and hindneck black or mouse gray, sharply defined against white of sides of head and neck; underparts, including sides of neck and head, satiny white; back and sides of body below brownish gray; wings the same except the secondaries, which are broadly margined with pale gray on outer webs for terminal half, forming a large white patch on spread wing. Bill greenish yellow, culmen black near base, feet pale yellow on inner surface, black on outer; iris orangy red or scarlet, the eyelids narrowly deep yellow. A separate color phase (or sibling species) exists in which the breeding plumage is generally lighter, the dark color of the crown does not reach the eyes and lores, and the bill is bright yellow rather than greenish yellow; this form has been called *clarkii* (Storer 1965).

WINTER PLUMAGE. Nearly the same as the breeding plumage, but slightly less contrasting. The head feathers are molted twice annually, whereas the body feathers are molted only once a year (Storer and Nuechterlein 1985).

JUVENILES. Crown and hindneck dark gray or dusky, the borders of the dark areas not sharply defined, and back feathers edged with grayish white.

DOWNY YOUNG. The lores and a triangular crown spot bare, the latter varying from orange to red; the plumage is nearly uniform in color, with the face and underparts nearly white to pale gray, the upperparts smoke gray, the bill black, the iris dark gray, and the legs and feet mostly slaty, with greenish lobes (Palmer 1962). The downy young of *clarkii* average paler than those of the typical form (Dickerman 1973; Ratti 1979).

Measurements and Weights

MEASUREMENTS. Wing: males 200–209 mm (average of 4, 205.5); females 187–97 mm (average of 5, 191). Culmen: males 74–78 mm (average of 4, 76.5); females 63–69 mm (average of 5, 65.2). Eggs: average of 20, 58.8 x 38.3 mm (Palmer 1962). There are apparently no mensural differences in the two morphs except for bill length and depth (Ratti, McCabe, and Smith 1983).

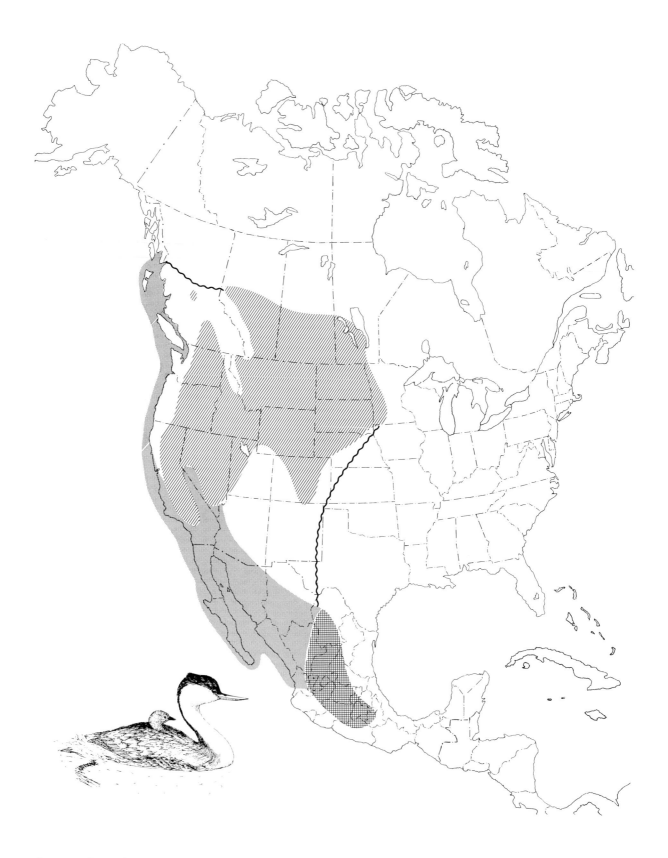

10. Current inclusive distribution of the western grebe, showing summer range (*hatched*), residential range (*crosshatched*),
wintering range (*shaded*), and limits of migratory corridors
(*wavy lines*).

WEIGHTS. All seasons, males 714–1,225 g (average of 9, 972); females 556–1,109 g (average of 16, 822) (museum specimens). Palmer (1962) and Storer, Siegfried, and Kinahan (1975) both reported substantially higher average weights for this species, the former giving an average of 1,474 g for 14 unsexed birds in winter and the latter reporting an average weight of 1,087 g for the species, sample size unstated. Estimated egg weight, 43.5 g (Schönwetter 1967).

Identification

IN THE FIELD. This large grebe is instantly identifiable on the basis of its extremely long, thin neck, which is entirely white on the sides and front, and the long, daggerlike yellowish to greenish bill. The head is strongly bicolored black and white in all adult plumages, and the eyes are a brilliant carmine. In the light-phase form the eyes are distinctly below the black crown, while in the dark-phase form the blackish color of the crown extends below and in front of the eyes to reach the base of the bill. The birds are associated with rather shallow and often alkaline prairie marshes during the breeding season, especially those having fairly extensive areas of open water for foraging and social display. During this period the advertising call, a loud whistled or bell-like note, is audible over great distances. This call may be either a single-noted *creet* or a double-syllable *cree-creet*, the latter being typical of dark-phase birds. In most parts of western North America the dark phase is more common, though unusually high numbers of light-phase birds occur at Goose Lake, California (95 percent), and Upper Klamath Lake, Oregon (49 percent) (Ratti 1981). In general there is a north-south cline in morph frequency, with nearly all birds at the northern end of the range of the dark phase, while about 10 percent of those breeding in northern Utah are light phase and most grebes breeding on the Mexican plateau of light phase. Only in the last-mentioned area have intermediate-phase birds been reported as fairly common.

IN THE HAND. The very large size (wing at least 187 mm), long neck, and very long (60–80 mm) bill, which is mostly yellowish to greenish, with a darker culmen, provide for easy in-hand identification. There are no major mensural differences between the two color phases; the bill color and the extent to which the black crown reaches downward in the area of the eyes and lores are the best distinguishing morphological traits. However, Ratti, McCabe, and Smith (1983) reported that light-phase females average statistically smaller than dark-phase females in culmen length, culmen depth, and length of bill from nostril to tip, whereas males differ statistically only in length of the bill from nostril to tip. Storer and Nuechterlein (1985) found that the two forms also overlap in all plumage characteristics except facial pattern, and birds from the Mexican population (both phases) averaged significantly smaller in all of four measurements than did more northerly populations.

Ecology and Habitats

BREEDING AND NONBREEDING HABITATS. Breeding habitats consist of fairly extensive marshes with rather large areas of open water supporting good fish populations. Stewart (1975) reported that in North Dakota the birds use permanent or deeper semipermanent ponds and lakes that range from slightly brackish to brackish, contain extensive areas of open water, and usually have an area of 50 acres (20 hectares) or more. While on their breeding areas the birds prefer foraging in well-lighted, unshaded water, often diving in water about 2 meters deep near the edges of tule beds (Lawrence 1950). Wintering birds favor salt and brackish waters, especially along coastal bays, but they also occur in freshwater sites, with the prime criterion the availability of small fish. They often extend out into deep offshore waters along the Pacific coast during winter, but they also use the bays and estuaries.

SOCIALITY AND DENSITIES. This is a relatively social species that typically nests in fairly dense colonies. Lindvall and Low (1982) reported that 95 percent of the nests they observed (nearly 400) were in colonies numbering from 5 to 88 nests. In smaller colonies the nests were spaced an average of 30 meters apart, and in colonies of more than 10 nests the average internest distance was 15 meters or less. Density information was not directly reported, but these authors stated that in 1974 an adult population of about 690 birds was present on the refuge, which then contained about 15,000 hectares of marsh and open water, representing about 22 hectares of habitat per adult. If only the open-water areas of the refuge are considered the density would be about 13 hectares per adult.

PREDATORS AND COMPETITORS. The major egg predators reported by Lindvall and Low (1982) were American coots (*Fulica americana*) and California gulls (*Larus californicus*). The latter species was also noted as an egg predator by Yocom, Harris, and Hansen (1958). Munro (1941) noted that American crows (*Corvus brachyrhynchus*) sometimes can drive grebes off their nests and steal their eggs. No doubt other predators exist and remain to be identified; Bent (1919) considered the mink (*Mustela vison*) a probable predator on adults.

143

The specialized foraging behavior of this species, with its concentration on small fish, places the western grebe in competition with various other fish eaters including loons, especially on coastal wintering areas. The strong sexual dimorphism in bill shape has been hypothesized as a means of reducing intersexual competition for food, and there is also a small but statistically significant difference in bill size between the light- and dark-phase adult birds (Ratti, McCabe, and Smith 1983). There are still only limited data suggesting that ecological niche partitioning between these two morphs might exist (Feerer 1977), and in many western areas both forms exist sympatrically (Ratti 1981).

General Biology

FOOD AND FORAGING BEHAVIOR. The early analysis by Wetmore (1924) of 19 stomachs taken throughout the year indicated a diet essentially 100 percent fish, the only other significant food remains being some aquatic insects (Corixidae). These fish were of various freshwater and marine genera (Mylocheilus, Cyprinus, Catostomus, Leuciscus, Atherinops, and Atheninopsis) and included individuals up to about 5 inches long. A larger sample of 24 birds was included in Lawrence's (1950) analysis, all collected on California breeding areas in summer months. There the food volume consisted of only 81 percent fish, mainly the bluegill (Lepomis macrochirus), while aquatic insects made up most of the rest. A similar study of 18 birds from the same general area but taken over a broader time period (March to December) confirmed the earlier finding that fish constitute the primary food source on the breeding grounds and that insects are eaten during spring and summer. In this case the fish included cyprinids and centrarchids 3.5–20.5 centimeters long, the centrarchids averaging 12 grams. A smaller sample of 13 stomachs from breeding grounds in Utah likewise indicated a concentration on fish, especially carp (Cyprinus carpio) (Lindvall 1976).

Foraging behavior has been documented by Lawrence (1950). He observed over 1,700 dives and determined a mean dive duration of 30.4 seconds, with most dives in water 4.5–9.0 feet deep. He found no correlation between dive duration and water depth. He noted that during diving the feet were stroked at an average rate of 1.4 strokes per second during unhurried progression and observed that during foraging the birds maintained a territorial spacing of at least 200 feet of open water between individuals. Foraging intensity was related to subsurface visibility, with highest rates of foraging when the sun's rays struck the water at a high angle, and foraging ceased abruptly in the evening. There was

also a decline in foraging during the middle of the day, apparently because warmer water temperatures and associated lower oxygen concentrations decreased the activity of fish and insects. Nuechterlein (1981a) suggested that light-phase birds tend to feed at shallower depths than dark-phase individuals and thus at different distances from shore. He believed this difference in behavior might help account for the puzzling geographic variations in color phase frequencies that have been reported (Ratti 1981).

MOVEMENTS AND MIGRATIONS. Very little specific information is available on migration in this species. As in other grebes, migratory flights are evidently nocturnal over land and partly diurnal along the coast. Furthermore, migration seems to be done in flocks rather than individually. Virtually the entire North American population winters along the Pacific coast, from south to central Mexico. Migrants begin leaving this latter area in late March or April, and by late April and early May they begin to move inland, crossing the Rocky Mountains at about this time and reaching the easternmost breeding areas of Manitoba and Minnesota. Fall migration is quite variable, but the first migrants reach salt water by early September, and some have arrived along the Pacific coast of Mexico by late October. The winter distribution pattern seems to vary from year to year, with maximum numbers during late December occurring as far north as Seattle in some years and as far south as San Diego in others (Palmer 1962). On the Atlantic and eastern Gulf coasts the species is rare to accidental in winter, but along the Texas Gulf coast it is fairly regular (Clapp et al. 1982). When on their breeding areas the birds become very sedentary; Nuechterlein (1982) has suggested that they might even be incapable of flight because breast muscles atrophy during that period.

Social Behavior

MATING SYSTEM AND TERRITORIALITY. Like other grebes, this species is at least seasonally monogamous. There is no information yet on mate retention or territorial site tenacity, but a substantial amount of data has accrued on mating preferences within color phases. Ratti (1979) reported a very high level of assortative mating within color phases in California and Oregon; only 97 of 606 pairs were of mixed color phases, and only 1 percent of 1,371 pairs observed at Bear River Refuge were of mixed type. In both places mixed pairing occurred at about 2 percent of the expected frequency assuming random choice of mates. Similarly, Nuechterlein (1981a) confirmed assortative mating in various localities in Manitoba, Oregon, and California and determined a very low (0–2.9 percent) level of participation of mixed-phase dyads during male/female courtship activities. He saw only one mixed pair among 91 pairs observed in April at Tule Lake Refuge and later saw another at Upper Klamath and two additional pairs at Tule Lake. Of 18 intermediate-plumaged birds observed at Upper Klamath and Tule Lake, only 3 were paired, and 2 of these were paired to one another. Intermediate-plumaged birds apparently have difficulty obtaining mates, though they are evidently fertile. Eggs of the single mixed pair that Nuechterlein observed in Manitoba hatched successfully. Intermediates may, however, be more frequent in the Mexican population (Feerer 1977). The basis for assortative mating is still not understood. The obvious possibility of imprinting on parental phenotypes has been suggested (Storer 1965), though Ratti (1977) was unable to confirm this with hand-raised chicks. Nuechterlein (1981a) believed that differences in the advertising call were critical for phase recognition and assortative mating preferences within color phase types. He suggested that imprinting might be involved or that mating preferences might be developed gradually through associative learning processes in adults of one or both sexes.

Territoriality is little studied in this species, although as noted earlier foraging territories may produce dispersal during feeding. Nest spacing suggests that territorial defense of the nest site must be very weak. Davis (1961) reported nests spaced as close as 6 feet apart, and Ratti (1979) observed a strong tendency for light-phase birds to nest close to other light-phase birds and the same for dark-phase individuals, with the average distances between nests of the same color phase 6.8 and 10.2 meters respectively. In a large North Dakota colony the average distance between nests was about 2 meters (Stewart 1975). Territorial encounters might be avoided under these close crowding conditions by the fact that the birds approach and leave their nests underwater, swimming submerged to open-water foraging areas.

VOICE AND DISPLAY. Vocalizations and display postures have been analyzed in detail by Nuechterlein and Storer (1982), whose work is the primary basis for this summary. The most significant vocalization in pair formation is the advertising call, which is performed by unpaired adults of both sexes and by paired birds toward one another. This call exhibits a high degree of individual variability and probably is significant in promoting individual recognition (Nuechterlein 1981b). In both color phases the call is lower pitched and more prolonged in males than in females, but light-phase birds usually utter a single *creet* note, whereas dark-phase individuals utter a double note. Ticking sounds uttered in irregular rhythms are produced by adults of both sexes during tick pointing, and similar ticking notes are uttered by alarmed parents with offspring. A harsh and wooden ratchetlike call is uttered during aggressive ratchet pointing displays, and a similar call is produced by nesting birds approached by intruders. Trilled sounds, sometimes approaching a teakettlelike whistle, are uttered during barging ("barge trilling"), and a similar call is uttered during neck stretching as well as during copulation. Finally, bursts of clucking sounds are produced during the "arch clucking" display at intervals of from 3 to 10 seconds. Calls of young include rather loud begging notes and perhaps other vocalizations as well.

Displays of the western grebe are numerous and often spectacular. Perhaps the simplest and most important, as well as the most common, is advertising. While swimming or resting in a normal posture, the crest is raised and the loud, single- or double-syllable call is uttered from 1 to 6 times, at intervals of 0.5–1.0 second. Unpaired birds of both sexes spontaneously perform this call, and it is also used for individual recognition among members of a pair or family at all times of the year. Tick pointing is usually used by two birds in visual contact, and while uttering ticking calls the neck is stretched, the bill is tilted upward slightly, and the crest is lowered as the bird quickly jerks its head from side to side. Frequently four or more birds may participate, but males predominate in such groups. Tick pointing may rarely directly precede rushing, but usually it tends to bring participating birds close together. When very close to each other (up to three body lengths apart) the birds perform ratchet pointing (fig. 38F), in which they lower their heads, stare at one another, and utter a loud call with bulging throats. This display may involve two males, a male and female, but only rarely two females. It may be alternated with quick bill dips and head shakes ("dip shaking"), may build to a rushing ("race") climax, or may gradually subside. The race or rushing display, the most spectacular of the western grebe's displays, consists of a rapid run over the water surface with the wings lifted but only partially raised to the sides, apparently providing an airfoil, while the feet patter rapidly over the water at a rate of 16–20 steps per second (fig. 38C), producing the only noise associated with the display. The race most often entails two birds (typically male and female) but is sometimes performed by lone males, by two males and a female, or very rarely by lone females or three males and a female. Racing by male groups may serve to attract the attention of females, and races by two males tend to be relatively ex-

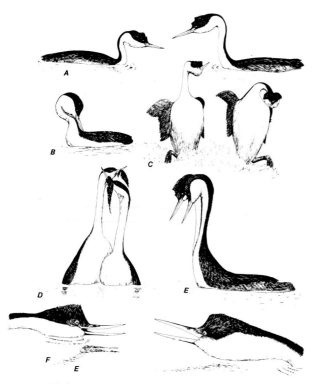

38. Social behavior of the western grebe (mainly after Nuechterlein and Storer 1982): A, greeting ceremony; B, bob preening; C, racing; D, weed dance; E, high arch; F, ratchet pointing.

tended. At the end of the race (of up to about 20 meters) the wings are lowered and the bird dives. At the end of the dive the birds usually emerge and perform neck stretching while uttering a trilled note as each attempts to regain eye contact with its partner. Following the race sequence, either of two displays is likely to occur. If the two racing birds are both males, they may follow the dive with barging ("barge trilling"), in which they move slowly forward in parallel, with wings folded, necks stretched, crests raised, and bills level, while turning their heads laterally and uttering a trilled call. Interested females sometimes join such groups, apparently to choose a male for further display interactions. When the race is done by a male and female it may be followed by separation, mutual ritualized preening, or the weed ceremony. This involves a mutual weed dive and approach while still carrying the weeds. When the birds meet they rise in a mutual weed dance (fig. 38D). After an extended weed dance the birds drop their weeds and begin mutual and often synchronized bob preening (fig. 38B), which is the equivalent of the "habit preening" of other species but alternates with a more exaggerated upward stretching of the neck and both be-

gins and ends in a distinctive high arch body posture (fig. 38E). This high arch posture is often associated with "arch clucking," performed either mutually between mates or by pair members at greater distances from each other. It also grades into a low arch posture that is identical to that of a bird soliciting copulation on a platform. Arch clucking is done only between paired birds and indicates that a pair bond has been formed. It is typically used as the final phase of the greeting ceremony of paired birds, which in its complete sequence includes mutual advertising calls as the birds approach each other (fig. 38A), followed by dip shaking and bob shaking and terminates with bob preening and arch clucking. Well-established pairs may reduce the sequence even further, to simple bob shaking or bob preening.

Copulatory behavior has not been described in detail, but it is preceded by soliciting on a mating platform, using a posture very similar to that of the low arch display posture (Nuechterlein and Storer 1982). While treading the male holds his head with the bill pointed downward near the female's, who holds her head level and slightly raised. The female apparently does not brush the male's breast with her crown, but after copulation she may raise her head and touch the male with her bill (Palmer 1962).

Reproductive Biology

BREEDING SEASON AND NESTING SUBSTRATE. The nesting season is fairly prolonged; in North Dakota there are egg records from May 15 to July 9, and dependent young have been observed as late as October 9. Manitoba and Saskatchewan records are for most of June, and those from Washington, Oregon, and California are from late May to early July (Bent 1919). In Utah there was a nearly three-month spread of nest initiation dates during a two-year period, but egg laying peaked in the first half of July (Lindvall and Low 1982). In Mexico the breeding period is apparently greatly extended, with eggs reported between May and October and downy young observed as late as December (Dickerman 1973; Williams 1982). Renesting efforts apparently are common and tend to extend the overall nesting period.

The nests are normally constructed over water, with materials built up from the substrate or anchored to submerged or emergent vegetation. Dry-land nest sites have been found (Nero, Lahrman, and Bard 1958) but are atypical. Normally the nests are in or very close to water deep enough to swim in; Lindvall and Low (1982) reported the average depth of the nearest open water as 21 centimeters for dry-land nests and noted that floating nests in open water extended 31 centimeters below the water surface. Nests built in emergent vegetation averaged 55 meters from land and only 0.4 meter from open water. The amount of emergent vegetation seemingly was insignificant compared with the need for ready access to open water for rapid escape from and probably also inconspicuous approaches to the nest. Nuechterlein (1975) considered that nest sites are selected on the basis of their relative safety from waves, water depth (at least 25 centimeters), and stem density (maximum densities preferred). The average depth of 315 nests was 41 centimeters, and 99 percent of the nests were in depths of at least 25 centimeters.

NEST BUILDING AND EGG LAYING. Nests can apparently be constructed rapidly; there are cases of "hastily improvised" new nests built over old nest bases shortly after thunderstorm damage to a nesting colony (Palmer 1962). The eggs are apparently laid approximately at daily intervals. Lindvall and Low (1982) reported an average clutch size of 2.6 eggs for 70 completed clutches representing both color phases in Utah. Ratti (1977) found no significant differences in clutch sizes of the two color phases; in 1975 a total of 111 nests of both phases had an average clutch of 2.47 eggs, while in 1976 the average clutch size for 154 nests of both phases was 2.36 eggs. These clutch sizes are low compared with those reported by Davis (1961) for Colorado, where 279 nests (some with incomplete clutches) averaged 3.41 eggs. Likewise, 12 North Dakota nests had as many as 7 eggs and averaged 4.2 (Stewart 1975), though it should be noted that "dump nesting" (joint or multiple laying in single nests) is prevalent in western grebe colonies and these large clutches are quite possibly influenced by this (Lindvall and Low 1982).

INCUBATION AND BROODING. Both sexes incubate, probably beginning with the laying of the first egg. Incubation averaged 24 days for 14 clutches and ranged from 21 to 28 days (Lindvall and Low 1982). Waves and varying water levels seem to be a frequent source of nesting losses in this species (Wolf 1955; Lindvall and Low 1982), and renesting efforts after nest failure are common. However, second nesting after successful hatching of a brood is still unreported. In one study 21 percent of the incubated clutches hatched at least one chick, and on average 0.35 class 2 (older downy) chick was produced per adult over a two-year period (Lindvall and Low 1982). Nuechterlein (1975) reported a two-year nesting success rate of 64 percent for 377 nests, but with substantial site-related variation in both years.

GROWTH AND SURVIVAL OF YOUNG. As with other grebes, the young are brooded on their parents' backs for the first few days after hatching, and adults have actually been observed to carry their chicks overland. Ratti (1979) reported that in 99 percent of all broods surveyed there were 1–3 chicks, with dark-phase adults having a slightly smaller brood size than light-phase parents during both of two years, although the differences were statistically significant for only one of the years. The overall average brood size Ratti observed for both summers was 1.76 chicks for 766 broods. Nuechterlein (1975) determined an early-August two-year adult/young ratio of 0.72, or 1.4 young per pair. There is still no information on growth rates in wild young, but in captivity light-phase chicks began to exhibit dark crown feathers at 10–15 days, whereas dark-phase young did not do so until they were 50–60 days old. Primary feathers began to emerge at about 40 days on young of both color phases, and the primaries had reached a length apparently suitable for flight by the time the young were approximately 70 days old (Ratti 1979). The upper asymptote of weight for 7 chicks was 1,106 grams.

BREEDING SUCCESS AND RECRUITMENT RATES. Judging from the limited data, nesting success in western grebes is quite variable, averaging only 21 percent in Lindvall and Low's (1982) study and ranging from 46 to 84 percent in Nuechterlein's (1975) study. This is likely to be increased somewhat by the still uncertain but apparently high incidence of renesting. If 3 eggs are considered the modal clutch size of western grebes, the average brood size of 1.76 young reported by Ratti (1979) suggests substantial (41 percent) mortality of eggs or of young chicks in successfully hatched clutches. The estimated production of 0.35 class 2 young per adult represents a maximum recruitment rate of 25.9 percent, even assuming no further chick mortality before fledging. This occurred in spite of an average class 2 brood size of 1.7 young per mated pair, suggesting that only about 40 percent of the mated pairs were successful in hatching broods. Ratti (1979) estimated a nearly identical brood size during two different years in the same area. He did not attempt to estimate overall reproductive success except to point out that light-phase birds seemed to be reproducing more efficiently in that population than were dark-phase individuals. The light-phase birds composed only 18.6 percent of the collective western grebe population but were associated with 33.6 percent of the total observed broods. Rudd and Herman (1972) estimated a production of 1.7 chicks per mated pair in what they believed was a normally reproducing California population, which would represent a maximum recruitment rate of 46 percent, assuming there were no further pre-fledging losses and that no nonbreeders exist in the population. Certainly neither of these assumptions is realistic, and thus the actual average recruitment rate probably lies somewhere between 25 and 45 percent, or somewhat less than is seemingly typical of other North American grebes.

Evolutionary History and Relationships

Nuechterlein and Storer (1982) concluded that based on its many distinctive display and ceremony differences as well as its unique morphological traits, the western grebe deserves generic distinction from *Podiceps*, its probable nearest relative. This relationship with *Podiceps* is indicated by its sharing the weed ceremony with members of that genus, as well as a similar though not necessarily homologous upright posture during advertising. Fjeldså (1982a) pointed out a substantial number of additional similarities between western grebes and *Podiceps* and suggested that it might well be included in that genus, since habit preening, barging, and weed ceremonies are all shared with *Podiceps*. The question of the most suitable way of dealing taxonomically with the two color phases is still perplexing; Ratti (1981) has pointed out that no other broadly sympatric polymorphic species of birds are known to exhibit such a low incidence of interbreeding, and yet it is also true that there may be no other case in which males of two distinct species regularly engage in the same displays to attract females.

Population Status and Conservation

There are no available estimates of western grebe populations for the entire continent, nor are there any good regional estimates. It is known that they have been nearly extirpated from many of the California breeding lakes (Feerer and Garrett 1977), and the effects of pesticides on western grebe reproduction are well documented (Rudd and Herman 1972). Because of their gregariousness and tendency to winter in coastal areas of California, they are highly susceptible to oiling mortality. Over half the oiled birds counted during one oil spill in San Francisco Bay were of this species, and perhaps as many as 11,000 western grebes were lost as a result of this single disaster (Smail, Ainley, and Strong 1972).

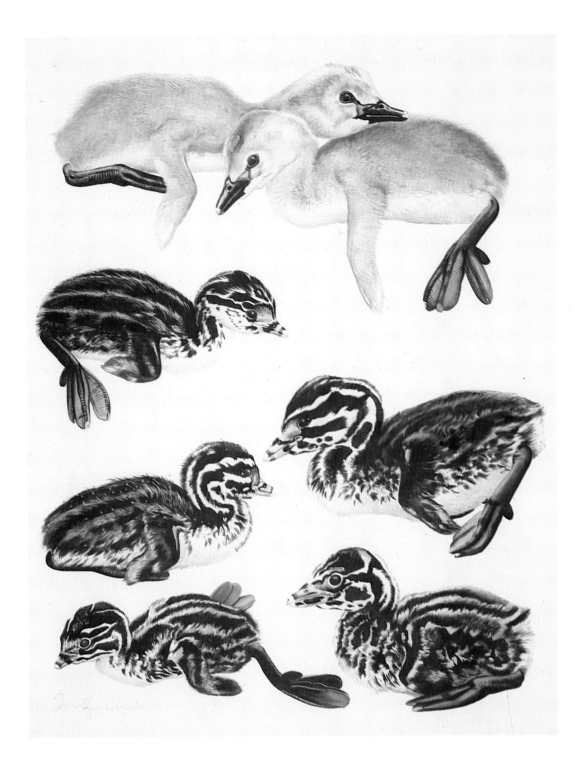

11. Downy young of grebes, including light (*top left*) and dark phases (*top right*) of western, black-necked (*upper left*), horned (*middle left*), least (*lower left*), red-necked (*middle right*), and pied-billed (*lower right*). Painting by Jon Fjeldså.

12. Dovekie, adult in breeding plumage.
Photo by Frank S. Todd.

13. Razorbill, adult in breeding
plumage. Photo by author.

14. Common murre, adult in winter
plumage. Photo by author.

15. Thick-billed murre, adult in breeding plumage. Photo by author.

16. Black guillemot, adult in breeding
plumage. Photo by Frank S. Todd.

17. Pigeon guillemot, breeding colony.
Photo by Frank S. Todd.

18. Pigeon guillemots, immature
plumage (*left*), rhinoceros auklet (*middle*) and common murre, breeding
plumage (*right*). Photo by author.

19. Marbled murrelet, incubating adult.
Photo by Stuart Johnson.

20. Kittlitz murrelet, incubating adult.
Photo by David G. Roseneau.

21. Cassin auklet, immature plumage.
Photo by author.

22. Parakeet auklet, adults in breeding
plumage. Photo by author.

Auks (*Alcidae*)

Dovekie

Alle alle (Linnaeus)

OTHER VERNACULAR NAMES: Little auk (British); skonge (Danish); mergule main (French); Krabbentaucher (German); agpaliarssuk (Greenland); haftyrdill (Icelandic); lyurik (Russian); alkekung (Swedish).

Distribution of North American Subspecies (See Map 11)

Alle alle alle (Linnaeus)
BREEDS from northwestern Greenland (Thule area) locally southward, and from eastern Greenland, Jan Mayen, Spitsbergen, Novaya Zemlya, and North Land (USSR) south to southeastern Greenland. Also breeds on Bear Island (Norway) and probably also on the New Siberian Islands. Probably breeds rarely in Bering Strait area, including Saint Lawrence and Little Diomede islands, and possibly on Ellesmere Island. Until recently also bred off northern Iceland (Grimsey Island).

WINTERS south from breeding range, in open water, to Southampton Island, Ungava Bay, along the Gulf of Saint Lawrence, southeastern Newfoundland, Nova Scotia, the Bay of Fundy, rarely to New Jersey, and to the Canary Islands, Azores, France, and the Baltic Sea.

Description (Adapted from Ridgway 1919)

ADULTS IN BREEDING PLUMAGE (sexes alike). Head, neck, and chest plain dark sooty brown (clove brown), becoming gradually darker on pileum and hindneck; a short white streak immediately above upper eyelid; rest of upperparts sooty black (slightly glossy), the second-aries sharply but narrowly tipped with white, the posterior scapulars narrowly streaked with the same; underparts of body immaculate white (abruptly defined against the dark brown of chest), the upper or outer portion of flanks broadly streaked with blackish; bill black; iris dark brown; legs and feet grayish, with dusky webs; inside of mouth light yellow.

WINTER PLUMAGE. Malar region, chin, throat, chest, and sides of nape white, the latter mottled with grayish and the feathers of chest with dusky bases; otherwise like the summer plumage. First-winter birds are more brownish than are older age groups (Kozlova 1961).

JUVENILES. Similar to the breeding plumage, but the head and upperparts browner and less glossy, scapular feathers unmarked, white markings above the eye very small, and throat flecked with brown and not strongly separated from the anterior breast (Glutz and Bauer 1982).

DOWNY YOUNG. Entirely sooty grayish brown, the underparts paler and more grayish. Iris blackish brown, bill blackish, legs and feet dusky.

Measurements and Weights

MEASUREMENTS. Wing: males 108–22 mm (average of 11, 115.8); females 106.5–118.0 mm (average of 5, 113.3). Exposed culmen: males 13–15 mm (average of 11, 13.9); females 12.5–14.5 mm (average of 5, 13.9) (Ridgway 1919). Eggs: average of 44, 48.2 x 33.0 mm (Bent 1919).

WEIGHTS. The average of 117 breeding-season males was 153.5 g, while 92 females averaged 146.8 g (Roby, Brink, and Nettleship 1981). Bradstreet (1982a) reported slightly higher averages for birds taken over a broader time span. Estimated egg weight, 28 g (Schönwetter

11. Current North American distribution of the dovekie, showing
colony locations, estimated breeding numbers, nonbreeding
range limits (*broken line*), and Eurasian range (*inset*).

1967). Newly hatched chicks average 21.5 g (Norderhaug 1980).

Identification

IN THE FIELD. This is only obviously small (dove-sized) alcid found on the Atlantic coast, and in the breeding season it has a sharply contrasting black-and-white pattern in which the black upperparts extend to the upper breast. In winter, white extends up to include the breast, throat, and lower cheeks and extends up around the ear coverts to a point level with the eyes, leaving a black semicollar in front of the wings. Calls are a series of chirping or piping notes, uttered on the breeding grounds. In flight the wing movement is very fast and resembles that of a chimney swift.

IN THE HAND. This is the only small alcid (wing under 125 mm) that has 12 rectrices and vertically aligned scutes on the lower front of the tarsus. Additionally, the bill is extremely short and is about as wide as deep at the base.

Ecology and Habitats

BREEDING AND NONBREEDING HABITATS. Breeding occurs along high arctic and, to a more limited degree, subarctic coastlines where the adjoining waters have surface temperatures in August of 0–6°C (Voous 1960). Dovekies typically nest in large colonies on mountain slopes, at altitudes of up to 300–500 meters, where talus slopes and scree provide abundant breeding sites. Most colonies are within easy flying distance of the sea, but the birds locally extend up valleys and fjords to a maximum reported distance of about 20 miles (32 kilometers) inland (Bateson 1961). However, Lovenskiold (1964) stated that in Greenland much of the most highly suitable scree breeding habitat is only a few hundred meters in width. The total altitudinal range of nesting birds there is from a few feet above sea level to about 600 meters, with maximum numbers associated with scree deposits below perpendicular rock walls. Stempniewicz (1981a) judged that the first requirement of breeding habitat is that it be near the sea, and he observed no nesting farther than 6 kilometers inland. Slope and exposure are also important, since they influence the rate of thawing of nesting sites. Sites having high humidity are avoided, since they retain water, and slopes of 25–35° are most suitable because they are fairly stable and allow for easy takeoffs.

During the nonbreeding season the birds are pelagic, tending to remain near the pack ice, foraging in the cold waters of ice leads, where small planktonic crustacean populations are most abundant, and in the western Atlantic the species is limited in its southern distribution by the warming influences of the Gulf Stream. Storm-driven birds sometimes occur well south of normal wintering limits, and occasionally "wrecks" occur when such birds are driven to coastlines or even well inland.

SOCIALITY AND DENSITIES. This is perhaps the most social of all the alcids; estimates of colony sizes are sometimes so large as to defy credibility, such as those at Thule, Greenland, which have been estimated at millions of individuals (Salomonsen 1967). However, Lovenskiold (1964) stated that early estimates of numbers in Spitsbergen colonies were greatly exaggerated and that a colony once estimated to have 10 million birds perhaps held as few as 1 or 2 million. In western Greenland Evans (1981) estimated a density of at least 0.25 nest per square meter, and on Spitsbergen the average nest density was judged by Stempniewicz (1981a) as 0.5–0.67 per square meter, while Norderhaug (1970) judged a nesting density of up to more than 1 nest per square meter on Spitsbergen. Roby, Brink, and Nettleship (1981) estimated that 7,000 pairs were present in a Greenland study area of 15,400 square meters, or 0.45 pair per square meter. As one proceeds south in the breeding range the numbers and breeding densities decline, and many of the southernmost colonies have disappeared in historical times as the surrounding ocean has gradually warmed. On wintering areas the birds tend to be highly dispersed but are associated primarily with offshore locations and avoid both ice-free and ice-covered areas, attaining their highest densities in areas having 75–99 percent ice cover (Renaud, McLaren, and Johnson 1982).

PREDATORS AND COMPETITORS. A considerable number of predators have been identified as enemies of dovekies, including white-tailed sea eagles (*Haliaeetus albicilla*), parasitic jaegers (*Stercorarius parasiticus*), peregrines (*Falco peregrinus*), snowy owls (*Nyctea scandiaca*), and common ravens (*Corvus corax*), but probably the most serious predators on adults and young are arctic foxes (*Alopex lagopus*) and glaucous gulls (*Larus hyperboreus*). Norderhaug (1970) stated that the parasitic jaeger and glaucous gulls are important predators of eggs and nestlings, and glaucous gulls often kill adult dovekies. Stempniewicz (1981a) agreed that the glaucous gull is an extremely serious predator of adults and chicks but found no evidence that the parasitic jaeger is a significant predator. According to Bateson (1961), glaucous gulls can catch adults in flight with relative ease, and arctic foxes often lie in wait at colonies to pounce on incoming birds or crawl into nesting holes after adults or young. In some areas they feed exclusively on these birds in summer and may even cache supplies of dead birds to use in winter. Roby, Brink, and Nettleship (1981) judged that glaucous gulls are serious predators of adults and young, and arctic foxes were found to take both eggs and adults. Stempniewicz (1981a) stated that the arctic fox prefers eggs and in trying to dig out nests sometimes causes them to cave in, killing the birds or destroying the eggs inside. Young and adult birds are mainly taken late in the breeding season. In Greenland, the Inuit take large numbers during summer, and some winter hunting occurs in southwestern Greenland as well (Salomonsen 1967). The white whale (*Delphinapterus leucas*) has been reported to take adult birds (Bent 1919), and it has been suggested that perhaps seals also feed on them to some extent.

General Biology

FOOD AND FORAGING BEHAVIOR. Norderhaug (1970) investigated the kinds and amounts of food adults brought to nestlings and found that it consisted of at least 95 percent planktonic crustaceans (mainly *Calanus*). During the 26–29 day nestling period each chick ate an average of 689.6 grams of plankton, which was converted into an auk biomass of 92.8 grams. Bradstreet (1982a) determined that during May and June adult and subadult dovekies ate nearly 100 percent copepod materials (especially *Calanus*), whereas during August amphipods became more important, composing 59 percent of adult foods and 90 percent of subadult foods. After fledging, adult males accompanied chicks to sea, where both fed almost entirely on amphipods (especially the hyperiid *Parathemisto*). Once abandoned by the adults, the young continued to concentrate on these amphipods, but a small proportion of the sample also comprised arctic cod (*Boreogadus*) and calanoid copepods. Golowkin, Selikman, and Georgiev (1972) found a similar high incidence of calanoid copepods in planktonic samples taken from adults transporting foods to young in August, with substantially lower quantities of Euphausiacea, Mysidacea, decapod larvae, and amphipods (Hyperiidae) as well as a very few fish. Roby, Brink, and Nettleship (1981) found that the average size of 204 "meals" (throat pouch contents) being brought to nestlings was 3.48 grams and that the food consisted mainly of copepods (*Calanus*) and amphipods (*Parathemisto* and *Aspherusa*).

Bradstreet (1982a) demonstrated that dovekies tended to select larger sizes of available prey species and also selected certain types of prey, in four of the cases analyzed taking large items in proportion to their abundance, while juvenile birds took amphipods at a rate 15–20 times as great as that group's abundance in the zooplankton population. Generally the birds took the largest copepods available in the upper 50 meters of water and also the largest life stage of each species. During the time that adults are feeding nestlings the pairs probably forage around the clock, feeding their young an average of 5.25 (Evans 1981) to 8.5 (Norderhaug 1970) times a day. Brown (1976) concluded it would be feasible for birds to forage as far as about 100 kilometers from their nesting sites and still allow each member of the pair to make four round trips a day to feed their young; he observed adults feeding at ranges of at least this far from Greenland nesting colonies. In contrast, Evans (1981) observed that most food was obtained within 2.5 kilometers of the colony and that the feeding rhythms closely matched the diel cycle of vertical migration in *Calanus* plankton (toward the surface at night, returning to deeper waters during daylight). Most observations on foraging dives indicate that the birds typically undertake rather short dives of 30 seconds or less, often less than 20 seconds (Glutz and Bauer 1982).

MOVEMENTS AND MIGRATIONS. Although not rapid fliers, these birds are highly mobile, tending to forage at considerable distances from their nesting colonies and undertaking migrations of substantial length. Renaud, McLaren, and Johnson (1982) studied spring migration in the area of Lancaster Sound (south of Devon Island) and western Baffin Bay and judged that a peak population of about 14 million birds may have been present in this area in mid-May of 1978 while en route to their Greenland nesting sites. Apparently northwestern Baffin Bay and adjacent Lancaster Sound constitute an important migratory corridor and staging area for most, if not all, the dovekies nesting in northwestern Greenland. In late summer (September) when the birds leave their colonies, they enter northern Baffin Bay and move down the high arctic (western) side of the bay, avoiding the West Greenland Shelf, and finally enter the Labrador Sea in October. They remain in the northwest Atlantic in the vicinity of Newfoundland until May (Brown et al. 1975). On the other hand, recoveries of birds banded in Spitsbergen indicate a movement to the southern tip of Greenland, followed by wintering on pack ice along its southwestern coastline. Birds wintering in this region may also include those nesting in East Greenland, Jan Mayen, and probably other breeding areas farther to the east (Salomonsen 1967). The time of arrival on breeding areas is quite variable, typically being during the first half of May in northern Greenland (Renaud, McLaren, and Johnson 1982; Ferdinand 1969), whereas in Spitsbergen they arrive during the first half of April (Bent 1919), and in USSR breeding areas such as Franz

Josef Land they may even arrive as early as late February or early March (Kozlova 1961). However, the laying period is more uniform, and at least in Greenland it seems to be strongly correlated with the timing of the maximum availability of *Calanus*. Similarly, the early fall departure from Greenland is associated with a rapid decline in the abundance of this source of food. Since the birds molt and become flightless immediately after breeding, they are apparently carried by currents across northern Baffin Bay in September to the east coast of Baffin Island (Evans 1981).

Social Behavior

MATING SYSTEM AND TERRITORIALITY. Norderhaug (1968) has proved that mate retention and nest site tenacity occur in this species. Based on information obtained from banding more than 11,000 dovekies in Spitsbergen, he determined that mating between the same male and female occurred for periods of up to at least 4 years and that the pair often returned to the same nest in subsequent years. Salomonsen (1967) also reported that nest site tenacity occurs in Greenland birds, and Stempniewicz (1981a) noted that 56 of 72 nest sites were used in the following year, presumably but not necessarily by the same birds. The age of initial sexual maturity is not known with certainty, but it is not likely to be earlier than the second year. Territorial behavior is apparently limited to the nest site. Considering the very high density of birds typically occurring in breeding colonies, a maximum territorial area of about a square meter per nest seemingly is normal. However, territorial fighting can be intense and appears to center on a stone immediately next to the nest crevice, which serves as a takeoff and landing site, an observation and resting post, and a place for courtship, including copulation. Preferred stones are those that are flattened at the top and protrude far enough from the surrounding surface to allow easy takeoff (Stempniewicz 1981a).

VOICE AND DISPLAY. Vocalizations of this species have been most completely analyzed by Ferdinand (1969), who recognized five calls of adults. One vocalization heard only one time was a single-noted call that probably represents a warning or distress note. Predominant during the breeding season is the trilling call, which is produced by both sitting and flying birds and is the call most frequently attributed to the species. It is a series of three separate motifs lasting a second or more, each motif being a series of rapidly rising and falling frequencies with many overtones, the second harmonic being the loudest. Probably each individual bird has an acous-

tically unique trilled call, with variations in call length, duration of the individual motifs, and substructure of the motifs. Massed calling by birds in flight produces "flock singing," which seems to exhibit coordinated variations in frequency and amplitude. Calls uttered by paired birds are of three recognized types. During aggressive encounters the birds simultaneously or alternately produce a hoarse, unmelodious call that lasts a few seconds. When sitting together on rock ledges pairs may utter weak clucking calls that, like the aggressive calls, sometimes take the form of an alternating duet. Finally, a snarling call was heard while a pair was searching for a nest site, when one of the birds lay down on the ground and made quivering movements, the other apparently uttering clucking calls. All these observed types of calls were constructed of only two (Ferdinand) to five (Rüppell 1969) different acoustic units that have been variously modified in length as well as in the number and relative strength of their harmonic elements.

Displays of the dovekie are still only very imperfectly understood, but they have been described by Rüppell (1969), Ferdinand (1979), and Evans (1981). The posturing is apparently quite simple. Resting dovekies assume a posture similar to that of singing birds or birds carrying food to their young (fig. 39A–C), although both singing and food-carrying birds have distinctly enlarged gular areas. Evans (1981) observed five major display postures, three of which involved head wagging. Head wagging, with the body low and the bill horizontal (fig. 39F), usually evoked the same response from the partner, especially if the two birds were at close quarters. If head wagging was done from an upright walking posture the response was less likely to be the same, and when performed with the bill pointing upward it was more apt to produce aggressive responses than when it was performed with the bill tilted downward. Likewise, simply walking upright with the bill pointed downward (fig. 39E) usually evoked no response and was often used by birds that had newly arrived in a group and were walking in close proximity to the others. Walking upright with the bill pointed upward was occasionally seen and evoked either aggression or no obvious response. Rüppell (1969) described this as a "parade" posture and observed that the crown feathers are distinctly raised at this time. Evans concluded that mutual head wagging is a courtship activity and noted that after hatching it became a common ceremony between pairs, replacing earlier mutual head bowing movements. At such times the birds would call and often touch bills during the ceremony, which lasted about 30 seconds, and the wings would sometimes be drooped and fluttered. Rüppell (1969) compared head wagging to billing in puffins, ex-

39. Social behavior of the dovekie (mainly after Ferdinand 1969): A, resting posture; B, carrying food; C, singing; D, carrying pebble; E, "parade" with bill down; F, mutual head wagging; G, H, copulation.

cept that in dovekies the birds' bills usually do not actually touch. Pair members also perform "butterfly flights" (Evans 1981), often preceded and followed by intense head wagging. These are downslope flights with very slow wingbeats (3–5 per second, instead of the normal 12–18 according to Rüppell), with associated calls of the same trilled type uttered during head wagging. Copulation is also preceded by head wagging. It occurs on land (often on the stones nearest the nest cavity), and during copulation the male droops his wings on either side of the female, who very briefly turns her head back to nearly touch his (fig. 39G,H). This posture is quite similar to that assumed by razorbills and murres during copulation.

Reproductive Biology

BREEDING SEASON AND NESTING SUBSTRATE. The nesting season is very short, and according to Evans (1981) the entire breeding season (including pair forma-

tion) takes only 3 to 3.5 months, distinctly shorter than in other North Atlantic alcids. There is a prelaying period of about a month, an egg-laying period of 16 days (mostly between June 20 and 27 in Evan's study area), and a combined incubation and fledging period of only 52 days. In Spitsbergen, Stempniewicz (1981a) reported a hatching period of 10 days in one summer and 14 days in the next breeding season, the latter being a year in which snow cover persisted unusually long and delayed the breeding cycle somewhat. Roby, Brink, and Nettleship (1981) estimated that the peak of egg laying at Robertson Fjord, in the Thule area of Greenland, was June 22–24 and the peak of hatching July 16–18.

The nest site is a rock crevice or the cavity formed by rock rubble in talus or scree. Typically the nest has a single entrance hole, often so small the bird can barely squeeze through it. In large colonies the individual nests may be interconnected by passages, allowing emergency escape routes when the entrance is blocked by a predator. The bottom of the nest site is of rock, typically with a pebble lining made up of small stones the adults gather in and around the nest (fig. 39D). These stones prevent the eggs from rolling during incubation. Sometimes nests are in holes under large rock blocks, in which case they may have earthen floors and be enlarged by the birds' clawing (Stempniewicz 1981a). Roby, Brink, and Nettleship (1981) found the highest nest densities in talus areas having rocks 0.5–1.0 meter in diameter.

NEST BUILDING AND EGG LAYING. Probably in most cases no nest is actually built, though pebble gathering probably requires some time. Certainly a good deal of time and effort is spent in finding and then defending a suitable nest site. Occupation of territories occurs as soon as the nesting sites become snow free, and fierce fighting may occur at that time (Stempniewicz 1981a). However, Evans (1981) observed little aggressive behavior, and perhaps differences in nest-site availability cause local differences in aggressiveness. Since only a single egg is laid, each bird's egg-laying period is very brief. As noted, entire colonies have highly synchronized egg-laying periods. Stempniewicz (1981a) reported that in the colony he studied (apparently totaling about 70 territorial pairs) the egg-laying period was 10 days in one year and 13 days the following year. Although two eggs or young have on occasion been found in a single nest, and adults exhibit a double brood patch, there is no good evidence that dovekies ever lay more than a single egg, according to Stempniewicz. However, their abundant food supply during the prelaying period not only allows dovekies to lay relatively large eggs, but also promotes rapid chick growth (Evans 1981).

INCUBATION AND BROODING. Both members of the pair incubate. During incubation the egg is placed on either side of the bird's axis, under one wing. When alarmed the incubating bird can thus move the egg to another part of the nesting cavity. The length of individual incubation sessions is quite variable, but on average there are four alternations per 24 hour period. Birds sometimes leave their egg unattended for several hours, perhaps when disturbed by gulls or foxes. Evidently both sexes participate about equally in incubation, and if one partner should die the other may continue to incubate for at least 3 weeks. The incubation period averaged 29 days (range 28–31) in Stempniewicz's study (1981a), longer than earlier unreliable reports of 24 day incubation periods. He also reported an average 4 day hatching period between pipping and emergence from the shell. He reported a 65.3 percent hatching rate from 98 eggs, while Evans (1981) reported a 65 percent hatching rate from 20 eggs. Nest abandonment was apparently a significant cause of egg losses in both studies.

GROWTH AND SURVIVAL OF YOUNG. During the first 24 hours after hatching the young bird is brooded almost constantly, and brooding gradually declines from 70 percent of the time on the 3d day to 10 percent by the 8th day, with some brooding occurring as long as 20 days after hatching. Feeding begins immediately, both inside the nest cavity and later (after about the 15th day) outside of it. By 3 days the chicks are able to run about through the maze of nesting chambers and most often defecate outside the cavity entrance. When they begin to leave the nest cavity they often climb the nearest stone, where they spend increasing amounts of time exercising their wings (Stempniewicz 1981a). Growth of the young is very rapid, and by 23 days the chicks reach their peak weight, then lose some weight. The number of parental feedings also begins to decline after 21 days. Down feathers are lost from the head, nape, and belly by 19–21 days and are virtually all gone by 25 days. Fledging occurs at 27–30 days (average 28.3 days) (Evans) but may occur as early as 23 days after hatching (Stempniewicz). Norderhaug (1970) reported an average fledging period of 27.1 days. As they fledge, young birds head out to sea singly, in groups of young, or as mixed groups of adults and young. At the same time or earlier, adults (both breeders and nonbreeders) begin leaving the colony (Stempniewicz 1981a,b). Bradstreet (1982a) reported that newly fledged chicks are attended by adult males. Nonbreeders and subadults reportedly leave Greenland colonies sooner than breeders, going to pack-ice areas where they molt during late July and August. By late August the adults from the Thule region abandon their chicks in northern Baffin Bay and leave the area, probably flying farther south before molting and thus reducing food competition with the chicks that remain behind (Bradstreet 1982a). Stempniewicz has made the interesting point that in this species the juvenal plumage of at least some populations closely resembles the adult breeding plumage, which might make it more difficult for predators to distinguish the younger and doubtless more vulnerable birds from older age-classes. A similar case seems to occur in razorbills and at least a small percentage of thick-billed murres.

BREEDING SUCCESS AND RECRUITMENT RATES. Stempniewicz (1981a) estimated that in his study area there was a 65.3 percent hatching success and a minimum 80 percent fledging success (51 of 64 hatched chicks, with 11 of uncertain fate). Ten of 13 hatched young (from 20 original eggs) fledged (77 percent) in Evans's (1981) study. Thus an overall breeding success of 50–52 percent is indicated by these two studies. Since the incidence of nonbreeders (subadults or nonnesting adults) is unknown, the actual recruitment rate cannot be accurately estimated, but it can be no higher than about 30 percent and is probably closer to half of that. There are apparently no estimates of adult mortality and survival rates, although Salomonsen (1967) noted that of 397 band recoveries of adults banded at Thule, 41 were made in the same summer. Of the remaining recoveries, 75 percent were in the first year, 17 percent in the second, 7 percent in the third, and 1 percent in the fourth. Banded adult birds from Disko Bay have been recovered far less often, but one was captured 8 years after banding. These few data suggest a higher adult mortality rate than seems typical of the larger auks, but it would probably be dangerous to speculate further.

Evolutionary History and Relationships

The dovekie has traditionally been placed in a monotypic genus, usually close to the great auk, razorbill, and murres. The American Ornithologists' Union (1983) considers *Alle* to constitute a monotypic tribe Allini, which is placed nearest the Alcini (*Uria* and *Alca*) in linear sequence. Voous (1973) inserted the genus between the murrelets and the auklets, producing a sequence that was later adopted by Glutz and Bauer (1982) and has also been recently adopted for *The Birds of the Western Palearctic* (Cramp and Simmons 1985). Kozlova (1961) described the dovekie's skull as relating it to the murre group and listed it next to *Alca* in linear sequence. I believe it is indeed closest to the typical murre/razorbill/great auk lineage, though perhaps its marked foraging divergence (and associated morphological adaptations) warrant tribal distinction. Its social displays (such as head wagging with billing, upward and downward bill tilting, and copulatory behavior) are very close to those of the razorbill, and it seems to me that *Alca* might be its nearest living relative.

Population Status and Conservation

As possibly the most abundant of all the alcids of the world, this species warrants no serious current concern from conservationists. Its major breeding areas are all remote high arctic sites, and though some peripheral southern breeding sites have disappeared in historical times this should cause no real concern and probably simply reflects climatic trends. Although dovekies are locally hunted and netted on their breeding areas in considerable numbers, such hunting does not appear to threaten their status (Salomonsen 1967). The birds are well dispersed in their wintering areas, most of which are well to the north of major oil shipping routes in pack-ice areas of the Atlantic.

Common Murre

Uria aalge (Pontoppidan)

OTHER VERNACULAR NAMES: Guillemot (British); lomvie (Danish); guillemot de troïl (French); agpa siggugtoq (Greenland); Trottellumme (German); langvia (Icelandic); umigarasu (Japanese); tonkoklyuvaya kayra (Russian); sillgrissia (Swedish).

Distribution of North American Subspecies (See Map 12)

Uria aalge aalge (Pontoppidan)
BREEDS from western Greenland (locally), Labrador, and Quebec south to Newfoundland and (formerly) Nova Scotia; around the coast of Iceland; from the Outer Hebrides, Shetlands, and Orkneys south to eastern and western Scotland; and central Norway from Lofoten to the vicinity of Bergen.

WINTERS offshore throughout its breeding range, extending farther south to Maine, casually to Massachusetts, New York, and New Jersey; and to northern Spain, France, Belgium, Netherlands, Denmark, and Germany.

Uria aalge californica (Bryant)
BREEDS from northern Washington south to California.

WINTERS offshore on adjacent seas. Casual south to Newport Beach, Orange County, California.

Uria aalge inornata Salomonsen
BREEDS from the Commander Islands, Saint Matthew Island, and northwestern Alaska to Kamchatka, the Kurile Islands, southern Sakhalin, eastern Korea, and Hokkaido, and through the Aleutian and the Pribilof Islands to southern British Columbia.

WINTERS offshore on adjacent seas north to the limit of open water.

Description (Modified from Ridgway 1919)

ADULTS IN BREEDING PLUMAGE (sexes alike). Head and neck plain olive brown or sepia to nearly clove brown, little if any darker on pileum and hindneck, but sometimes slightly more grayish on crown; rest of upperparts plain dark grayish brown (nearest chaetura drab or fuscous, but more grayish than the latter), the secondaries narrowly but sharply tipped with white; underparts, including median portion of lower foreneck, immaculate white except on outer portion of sides and flanks, where broadly streaked with grayish brown; bill black; inside of mouth yellow; iris dark brown; legs and feet dull black or dusky. In the "bridled" color phase a narrow white eye ring and postocular stripe are present.

WINTER PLUMAGE. Whole underside of head, foreneck, malar, suborbital, and auricular regions, and stripe on each side of occiput white, the latero-occipital area separated from the white below it, except posteriorly, by a postocular stripe of dark smoky brown, extending along upper edge of auriculars; bill and feet more brownish. First-winter birds have fewer dark flank stripes and more white nape stripes (Kozlova 1961).

JUVENILES. Similar to winter adults but without white on sides of occiput, and white of foreneck faintly mottled with grayish brown or dusky; bill smaller. There are no black flank stripes, and the upper body feathers are edged with brown (Kozlova 1961).

DOWNY YOUNG. Head and neck sooty black, finely streaked with dull grayish white; upperparts plain deep grayish brown or brownish gray, the sides and flanks similar but paler; chest, breast, abdomen, and vent region immaculate white. Iris brown, bill bluish gray, mouth pale flesh color, and legs and feet yellowish with blackish markings (Harrison 1978).

Measurements and Weights

MEASUREMENTS (of *aalge*). Wing: males 191.0–204.5 mm (average of 9, 197); females 185–208 mm (average

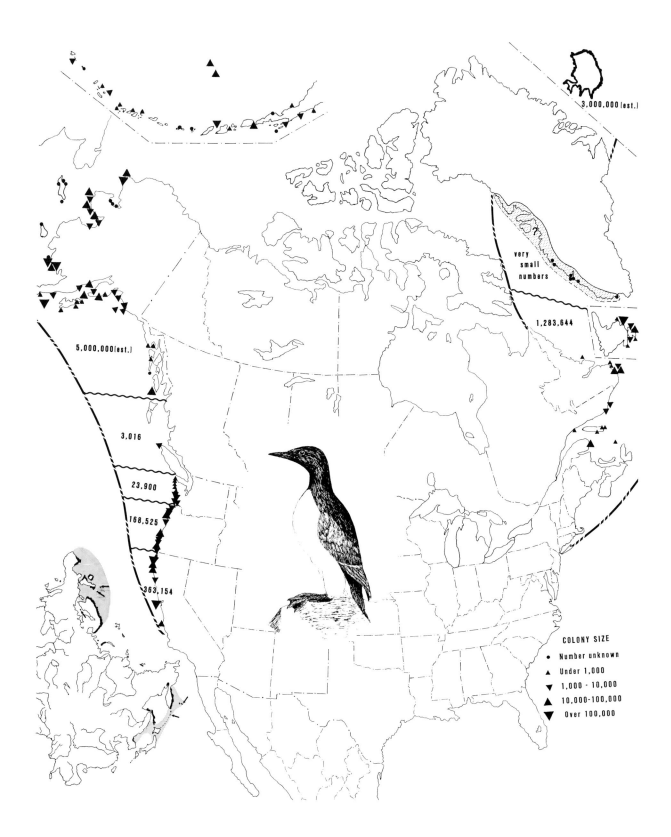

3,000,000 (est.)

very
small
numbers

1,283,644

5,000,000 (est.)

3,016

23,900

168,525

363,154

COLONY SIZE

• Number unknown

▲ Under 1,000

▼ 1,000 - 10,000

▲ 10,000-100,000

▼ Over 100,000

12. Current North American distribution of the common murre; symbols as in map 11.

of 16, 190.7). Exposed culmen: males 40.0–49.5 mm (average of 9, 46.8); females 40–50 mm (average of 16, 43.5) (Ridgway 1919). Eggs: average of 64, 81 x 50.5 mm (Bent 1919).

WEIGHTS. The average of 41 summer males was 980.0 g and that of 37 females was 988.9 g (Swartz 1966) at Cape Thompson, Alaska. A sample of 121 males and 117 females from Newfoundland breeding areas averaged 1,006 (775–1,202) g and 979 (815–1,187) g respectively (Threlfall and Mahoney 1980). The calculated egg weight is 108–14 g for various North American races (Schönwetter 1967). Newly hatched young weigh 66–79 g (Fjeldså 1977).

Identification

IN THE FIELD. The strongly black-and-white plumage and moderately large body (size of a scoter or an eider) separate this species from all other alcids except the thick-billed murre and the razorbill. The razorbill has a heavier and less pointed bill, and the thick-billed murre has a thin white mandibular stripe in breeding plumage and has more black on the upper face in the winter plumage, producing a smudgy black area behind and below the eye. Both species utter hoarse, moaning calls. The common murre has a rare "bridled" color phase in which a narrow white eye ring and postocular stripe appear in the breeding plumage. This variation does not occur in the thick-billed murre.

IN THE HAND. The combination of relatively large size (wing length at least 185 mm) and a fairly long (culmen at least 40 mm) and tapered black bill distinguishes this from all other alcids except the thick-billed murre. Apart from the absence of a pale whitish bill stripe, the common murre also has a more brownish crown, and its bill depth at the nostrils is less than a third the length of the exposed culmen.

Ecology and Habitats

BREEDING AND NONBREEDING HABITATS. Breeding colonies of common murres are largely restricted to subarctic and temperate coastlines having surface water temperatures in August ranging from 4°C in the north to 19°C in the south, with the northern limit corresponding fairly well to the southern edge of the pack ice in March. In other words, it breeds north to those coastal areas that remain free of pack ice throughout the year, whereas the thick-billed murre breeds north to those areas that remain open only throughout the summer (Voous 1960). Within these limits, murres of both species breed mostly on rocky coasts that usually have steep seaward cliffs, though low-lying coasts may also be used if they are remote and predator-free. Stratified rock layers providing nesting ledges, or weathered pinnacles and similar promontories, are important habitat components (Tuck 1960). Where both species of murres occur, the thick-billed murre is likely to occur on narrower cliff ledges and smaller promontories than the common murre (Voous 1960). At least in the North Atlantic the common murre prefers flat places such as rock ledges for nesting and avoids crevices, whereas thick-billed murres are less rigid in their requirements. In high-arctic areas where there are no common murres, thick-bills occupy all kinds of sites, but where both occur together the thick-bills are forced into suboptimal nesting areas, perhaps because the common murres arrive earlier and occupy favored nesting ledges, leaving only the marginal sites for the tardy thick-billed murres. Bedard (1969c) illustrated a rather broad range of habitats used as nesting sites for common murres on the Saint Mary Islands of the Gulf of Saint Lawrence, rarely including deep rock fissures with debris present and even more rarely rock fields, where the birds occupy crevice habitats primarily used by razorbills. Non-breeding habitats are coastal and pelagic areas extending as far south as the 15°C February isotherm, and probably north to the limits of pack ice. Typically they are found in the offshore zone (at least 8 kilometers out to sea), and no more than a few hundred kilometers offshore at their southernmost breeding limits (Tuck 1960).

SOCIALITY AND DENSITIES. These are highly social birds on the breeding areas, with maximum densities of 28–34 birds per square meter reported by Tuck (1960) in three 1 meter plots in the densely occupied core of one colony, with some birds occupying no more than 0.5 square foot of ledge. Apparently the birds prefer such areas to less densely crowded ones, and especially after incubation has begun the birds are not aggressive toward neighboring pairs. Fairly high densities sometimes occur among wintering flocks of murres (either or both species) as well; Tuck reported maximum densities of about 10,000 birds per square mile (3,900 per square kilometer) on the Grand Banks area off Newfoundland in February. However, the birds tend to be well spaced while foraging, though still gaining the values of gregariousness in searching for shoaling fish.

PREDATORS AND COMPETITORS. Certainly among the major predators of murres are the larger gulls, which often breed close to murres' nesting colonies. On the Atlantic coast this includes such species as the great black-backed gull (*Larus marinus*) (Johnson 1938), while on the Pacific coast the glaucous-winged gull (*L.*

glaucescens) is similarly important. Other gulls that have been noted as egg or chick predators of common or thick-billed murres, or both, include the glaucous gull (*L. hyperboreus*) and western gull (*L. occidentalis*), while common ravens (*Corvus corax*) and, in Europe, carrion crows (*C. corone*) have been implicated (Tuck 1960; Glutz and Bauer 1982). The gulls, ravens, and crows are more often scavengers than predators, since the presence of one of the adults at the nest will keep such birds away, and only when the colony is disturbed can they easily obtain actively tended eggs or young. Likewise, most murre nests are fairly inaccessible to foxes (*Alopex lagopus*), and possibly even the snowy owl (*Nyctea scandiaca*) is more often a scavenger than an active predator, in Tuck's opinion. He did note that rough-legged hawks (*Buteo lagopus*), goshawks (*Accipiter gentilis*), peregrines (*Falco peregrinus*), and gyrfalcons (*F. rusticolus*) sometimes prey on murres, but even these may largely content themselves with injured birds.

Certainly the most serious competitor of the common murre is the thick-billed murre, whose nesting and wintering ranges both partially overlap with those of the common murre. Tuck (1960) mapped a total of more than 100 murre breeding colonies, of which about a third were of mixed species. Data on foods taken during the breeding season (Swartz 1966) indicate a substantially lower proportion of invertebrates in the diets of adult common murres, and to a lesser extent in chicks, while comparable studies by Hunt, Burgeson, and Sanger (1981) show similar differences in adult samples. On the other hand, no major differences in the invertebrate component were evident in the data of Belopolskii (1957). Spring (1971) has correlated differences in the morphology of these two species with differences in locomotor abilities and suggested that the thick-billed murre is better adapted to feeding on invertebrates and bottom-living fish, while the common murre is better at chasing pelagic fish. The common murre's taller stance and better walking ability also help it attain agonistic superiority during competition for nesting ledges, while the thick-billed murre's greater flying efficiency may be correlated with longer migratory flights and long foraging flights during the breeding season.

General Biology

FOOD AND FORAGING BEHAVIOR. The common murre feeds predominantly on schooling fish throughout the year. A sample of 14 wintering birds from coastal waters off Denmark was found to have been feeding entirely on fish, primarily herring (*Clupea harengus*), mainly small individuals up to about 6 centimeters. Other fish that

have been reported as prey include (*Sprattus sprattus*) and sand launce (*Ammodytes* sp.), and all three of these food sources have been found to be important for nestlings, particularly sprats (Hedgren 1976). Pearson (1968) reported that in the Farne Islands, at least, this species specializes on midwater fishes averaging about 8 grams, with heavy use of sand launce 100–155 millimeters long. Swennen and Duiven (1977) found that the species tended (under experimental conditions) to take larger and heavier prey fish than did either razorbills or Atlantic puffins, with the preferred weight of two prey species being 14–16 grams. Overlapping diets between common and thick-billed murres have already been mentioned, including the greater preponderance of fish in the diet of the former species. Swartz (1966) judged that in addition to a greater reliance on fish the common murre seems to have a specific preference for sand launce. There were fewer polar cod (*Boreogadus saida*) in samples from common murres than from thick-billed murres, though in both cases this was the most frequent item in adult stomach samples.

Prey are captured by extended dives, mostly at depths of 4–5 meters, but sometimes by bottom feeding at 8 meters (Madsen 1957). Under pelagic conditions the birds may dive even deeper, rarely as much as 110–30 meters, at which depths it has been captured in crab pots off coastal Alaska (Forsell and Gould 1981). Most dives last less than 30 seconds, although dives of up to 74 seconds have been reported (Glutz and Bauer 1982). Foraging tends to occur in flocks early in the breeding season, but as the year progresses murres increasingly forage individually. When bringing food to chicks they nearly always carry only one fish at a time, holding it lengthwise with the head inward (fig. 40G). Fish up to 15 centimeters long can be swallowed by murre chicks only about a week old, and the chicks require roughly half their weight in food daily, tripling their hatching weight (of about 75 grams) in 3 weeks (Tuck 1960).

MOVEMENTS AND MIGRATIONS. Understanding the migrations of common murres in most areas is complicated by their great similarity to thick-billed murres. The relatively few birds breeding in western Greenland are only partially migratory, wintering mainly in Sukkertoppen and Godthab districts and in small numbers farther south (Salomonsen 1967). Based on banding information, it is known that the offspring of common murres nesting at Funk Island (off Newfoundland) swim northwest against the Labrador Current in company with adults, reaching the Labrador coast by early August. By early October most of them begin to move south, with some going through the Strait of Belle Island and most passing along the northeast coast of

40. Social behavior of the common and thick-billed murres (after Norrevang 1957 and photos of author): A, side preening, B, greeting, C, mutual bowing, D, preen solicitation, and E, allopreening in common murre; F, preflight posture, G, fish carrying, and H, food presentation by mate in thick-billed murre.

Newfoundland and apparently largely wintering with thick-billed murres off the southeast coast. First- and second-year birds apparently also summer off the south and southeast coasts of Newfoundland, with some returning to breeding colonies in the second year and perhaps nearly all of them doing so by their third year (Tuck 1960), although initial breeding may not actually begin at that time. Movements off the Pacific coast are still very poorly known, but during winter and spring the distribution of both murre species is greatly influenced by the location of pack ice, the birds often feeding near its edge. Most of the birds winter over the continental shelf, especially in the vicinity of large colonies (Gould, Forsell, and Lensink 1982). A very important wintering area for Bering Sea and northern Pacific coast murres is the Kodiak archipelago, where common murres are the most abundant seabird species, outnumbering the thick-billed murre about 30 to 1 in bays. Murres also occur in offshore areas in slightly smaller numbers, but in deeper waters the thick-billed murre

occurs in higher proportions, suggesting that it may prefer to winter somewhat farther offshore than does the common murre (Gould, Forsell, and Lensink 1982). Most of the wintering, migrating, or nonbreeding murres are organized in flocks of from fewer than 5 birds to more than 1,000, and collectively perhaps over a million common murres winter throughout the Kodiak area. This would represent a large part of the total nesting population of the Bering Sea and Gulf of Alaska (Forsell and Gould 1981). As for other migrating seabirds using this area, Unimak Pass is probably the most important corridor to and from the eastern Bering Sea (Gould, Forsell, and Lensink 1982). Probably in most regions breeding birds winter as close to their nesting colonies as the climate permits, to allow early return and occupation of favorable nest sites. Thus in some regions the birds may return to nesting areas shortly after completing their postnuptial molt (Birkhead 1978).

Social Behavior

MATING SYSTEM AND TERRITORIALITY. This species exhibits both mate retention and nest site tenacity; 3 out of 6 pairs of common murres at Funk Island comprised the same birds in two successive years, and banded murres have returned to the same colony, and usually the same nest site, for at least as many as 5 years (Tuck 1960). Birkhead (1977a) observed a 95 percent rate of site tenacity among 74 marked birds the following year. Since the sexes often arrive separately, with males probably the first to occupy potential nest sites, mate fidelity is probably achieved by the strong site fidelity of the species. It is questionable whether murres can in any way be called "territorial" (Norrevang 1957), though when the birds first arrive ashore they exhibit individual distance characteristics and a good deal of bickering and jostling, which breaks down as pair bonds are established or reestablished (Tuck 1960). Swartz (1966) noted that in fighting associated with nest-ledge selection birds sometimes left the ledge and continued fights in the water. Five such sets of fighting birds were found to be all males. Swartz has also suggested that at least some pairing might occur before occupation of the nesting ledges, since he observed copulation as early as the day of arrival at cliff sites. Tschanz (1959) stated that only the actual incubation site and a narrow strip leading to the landing and takeoff site are defended by breeding birds.

VOICE AND DISPLAY. Vocalizations of this species have been summarized and illustrated with sonograms by Glutz and Bauer (1982). Adult vocalizations include a general "nodding" call that occurs during ordinary colony activities and a short or two-part call that indicates slight excitement. This grades into a much more prolonged excitement crow or fighting crow, the latter sometimes lasting over 5 seconds. Another prolonged call is the "disapproval" call, a defensive call that may be directed at intruding birds. Calls to the mate or chicks include a contact or greeting bark, a copulation call, and an attraction call. Calls of chicks include a contact call, a whining note, and a loud position call that helps adults recognize and locate their separated chicks.

Displays of the common murre have been described by Birkhead (1976, 1978). He recognized six appeasement postures, including three passive (aggression-avoidance signals) and three active ones (aggression-termination signals). Active appeasement displays include side preening (fig. 40A), which serves as both an active and a passive signal, a stretch away posture (a rapid in-and-out movement of the neck) performed mainly by incubating birds, and a turn away posture that was usually assumed during or after fights. These three signals also serve as passive appeasement signals, as do two forms of ritualized walking (a posture with head up and neck stretched diagonally upward very similar to that assumed immediately after landing, or a posture with head down and neck stretched forward, usually with the wings both raised) and the postlanding display. In this posture the bird walks while maintaining the heraldic posture assumed immediately after landing near conspecifics, with the wings extended upward and the head, outstretched neck, and body all in nearly vertical alignment. The postlanding display is probably a combination or mosaic posture, involving both postural recovery after landing and preparation for agonistic interactions with nearby birds (Mahoney and Threlfall 1982). Threat display consists of a vertically erect posture, usually with the wings somewhat extended, combined with bill pointing or jabbing movements toward the opponent. Display interactions between paired birds consist of protracted greeting ceremonies (fig. 40B), including mutual billing or nibbling and mutual preening (fig. 40E). Preening between paired birds is often preceded by a preen solicitation posture, with the bill tilted upward and the eyes nearly closed (fig. 40D). This often results in the partner's preening the neck or head region of the soliciting bird. Two displays of paired birds are associated with site ownership. One of these is mutual bowing (fig. 40C), and the other is the postlanding display posture, which is performed more frequently at a nest site than when landing at a loafing area. Mutual bowing takes a form fairly similar to the preflight bowing movements of thick-billed murres (fig. 40F) and ex-

cept for the context might be easily confused with them. Thick-billed murres also perform an "alarm bowing," a "head vertical, bill vibrate" posture comparable to the corresponding "ecstatic" posture in the razorbill, and a fish presentation display (fig. 40H) (Birkhead 1976). Some authors have questioned whether the last is actually a functional behavior, for though birds often carry fish in the bill for extended periods before the young have hatched, they rarely actually present these fish to their mates. Slow "butterfly flights" are sometimes also performed by murres; Tuck (1960) believed these might be performed by birds that had been forcibly relieved of nesting duties by their mates. Birkhead (1976, 1978) did not observe any aerial displays in murres but indicated that one is present in razorbills.

Copulation is initiated by the female's falling or leaning forward and uttering a call (Birkhead 1978). The male mounts her from the side, using his wings to maintain balance and usually drooping them over her sides during treading (fig. 41G). At that time the female throws her head back, gapes, and utters a hoarse call, then the head returns to the plane of the body. The male may also call during treading. After copulation the female rises and the male glides off her back (Tuck 1960). Although monogamy is maintained, males also exploit any available opportunities for copulation with other females. However, females rarely accept the advances of strange males, which are very prevalent immediately before egg laying (Birkhead 1978).

Reproductive Biology

BREEDING SEASON AND NESTING SUBSTRATE. Egg records from the Gulf of Saint Lawrence are from late May to late July, with a peak during the second half of June. Records from the Farallon Islands of California are from early March to late July, peaking in the second half of June (Bent 1919). At the northern edge of the species' range at Cape Thompson, Alaska, egg records are from late June to early September, with an approximate month-long spread in hatching dates (Swartz 1966). Even at this latitude there is a fairly high rate of egg replacement, especially if the loss occurs shortly after laying, with an average egg-replacement interval of about 2 weeks. Repeated egg replacement is also typical; a single female might thus produce several eggs in a breeding season (Glutz and Bauer 1982), though only one offspring is raised per season. Birkhead (1980) reported that only about 10 percent of the eggs in his study area were replacement eggs and that egg-laying synchrony existed at two levels, including a colony-level effect and an intracolony effect. Although level ledges on precipitous slopes are the favored nest sites of this species, Bedard

(1969c) noted that the birds also use crevice and cavity sites to a limited degree, so long as they are able to stand and copulate there. Birkhead (1977a) reported that the average width of ledges used by nesting birds on medium-density sites of his study area was only 0.29 meter; birds nesting in higher-density sites had a higher breeding success rate than those occupying less dense sites, apparently because sparsely spaced birds had poorer antipredator defenses against gulls and also exhibited poorer breeding synchrony than did those nesting close to one another.

NEST BUILDING AND EGG LAYING. No nest is built; the egg is laid on the rocky substrate. A few pebbles or other materials may be dropped at the nest site, perhaps to reduce rolling of eggs, especially early in incubation before the egg has become cemented to the substrate by excrement and sediment. Although individual murre eggs are highly variable in color and spotting pattern, the birds have only limited egg recognition capabilities and often will accept eggs laid by other murres or even egg-shaped rocks (Tuck 1960). However, Tschanz (1959) concluded that common murres are unlikely to retrieve an egg laid by another bird unless it strongly resembles their own. The birds also recognize and respond to their own specific nest sites and are unlikely to retrieve their eggs when they are substantially displaced, such as into the common activity area. At times, however, a bird will defend a new territory around an egg that has rolled away from its original site (Norrevang 1957).

INCUBATION AND BROODING. Both sexes incubate with similar intensity; Tuck (1960) believed that males might even be more assiduous than females. Incubation begins with the laying of the first egg. The bird places the egg between its legs on the outstretched toe membranes; it normally maintains a semiupright posture with its head toward the cliff ledge and the large end of the egg also toward the cliff. Unlike razorbills, the birds keep the incubating eggs constantly at the same angle relative to the cliff, and there are fewer changeovers between pairs during incubation. Further, the shape of the egg allows it to turn in a smaller circle than those of razorbills (Ingold 1980). The incubation period is somewhat variable but averages about 32 days (Tschanz 1968; Hedgren 1980; Birkhead 1980), with reported extremes of 30.5 to 35 days.

GROWTH AND SURVIVAL OF YOUNG. Because of the high rate of egg loss in murres and the associated need for renesting, the hatching period is relatively prolonged, even in northernmost areas. Swartz (1966) observed that although the peak of hatching of murres (both species) at Cape Thompson occurred during a 10

day period, the entire hatching period extended more than a month, with most late hatching the result of egg replacement. Both adults feed the chick, which is rarely left unattended. Nonetheless, there is often a fairly high loss of chicks to exposure or falls during the first 6 days after hatching, after which clinging, hiding, and thermoregulation abilities have become better developed. Thermoregulation may not be completely developed until shortly before the chick fledges (Tuck 1960). Tschanz and Hirsbrunner-Scharf (1975) compared behavioral adaptations of murre and razorbill chicks in terms of relative capabilities for survival on cliff ledges and, as noted in the razorbill account, found numerous significant behavioral differences. Birkhead (1977a) reported that the young are fed an average of 3.2 times a day (maximum of 7), with the adult carrying back a single fish on each trip (average fish weight 8.8 grams) and requiring an average of 82 minutes per foraging trip. Plumage development and weight gain nevertheless proceed fairly rapidly, and fledging occurs about 3 weeks after hatching. At that time the chick is approximately three times its hatching weight but only about 25 percent of adult weight, and it is still unable to fly. It leaves the colony by scrambling, flying, or gliding down to the sea in company with one of the adults, nearly always after dusk (Greenwood 1964). The young birds immediately leave the vicinity of the colony, and for the first few weeks each chick is evidently tended by a parent. Although Tuck (1960) believed such postfledging parental accompaniment is rare, Greenwood (1964) considered it normal and important for chick survival. Early fledging probably has several advantages; it reduces the time parents must transport food long distances to the nestling, lets the chick learn to forage effectively while it is still in the care of a parent, and allows an early postnuptial molt by adults (Birkhead 1977a).

BREEDING SUCCESS AND RECRUITMENT RATES. Birkhead (1977a) reported that of 486 eggs laid in a three-year period, 392 chicks hatched and 349 were fledged, representing a breeding success of 0.72 fledged young per pair. During a six-year study Hedgren (1980) found a high rate of breeding success in all years, with an average production of 0.8 young fledged per breeding pair. Eggs laid late in the season, either by inexperienced birds or as replacements, suffered higher egg or chick losses than did those produced earlier. Postfledging mortality rates of juveniles are still largely unknown, although Tuck (1960) noted that of 568 recoveries of murres (both species) that had been banded in Canada, predominantly when chicks, 70 percent occurred during the year following banding. Birkhead (1974) estimated an adult annual mortality rate of 12.1

percent based on recoveries of birds banded as adults, and a similar estimate of 13 percent annual mortality for adults was made by Southern, Carrick, and Potter (1965). Later, Birkhead and Hudson (1977) estimated a slightly lower (9.5 percent) adult mortality rate and stated that most birds probably do not begin breeding until their 5th year. Because of band-loss problems in long-lived birds it is very difficult to estimate mortality rates, and 6 percent annual adult mortality may be closer to the actual case (Birkhead 1974). This would result in an average life expectancy for adults of 16 years; banded birds have been known to survive as long as 32 years (Glutz and Bauer 1982).

Evolutionary History and Relationships

Clearly the common murre and thick-billed murre constitute a superspecies, a group with a common ancestry that probably became isolated as recently as early Pleistocene times. Storer (1952) judged that the genus *Uria* probably had an Atlantic Ocean origin and that the thick-billed murre may have speciated in an area north of Siberia, later moving south and encountering the more temperate-adapted common murre. Only a single convincing case of hybridization has been reported (Tschanz and Wehrlin 1968), which is rather surprising in view of the species' overlapping ranges, similar ecologies, and very similar vocalizations and displays.

Population Status and Conservation

Tuck (1960) described changes in murre populations of eastern Canada: in the Labrador Current region there has been a general increase in numbers, whereas in the Gulf of Saint Lawrence the murres have not fully recovered from a crash that occurred in the 1880s. Nettleship (1977) estimated an eastern Canadian population in excess of 1,200,000 birds, which he believed to be declining in the Gulf of Saint Lawrence and stable or increasing in Newfoundland and Labrador. In Alaska, where the two murres are perhaps the most numerous of all pelagic breeding birds, they may have a combined population of about 10,000,000, with the two forms probably fairly similar in overall abundance (Sowls, Hatch, and Lensink 1978). In California the species is probably increasing, although the large colony on the Farallon Islands is still only a fraction of its original historical levels (Sowls et al. 1980). Surveys in Great Britain have indicated a general decline in southern England and Wales and varied population trends in Scotland. Oil spills represent a constant threat to this species in many parts of its breeding and wintering range, and the birds are also highly sensitive to losses resulting from

human disturbance during the breeding season. Finally, losses from fishing nets are sometimes substantial; Piatt, Nettleship, and Threlfall (1984) reported that such losses may have represented from 3 to 13 percent of the breeding stock of some Newfoundland murre colonies in recent years, or potentially more than the species' annual recruitment rates. This mortality is particularly significant inasmuch as it typically occurs during the peak of the breeding season and so probably leads to nestling mortality as well.

Thick-billed Murre

Uria lomvia (Linnaeus)

OTHER VERNACULAR NAMES: Brünnich's murre; Brünnich's guillemot (British); Pallas's murre; kortaebbet lomvie (Danish); guillemot de Brünnich (French); Dickschnabellumme (German); agpa (Greenland); stuttnefja (Icelandic); hashibuto umigarasu (Japanese); tolstoklyuvaya kayra (Russian); spetsbergsgrissla (Swedish).

Distribution of North American Subspecies (See Map 13)

Uria lomvia lomvia (Linnaeus)
BREEDS from Somerset Island, northwestern Greenland, Iceland, Jan Mayen, Spitsbergen, and Novaya Zemlya south to northern Hudson Bay, northern Quebec, Labrador, and islands off the coast of Newfoundland; formerly to Maine and to northern Russia.

WINTERS in open waters within the breeding range from Greenland south into Hudson Bay and on the Atlantic coast to New York, New Jersey, and sometimes South Carolina; casually to Lake Huron, Lake Erie, Lake Ontario, and Lake Champlain; and from Iceland, northern Norway, and the Kara Sea to northern France, Denmark, northwestern Germany, and western Sweden.

Uria lomvia arra (Pallas)
BREEDS along the coast of northeastern Siberia to the Diomede Islands, Kotzebue Sound, and northern Alaska; south to the east coast of Kamchatka, the Commander Islands, the Kurile Islands, the Pribilof Islands, the Aleutian Islands, and Kodiak. Has bred in small numbers on Triangle Island, British Columbia, since 1981.

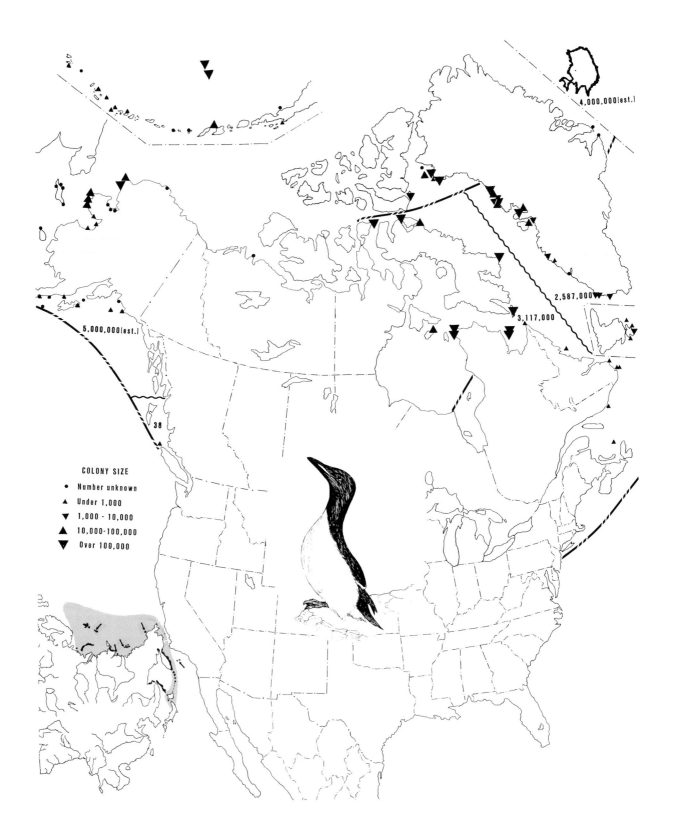

COLONY SIZE
• Number unknown
▲ Under 1,000
▼ 1,000 - 10,000
▲ 10,000-100,000
▼ Over 100,000

4,000,000(est.)

2,587,000

3,117,000

5,000,000(est.)

38

13. Current North American distribution of the thick-billed murre; symbols as in map 11.

WINTERS from Bering Sea south to Sakhalin and Honshu, and to southeastern Alaska, less frequently to British Columbia.

Description (Modified from Ridgway 1919)

ADULTS IN BREEDING PLUMAGE (sexes alike). Sides of head and neck, chin, throat, and foreneck uniform clove brown, passing into sooty slate blackish on pileum and hindneck; upperparts plain sooty slate blackish (similar to but rather more grayish than color of hindneck and pileum), the secondaries narrowly but sharply tipped with white; underparts, including median portion of lower foreneck, immaculate white, the exterior feathers of sides and flanks broadly edged on outer webs with sooty blackish; bill black, the basal half (approximately) of maxillary tomium bluish gray, sometimes conspicuously light colored; iris dark brown; legs and feet blackish, tinged with reddish.

WINTER PLUMAGE. Whole throat, foreneck, malar, subocular, and lower auricular region white, but not extending above eye stripe as in *aalge,* and the light-colored area of maxilla less conspicuous than in summer; the lower part of foreneck faintly mottled transversely with dusky; otherwise as in summer. First-winter birds exhibit mottling on the sides of the head (Kozlova 1961) and have smaller bills than older age-classes (Gaston 1984).

JUVENILES. Dimorphic, some birds resembling summer adults, with black chin and throat, and others similar to adult winter birds (Gaston and Nettleship 1981).

DOWNY YOUNG. Above dusky grayish brown or sooty, the head and neck finely streaked with pale buffy grayish; throat, foreneck, sides, and posterior underparts pale brownish gray, chest, breast, and abdomen dull white. Closely resembles the common murre, but the light barbs of the head are more buffy, the dorsal part of the body more brownish and sometimes tinged with cinnamon (Fjeldså 1977).

Measurements and Weights

MEASUREMENTS. Wing: males, 198–215 mm (average of 5, 208.2); females 198–209 mm (average of 4, 203.5). Exposed culmen: males 30.5–36.5 mm (average of 5, 34.3); females 31.0–37.5 mm (average of 4, 35.1) (Ridgway 1919). Eggs: average of 41, 80 x 50 mm (Bent 1919).

WEIGHTS. A sample of 79 males in summer from Cape Thompson, Alaska, averaged 975.5 g, while 60 females averaged 949.8 g (Swartz 1966). Gaston and Nettleship (1981) reported smaller weights from Prince Leopold Is-

land. Estimated egg weight, 106 g (Schönwetter 1967). Newly hatched chicks weigh about 70 g (Gaston and Nettleship 1981).

Identification

IN THE FIELD. Like the common murre, this species is about the size of a sea duck, but it is strongly black and white, with a sharply pointed bill. The bill is more tapered than in the razorbill but is heavier than in the common murre, and in any adult-stage plumage a narrow white mandibular stripe extends back along the base of the upper mandible, which is lacking in the common murre. Further, in breeding plumage the white of the breast meets the black foreneck in a distinctly acute peak rather than a nearly straight line. In winter plumage the thick-billed murre has less white on the sides of the face, with blackish extending below and behind the eye to the upper ear coverts. The white mandibular stripe is present as well but is less conspicuous in winter. Both species utter hoarse moaning calls while on the breeding grounds, sounding like a repeated *arr, awk,* or *uggah.*

IN THE HAND. The combination of a wing length of at least 185 mm and a culmen length of at least 40 mm, together with a tapered blackish bill, eliminates all other living alcids except the common murre. The thick-billed murre has a whitish bill stripe (most evident in summer; paler in winter and lacking in juveniles), and its bill is slightly heavier basally, so that the bill's depth at the nostrils is more than a third the length of the exposed culmen.

Ecology and Habitats

BREEDING AND NONBREEDING HABITATS. Thick-billed murres are similar to the common murre in their choice of breeding habitats, but they occur at considerably higher latitudes, occupying arctic and subarctic coastal areas with August surface water temperatures of from 0°C to about 10°C and overlapping with the common murre in the zone of 4.5–10°C (Voous 1960). Almost all the largest eastern Canadian and Greenland colonies of thick-billed murres (and all the pure thick-bill colonies) occur in an area north of the line of 10 percent sea ice cover during late July and the first third of August (Gaston and Nettleship 1981). The birds are generally more abundant on islands than on the mainland coasts, and nesting birds favor areas where foxes are rare or absent. In the winter the species is somewhat more pelagic than the common murre and tends to occupy deeper waters and more offshore zones (Forsell and Gould 1981).

SOCIALITY AND DENSITIES. Thick-billed murres are sometimes as sociable as common murres, often nesting in direct bodily contact with neighbors, especially on steep cliffs. Perhaps they do not normally occur in such great densities as common murres when both species are nesting in the same areas, but at least in part this is because common murres tend to displace thick-billed murres from the most favorable nesting locations, forcing them into peripheral areas where suitable nest sites are more scattered. Thick-billed murres are very slightly smaller than common murres (contrary to Bergmann's rule), and so it is possible that the common murre may obtain the choicest nesting locations because it tends to arrive first and begin nesting sooner on jointly used sites (Tuck 1960). In an area used only by thick-billed murres, Gaston and Nettleship (1981) found that the birds occupied the entire cliffside from about 6 meters above sea level, with the distribution determined by the locations of suitable ledges of hard strata.

PREDATORS AND COMPETITORS. Almost certainly the same general predators that affect the common murre also prey on the thick-bills, although these would include the more arctic-adapted species of gulls such as glaucous gulls (*Larus hyperboreus*). In spite of earlier observations to the contrary, Pennycuick (1956) did not consider glaucous gulls a serious problem in his study area, and he also found no direct evidence of predation by arctic fox (*Alopex lagopus*). A small percentage of eggs were likewise lost to gulls in the colony studied by Gaston and Nettleship (1981). However, the glaucous gull locally not only takes eggs but also sometimes is a serious predator on chicks leaving their nests at the time of fledging (Daan and Tinbergen 1979; Williams 1975). Swartz (1966) noted that glaucous gulls and common ravens (*Corvus corax*) were the second most important cause of egg losses (after falls from ledges), and foxes may also have caused some egg losses. Competition between the common murre and thick-billed murre has been discussed in the account of the former species, and it includes both nest site and foraging similarities. Sergeant (1951) suggested that in mixed colonies on Bear Island (USSR) the thick-billed murre is the dominant nesting form in smaller, more irregular sandstone cliffs and at the very edge of the clifftops, whereas the common murre occupies flat-topped areas and both species occur on long dolomite ledges. The common murre always selects open and flat situations for nesting, while the thick-bill is less rigid in its requirements. In Sergeant's view these differences relate to minor variations in postures assumed by the two species during incubation. Williams (1974) has also analyzed nest site characteristics of these two species on Bear Island.

General Biology

FOOD AND FORAGING BEHAVIOR. As noted in the account of the common murre, the foods of these two species are very similar except for an apparently higher concentration by the thick-billed murre on crustaceans and other invertebrates, at least during some periods. Swartz (1966) noted that only 63.9 percent of the 176 thick-billed murres examined contained fish remains, whereas 95.5 percent of the common murres did. On the other hand, invertebrates were present in 33.8 percent of the thick-bill samples, compared with only 6.1 percent of the common murres. Hunt, Burgeson, and Sanger (1981) judged that invertebrates might be more important to thick-billed murres both before the eggs are laid and after the young have fledged, whereas during incubation and the chick-raising period there is a heavy use of fish. Gaston and Nettleship (1981) determined that over 99 percent of the food provided to chicks was fish (134 samples), nearly all arctic cod (88.2 percent) or sculpins (*Triglops* spp.). They found fish in 96 percent of the digestive tracts of 80 adults examined (and containing food remains) between May and August and found crustaceans in 21 percent.

Bradstreet (1982b) examined foraging behavior of birds and mammals at ice edges during late spring and concluded that thick-billed murres and other species favor ice-edge habitats over open ocean because they provide greater access to such favored foods as arctic cod (*Boreogadus saida*). During two of three years of study thick-billed murres fed largely (86–96 percent dry weight) on this species, while in the other year they foraged proportionately more (50 percent dry weight) on a pelagic amphipod (*Parathemisto*) that is not associated with ice edges. In that year the birds were apparently unable to obtain enough cod and thus ate large numbers of the much smaller amphipods. This was during the murres' prelaying to early incubation period. Foraging depths are known to extend to 73 meters, and dives may cover 30–40 meters in horizontal distance. While foraging the birds typically remain underwater 54–71 seconds, but they may remain submerged as long as 98 seconds when frightened (Glutz and Bauer 1982). Gaston and Nettleship (1981) judged that the thick-billed murre has a potential maximum foraging radius of 100 kilometers during the chick-rearing season (200 kilometers during the prelaying period) and an effective diving depth of 20 meters. This mobility, together with the species' ability to shift from fish to crustaceans when necessary, helps explain why it is the most abundant of the arctic seabirds.

MOVEMENTS AND MIGRATIONS. The best information on migration comes from the analysis by Gaston (1980).

He judged that the murres of the Lancaster Sound area leave their breeding areas shortly after breeding terminates, reaching Greenland in early September after apparently crossing northern Baffin Bay and there joining birds from the northwestern Greenland colonies. Some of these birds migrate down the eastern coast of Baffin Island, probably mingling there with birds from Reid Bay in Davis Strait. South of Davis Bay either all the birds may move to the coast of Greenland, or some may go directly to Labrador. From November onward there is a movement from Greenland to Newfoundland that may continue until at least January. Birds from Hudson Strait pass eastward from there during September and early October, passing the coast of Labrador in October and arriving off the Newfoundland coast in November. The return begins in late March, with the time of migration dependent upon pack-ice distribution. Those heading for Lancaster Sound move first to Greenland and then cross over to Lancaster Sound, arriving there in late May or early June. The passage of Davis Strait lasts about a month and involves at least 3.5 million birds. Movements in the Pacific area are much more poorly known, but as noted in the account of the common murre, there is a very large population of more than a million murres wintering in the Kodiak archipelago, some of which are thick-bills. The total thick-bill population in Alaska may number close to 5 million birds, and it is likely that most of these winter in the Bering Sea, inasmuch as this species has been found to be many times more numerous there than in the Gulf of Alaska during pelagic surveys (Gould, Forsell, and Lensink 1982). Probably at least most of the breeding birds winter as close to their nesting colonies as pack-ice patterns allow, which would mean that many winter in the central Bering Sea from waters north of the central and western Aleutian Islands north roughly to the vicinity of the Pribilofs.

Social Behavior

MATING SYSTEM AND TERRITORIALITY. Seasonal monogamy, with regular remating with the previous mate and a return to the same nest site, is typical of this species (Tuck 1960). A small number of birds marked by Gaston and Nettleship (1981) exhibited strong site tenacity and presumed mate retention. Overall breeding success at sites occupied both years was substantially higher (74.4 percent) than for sites used in only one of the two years (57–66 percent), suggesting a distinct biological advantage of site fidelity, though the use of the sites by the same birds in both years remained unproved. As in the common murre, territorial defense is probably limited to the immediate nest site, a few

square decimeters in area. Fights, probably over nest sites, are often intense in the early part of the breeding cycle, but after the beginning of egg laying those that include bill grappling are rare. At least in one study plot the highest intensity of fighting occurred during the last 10 days before egg laying, with a maximum observed rate of 25.2 fights per 100 birds per hour (Gaston and Nettleship 1981).

VOICE AND DISPLAY. Tuck (1960) stated that the calls of thick-billed and common murres appear to be remarkably similar, and more detailed comments have been made by Pennycuick (1956) and Tschanz (1972). Like those of the common murre, the calls tend to intergrade and to vary considerably in duration, loudness, and tonal characteristics. Chicks have at least four distinct calls, including a "cheeping" call uttered by cold or hungry chicks, a whining note, a bell-like call uttered on actual contact with an adult, and a squealing three- or four-note "water call" that is most often given when the chick is greatly frightened or about to fledge. This last vocalization produces a strong agitated response from adults. Representative sonograms of this species' calls are provided by Glutz and Bauer (1982).

Displays of the thick-billed murre (see fig. 40) are extremely similar to those of the common murre. Major differences between the species seem to consist of their vocalizations as well as the obvious visual differences associated with bill shape and color. The thick-billed murre also utters a contact bark that is lacking in the common murre, usually produced with sideways neck bending and shaking of the whole body. Afterward the bird bows so deeply that its bill may touch the substrate (Williams 1974; Glutz and Bauer 1982).

Reproductive Biology

BREEDING SEASON AND NESTING SUBSTRATE. Records of eggs from the Gulf of Saint Lawrence and eastern Labrador are from 5 June to 25 July, with a probable peak during the latter half of June (Bent 1919). Likewise, most egg laying in Greenland takes place during the latter half of June (Salomonsen 1967). On Prince Leopold Island in arctic Canada the egg-laying period extended from about June 20 to late July, with a peak about the first of July, and ranged from 35 to 45 days during three breeding seasons. In spite of the high latitude, about 30 percent of the eggs lost were replaced, though pairs losing their eggs after June 20 did not lay again. The replacement interval was 13–15 days (Gaston and Nettleship 1981). Egg records in the Bering Sea region are from June 2 to September 1, with a peak from mid-June to mid-July, while records from north of the Bering

Strait are from July 3 to August 1 and a few records from south of the Alaskan Peninsula are all for June (Bent 1919). In a mixed murre colony at Cape Thompson, Alaska, the major period of egg laying extended about a month, from late June to late July, with a peak of initial laying lasting about 10 days and very few replacement eggs laid after early August (Swartz 1966).

The preferred nest substrate consists of narrow ledges on cliff faces; ledges as narrow as the length of the bird's foot plus its tarsus are used. Since the birds typically incubate by raising the sternum and having it externally supported, they can push against the cliff face and incubate their eggs in a nearly upright position. Gaston and Nettleship (1981) analyzed various aspects of nest sites with respect to breeding success. Sites with a single immediate neighboring pair were most common, but breeding success increased with the number of neighbors, up to two. Ledge slope (level or sloping toward or away from the sea) had relatively little effect, with level sites having the highest breeding success. Ledge width had little effect on breeding success, but most nests were on narrow ledges (so narrow that the tail hung over the edge). Most sites were protected by a rock wall on at least one side; generally breeding success increased with the degree of such protection. A small percentage of the nest sites had cracks large enough for chicks to hide in; such sites had higher average (but not statistically significant) breeding success. Most nest sites lacked rooflike overhead protections, and this did not seem to influence breeding success. Some sites still had snow cover at the time of initial egg laying; this seemed to have little direct effect on breeding success.

NEST BUILDING AND EGG LAYING. No nest as such is built, but like common murres the birds sometimes deposit pebbles in the vicinity of their nest sites. Pennycuick (1956) considered this a kind of "vestigial" nest building. Gaston and Nettleship (1981) observed that in spite of varying weather conditions, the egg-laying period at Prince Leopold Island varied only slightly in three different years, and they judged that its timing might be fixed in relation to certain events in the marine environment (such as timing of peak abundance of foods for chicks) rather than occurring as soon as the birds are physiologically capable of producing eggs. They observed no instances of more than one egg replacement, although Tuck (1960) reported that at Cape Hay two replacement eggs might be produced. He suspected that most egg losses he saw at Cape Hay were caused by eggs' falling off their ledges, while some others resulted from rock falls or eggs' rolling into crevices

and a very few were taken by gulls. Gaston and Nettleship (1981) reported a rather low rate of egg losses, with most of the observed hatching failures apparently resulting from infertility or dead embryos and most egg disappearance occurring very shortly after laying.

INCUBATION AND BROODING. Incubation begins immediately after the egg is laid and, as noted earlier, is typically done in a semierect posture. Both sexes participate about equally in incubation; Tuck (1960) believed the male might actually take the larger role. Gaston and Nettleship (1981) estimated that the hatching success of initial eggs was from 76 to 84 percent during three years of study, and eggs laid early in the season had a greater hatching success rate than did those laid later. There is some variation in observed incubation periods; Tuck (1960) noted that 60 percent of 100 marked eggs hatched 34 days after laying, 30 percent after 33 days, and 10 percent after 32 days. Gaston and Nettleship (1981) observed a maximum range of 25–40 days, with modal lengths of 32–33 days in three different years. They observed a hatching success rate of 75.7–84.0 percent for initial eggs and a slightly lower (69 percent) rate for second nesting efforts. Tuck (1960) observed a much higher rate of egg loss at Cape Hay and judged that no more than 60 percent of the birds attempting to breed actually succeeded in hatching an egg.

GROWTH AND SURVIVAL OF YOUNG. As soon as the chick has hatched the adults begin to feed it, occasionally even bringing fish before the chick has escaped from its shell. The average weight at hatching is about 70 grams (Uspenski 1958; Gaston and Nettleship 1981). Typically, weight increases for the first 8–10 days of life are at the apparent maximum possible rate, suggesting that the adults provide all the food the young can handle. After 10 days there are divergences in weight that seem to reflect variations in the parents' ability to provision their young, and apparently after about 16 days the adults are unable to feed their chicks at a rate that covers their metabolic needs. Thus weight gain tends to decrease, and plateaus or even slight declines occur just before fledging, probably because of the chicks' increased physical exercise. However, there are substantial individual differences in average weights at fledging that could have strong effects on postfledging survival probabilities (Gaston and Nettleship 1981). The actual fledging age varies somewhat between individuals, in part because of a tendency for synchronized fledging throughout a colony, but it averages about 20 days. Williams (1975) judged that synchronized fledging is most likely to serve as an antipredator device (by "swamping" the predators with available prey) when access to the sea is hampered by difficult terrain, and that the risk of predation is greatest before and after the primary fledging hours of late evening. Typically the chick leaps from its nest site and glides down the cliffside to the sea, followed closely by one of its parents. By calling to one another, adult and chick meet on the water and then swim out to sea together. Williams's observations suggest that thick-billed murres may protect their young more intensely than common murres do, possibly because of their more restricted breeding sites. Hatch (1983) studied initial postfledging survival in both species of murres in Alaska and observed a very high (nearly 50 percent) mortality rate in thick-billed murre chicks during the first 5 days after fledging, at least among those birds that remained in freshwater ponds during this period. He found a very high positive correlation between weight at fledging and immediate postfledging survival, attributing the probable advantages of heavy fledging weight to ability to survive unfavorable weather and limited food supplies during this critical period.

BREEDING SUCCESS AND RECRUITMENT RATES. During their three years of study, the overall breeding success rate (percentage fledged young relative to total initial eggs) reported by Gaston and Nettleship (1981) ranged from 68.4 to 78.8 percent, averaging approximately 0.73 fledged young per breeding pair. They judged that only about 40 percent of the fledged chicks survived the fall journey to their wintering areas, and thus the actual maximum annual recruitment rate (addition of fully grown young to the population) is approximately 0.25 young per pair. The difference between this estimate and the actual recruitment rate depends on the incidence of nonbreeders in the population. If the species has a normal annual adult survival rate of 91 percent as estimated by Birkhead and Hudson (1977), and if only 4–6 percent of the banded chicks survive to their 5th year (when most initial breeding probably occurs), then is it quite possible that at least half of the general wintering population is made up of nonbreeding birds. Obviously a minimum recruitment rate of 9 percent would be needed to maintain the population over long periods.

Evolutionary History and Relationships

This has been discussed in the account of the common murre. The incidence of possible hybridization between these two species in the wild is still uncertain but apparently is extremely small (Tschanz and Wehrlin 1968; Cairns and deYoung 1981) considering the rather large area of sympatric contact between them.

Population Status and Conservation

Nettleship (1977) judged that the breeding population of thick-billed murres in eastern Canada numbered more than 5 million birds, with probable declines occurring in the eastern Canadian arctic and the Gulf of Saint Lawrence and uncertain trends in eastern Newfoundland and Labrador. Much less is known of actual numbers elsewhere in the species' North American range, but in Alaska the total numbers probably also are in the vicinity of 5 million birds (Sowls, Hatch, and Lensink 1979). Gaston and Nettleship (1981) stated that these current very large populations are no guarantee of the species' continued survival, especially since the birds are concentrated in a few massive colonies that may be the relicts of consolidations of earlier smaller ones. These vast colonies depend on massive food supplies, including a single species of fish (arctic cod) in the Prince Leopold Island area, and a change in availability could have calamitous consequences for millions of birds. Murres are also highly vulnerable to oil spills, increased losses of eggs and chicks associated with human disturbance to nesting colonies, loss of nesting sites through erosion or earthquakes, and similar environmental problems (Tuck 1960). King and Sanger (1979) assigned an oil vulnerability index of 70 to both species of murres, one of the lowest indexes for the entire family but well above the overall average (51) for marine-oriented birds.

Razorbill

Alca torda Linnaeus

OTHER VERNACULAR NAMES: Razor-billed auk; alk (Danish); petit pingouin (French); Tordalk (German); álka (Icelandic); gagarka (Russian); tordmule (Swedish).

Distribution of North American Subspecies (See Map 14)

Alca torda torda Linnaeus
BREEDS from western Greenland and the Labrador coast to southeastern Quebec, eastern Newfoundland, southern New Brunswick, and eastern Maine; and from Norway and northern Russia south to southern Norway, southern Sweden, and southern Finland.

WINTERS from southwestern Greenland (in small numbers) south to New York, rarely to New Jersey, casually

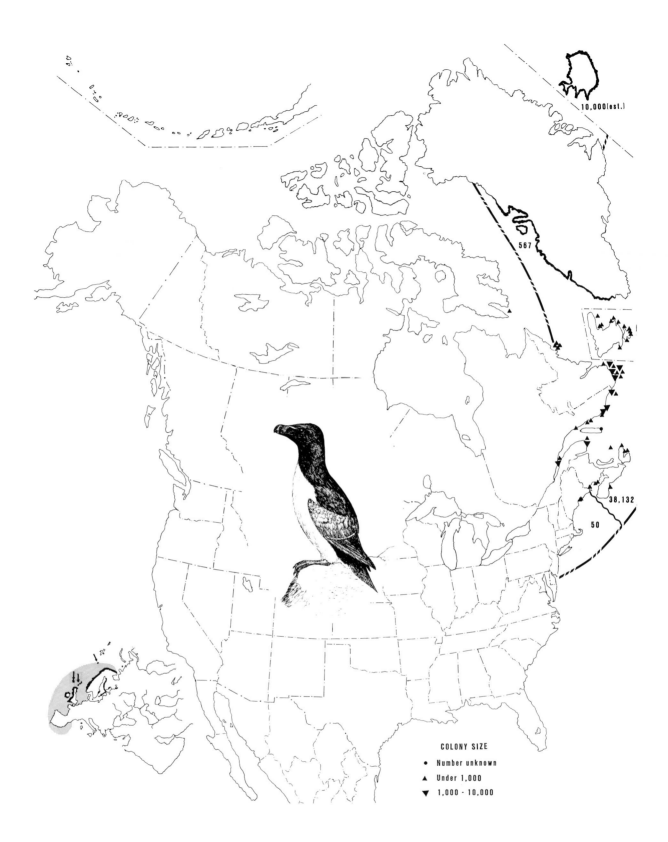

14. Current North American distribution of the razorbill; symbols as in map 11.

to Virginia and South Carolina; and from southern Norway and the Baltic to Portugal and the western Mediterranean Sea.

Description (Modified from Ridgway 1919)

ADULTS IN BREEDING PLUMAGE (sexes alike). Head and upper neck plain dark brown (bright clove brown or deep olive brown), becoming much darker on pileum and gradually darkening into slate black on hindneck and rest of upperparts; secondaries narrowly but sharply tipped with white; a narrow white line extending from anterior angle of eye to near base of culmen; underparts, including axillaries and under wing coverts, immaculate white, this extending forward to and including the lower foreneck; bill black, with one or more of the transverse grooves whitish; interior of mouth yellow; iris dark brown; legs and feet dull black.

WINTER PLUMAGE. Whole under portion of head and neck and space behind auricular region white; no white line between bill and eye; bill without the basal lamina; otherwise as in summer.

FIRST WINTER. Similar in coloration to the winter adult, but bill smaller and without grooves. The upperparts are more brownish, and there is only a pale stripe from the eye to the bill (Kozlova 1961).

JUVENILES. Sometimes similar to the adult breeding plumage; the entire head and neck black, with a narrow white stripe from the eye to the culmen as in the adult breeding plumage. This in turn is replaced by the first-winter plumage when the bird is about 4 months old (Heinroth and Heinroth 1931). Plumage dimorphism occurs, with some young having juvenal plumages resembling the adult winter plumage (Gaston and Nettleship 1981).

DOWNY YOUNG. Head, neck, and underparts plain dull whitish, usually more or less tinged above with brownish buff; back, rump, and flanks varying from pale brownish buff, more decidedly brownish posteriorly, to dark sooty brown, the down dusky immediately beneath the surface; posterior and lateral underparts more or less tinged with brownish buff or sooty brownish. This plumage is lost by about 3 weeks of age (Heinroth and Heinroth 1931). Iris brown, bill, legs and feet black, mouth pale yellow (Harrison 1978).

Measurements and Weights

MEASUREMENTS. Wing: males 188–201 mm (average of 7, 195.6); females 194–98 mm (average of 2, 196). Culmen: males 32.0–35.5 mm (average of 7, 34.1); females 33.5–35.0 (average of 2, 34.2) (Ridgway 1919). Eggs: average of 80, 75.9 x 47.9 mm (Bent 1919).

WEIGHTS. The average of 81 adult males was 734.0 g (range 524–890), and that of 61 females was 700 g (range 620–800) (Belopolskii 1957). Bianki (1977) noted that 50 birds of both sexes averaged 701 g, and Spring (1971) reported an average of 719 g for 1,442 birds. Estimated egg weight of nominate *torda*, 90 g (Schönwetter 1967). Newly hatched young weigh 54–74 g, averaging about 63 g (Dementiev and Gladkov 1968).

Identification

IN THE FIELD. This is the only fairly large (over 12 inches long) alcid that is entirely white below and black above, with a heavy black-banded bill that is distinctly blunt tipped. When swimming, its bill and tail are often tilted upward. At close range a narrow white line can be seen extending from the eye to the tip of the forehead. In winter the color pattern is similar, but the throat, cheeks, and ear coverts are also white. On the breeding ground, low gutteral or croaking sounds are uttered, sounding like repeated *arr, ood,* or *hurr-ray* notes.

IN THE HAND. This is the only living species of alcid with a bill that is both black and strongly compressed, blunt-tipped, and with a depth at the base nearly as great as the length of the exposed culmen. Young birds have somewhat less massive bills than adults, but even in these the upper mandible is strongly decurved near the tip, and the dorsal and ventral bill profile is nearly parallel rather than tapered.

Ecology and Habitats

BREEDING AND NONBREEDING HABITATS. This species breeds along coastlines of temperate and subarctic seas where there are August surface water temperatures of between about 4°C and 15°C and it also very rarely penetrates brackish water (in the Gulfs of Bothnia and of Finland) (Voous 1960). Reported rare breeding on fresh water (in Finland near Lake Ladoga and the USSR border) is apparently erroneous (Glutz and Bauer 1982). In general its breeding habitats are very similar to those of common murres, with which it often shares breeding colonies. However, it seeks out deeper nesting sites and occurs in large colonies only to about 300 meters above sea level (Glutz and Bauer 1982). In the western Atlantic the northern limit of the breeding range occurs in temperate areas of eastern Canada and adjacent New England, and the species occurs up the Low Arctic coastline of western Greenland in small but generally undocumented numbers (Brown et al. 1975). In Greenland the birds nest in rock crevices, in coarse boulders of rough talus, and in holes in rock, the sites always being sheltered and under cover. There they occur up to about 300 meters elevation, but most nests are near sea level, mainly on the small islands in the offshore fringe of skerries (Salomonsen 1967). Suitable nesting sites have foraging areas within a 15–20 kilometer radius, which are seldom in the littoral zone (Glutz and Bauer 1982). Bedard (1969c) compared nest site preferences of razorbills with those of other alcids breeding in the Gulf of Saint Lawrence and reported nesting habitat overlaps with common murre, black guillemot, and Atlantic puffin. However, 82 percent of the nesting of razorbills occurred in two habitat types, deep rock fissures with associated debris, and boulder fields. A small amount of nesting was also done in deep rock fissures lacking debris and in shallow vertical fissures. Most colonies were within 60 meters of the sea, and none were more than 80 meters away. At Kandalaksha Bay, USSR, the birds avoid forest-covered islands, and most nesting occurs in areas of lichen-covered rocks.

During the nonbreeding season the birds are pelagic, wintering mainly in cold-temperate waters and concentrating in offshore areas similar to those used by murres, but apparently at least locally foraging on rather different food types (Tuck 1960).

SOCIALITY AND DENSITIES. This is a colonial nesting species, with the size of the colony no doubt reflecting both the relative availability of suitable food resources and the abundance of nesting sites. Bedard (1969c) and Brown et al. (1975) have reviewed the locations and densities of Canadian and western Greenland breeding sites, which typically support from a few pairs to as many as 5,000 birds. Nettleship (1977) indicated that 44 known Canadian colonies had summer populations totaling over 38,000 birds, an average of 866 birds per colony. In the most highly favored nesting sites (boulder fields and deep rock fissures with associated debris) the nesting density may range from 0.25 to 4 pairs per square meter of surface area (average of 9 sites, 1.8 per square meter) (Bedard 1969c).

PREDATORS AND COMPETITORS. The well-hidden nest sites probably substantially reduce the predation rate, though evidently some eggs are regularly taken by egg predators, including gulls (Bedard 1969c) and some disappear from nests for uncertain reasons but presumably taken by jackdaws (Corvus monedula), herring gulls (Larus argentatus), or unknown egg predators (Plumb 1965). Hudson (1982) reported egg losses to herring gulls, great black-backed gulls, jackdaws, and carrion crows (Corvus corone). There is no doubt a much higher predation rate on the vulnerable chicks; Lloyd (1979)

judged that most of those disappearing from nest sites were taken by herring gulls, and other large gulls such as great black-backed gulls (Larus marinus) have been implicated elsewhere in chick losses. In the Murmansk coast area the largest numbers of eggs and young are apparently taken by great black-backed gulls and herring gulls, while avian predators of adults in general have been identified as sea eagles, peregrines, snowy owls, and goshawks. Probable mammalian predators on nesting colonies include weasels (Mustela erminea), otters (Lutra lutra), red foxes (Vulpes vulpes), and arctic foxes (Alopex lagopus) (Kartashev 1960).

Competitors are most likely to be the two murres, particularly the common murre, whose breeding range and nesting site requirements broadly overlap with those of the razorbill and whose foraging behavior is also quite similar (Tuck 1960). The common murre is substantially larger than the razorbill and thus has an advantage when the two species are competing for the same nest site (Bedard 1969c). The Atlantic puffin also overlaps to some degree in both nest site usage and foods eaten, but it is smaller than the razorbill and should be at a corresponding competitive disadvantage. The three species take prey fishes that differ in maximum and preferred size and weight, with the razorbill closer to the puffin than to the common murre in these respects (Swennen and Duiven 1977).

General Biology

FOOD AND FORAGING BEHAVIOR. One of the most complete food analyses available for the razorbill is that of Madsen (1957), which is based on a sample of 120 birds (71 with food remains present) taken in Danish coastal waters between November and February. Of those with food present, 97 percent had remains of fish and 83 percent had fed exclusively on fish, the remainder having also eaten crustaceans. One stomach contained only crustacean remains, and another only the remains of polychaetes. Of the fish, sticklebacks (Gasterosteus spp.) had been eaten by 40 percent of the birds, herring (Clupea harengus) by 34 percent, gobies (Gobiidae) by 32 percent, cod (Gadidae) by 10 percent, and other types in smaller amounts. Generally only a single type of fish was present, but as many as four different types of fish prey were found in individual samples. Three types of fish (herring, sticklebacks, and gobies) composed about 80 percent of the total food remains, while crustaceans (almost entirely Mysidae and amphipods, probably Gammarus) made up about 10 percent of the remains. Lloyd (1979) reported that the primary food given chicks on Skokholm Island was sand launce (Ammodytes), and other studies such as those from the USSR have generally supported the view that sand launce, herring, and capelin (*Mallotus*) are major foods during chick raising (Kartaschev 1960; Bianki 1977).

Fish and other prey are captured by extended dives in shallow to fairly deep water. Madsen (1957) stated that although the razorbill may sometimes dive to depths of 5 meters, it more often prefers depths of 2–3 meters for bottom foraging. It also dives pelagically in water of great depths, especially during winter, and may remain submerged up to about 40 seconds (see tables 10 and 11). A possible maximum diving depth of 120 meters was established by Piatt and Nettleship (1985), and a dive duration of 52 seconds has been noted by Bianki (1977). When carrying sand launce or similar-sized fish back to their young, adults may handle anywhere from 1 to 9 average-sized fish simultaneously, but 5 or 6 are most frequent (Perry 1975). In the case of very small fish, up to 20 might be carried, and the young are fed sand launce from 76 to 158 millimeters long, depending on their age (Lloyd 1976a; Perry 1975). The rate of feeding young is quite variable, from about 1 to 7 times a day, but typically from 2 to 5 times. Lloyd (1977) observed an average feeding rate of 4.7 times a day for control (unmodified) nests but an average rate of 9 times a day for experimental nests in which a second chick had been added. The total weight of food provided per day averaged 22 grams in control nests.

MOVEMENTS AND MIGRATIONS. Very little is known of the migrations of birds breeding in western Greenland, but they are believed to follow the Labrador Current southward to winter off the Labrador coast (Salomonsen 1967). At that time of year they are easily confused with murres, and so wintering distribution patterns are very incomplete (Brown et al. 1975). However, data from birds banded in Great Britain provide some indication of movements and migrations in this species (Lloyd 1974). In general, younger (first-year) birds have been found to winter farthest from their natal colony (averaging 984 kilometers for 99 birds banded in the South Irish Sea), while second-year birds have somewhat reduced migratory tendencies (936 kilometers for 30 birds), third- and fourth-year birds even less (715 kilometers for 18 birds), and older age-classes the least (529 kilometers for 39 birds). Lloyd concluded that razorbills banded in Britain are truly migratory during their first two years of life, but in later years their behavior is more nearly one of random dispersal away from their breeding colonies. The birds typically return to their natal colony at the end of their third year, although few birds breed before the fifth year of life. After breeding, they return to the same site year after year. One bird that was initially trapped as an adult was retrapped while breeding on Skokholm eighteen years later.

Social Behavior

MATING SYSTEM AND TERRITORIALITY. The mating system is one of seasonally renewed monogamy. Mate retention is the rule; on Skokholm 72 percent of color-marked birds paired with the same partners for two or more years. Territorial defense must be limited to the area of the nest site itself, considering the high breeding densities reported by Bedard (1969c). Nest site tenacity is also very well developed in this species (Bianki 1977). Perry (1975) has described the violent fights that sometimes occur at the mouths of nesting crannies and probably are associated with nest site defense. Sometimes two nests may occur under the same rock, but typically they are on opposite sides and separated either by natural partitions or by a certain distance if the birds use a single roomy hollow. According to Bianki (1977), breeding by birds on the Murmansk coast of the White Sea, USSR, may begin as early as the second year of life, although the youngest age-class for which nesting was proved was of a 3-year-old. Lloyd and Perrins's (1977) evidence from Great Britain indicates a much more deferred period of four or five years to initial breeding.

VOICE AND DISPLAY. General vocalizations of razorbills have been studied and classified by Paludin (1960) and Bedard (1969c), while Ingold (1973) has investigated vocal signaling between parents and chicks and a comparative summary has been provided by Glutz and Bauer (1982). Adult sounds are all rather similar in their growling aspect, but they do differ in minor characteristics of duration, tone, and cadence. On their breeding areas razorbills vocalize toward their eggs, toward their chicks, toward their mates, and toward other razorbills or enemies. Bedard (1969c) recognized seven different situations eliciting calls by adults, including a female copulation call, a "salutation" call associated with mutual billing and preening, a call occurring during the initiation of incubation, a call uttered during hostile interactions (the "ruee" posture and perhaps also during beak lowering while gaping), a call directed toward the nestling, an "alert" call, and a call uttered during the "ecstatic" posture. Chicks have four types of calls, of which the "leap call" is the loudest and carries farthest. Parents learn to recognize their chicks on the basis of acoustic characteristics of their calls during the first 10 days after hatching; they are able to find their own chicks after the frequent separations occurring at their "jumping off" age of 16–23 days (Ingold 1973).

Display postures of razorbills are largely derived

from agonistic sources that have been variably ritualized. Birkhead (1976, 1978) has identified both "threat" and "fighting" displays in razorbills but did not observe any specific appeasement gestures comparable to those he found in the more gregarious common murre. Like murres and dovekies, razorbills perform a "butterfly flight" immediately after taking off from ledges, characterized by unusually slow wingbeats and of doubtful display significance. Paired birds spend much time in mutual billing, "nebbing," and caressing (fig. 41C,D) that seem to variably grade into one another and head rubbing as well. These activities are used by paired birds as greeting ceremonies and probably play important roles in both pair formation and pair maintenance. Actual fighting is also done primarily by beak stabbing and biting (Perry 1975), and thus these billing activities probably derive from ritualized aggressive antecedents. Bedard (1969c) recognized two phases of these activities, including "salutations" (billing and nibbling) and "mutual caresses." Intraspecific agonistic

postures include a beak lowering and associated gaping (fig. 41A), together with a growling call, performed toward other intruding birds. This posture somewhat resembles "foot staring" behavior (fig. 41B), a posture of uncertain social significance often performed by grouped adults and at times also performed by paired birds during greeting ceremonies. Another posture of special interest is the "ecstatic" (Bedard) or "head vertical, bill vibrate" (Birkhead) posture (fig. 41A), which consists initially of laying the head back in a manner similar to but distinct from that of a mated bird soliciting preening (fig. 41C) and, with a vibrating throat, uttering a growling call. At times the head may be retracted until it almost touches the bird's back. In the second phase the bird lowers its head, with open bill, to its breast. This display is performed most frequently during pair formation and chick raising; in the former case it is a male display and in the latter is performed by both sexes. Its function is still uncertain, but it may serve as an advertisement signal for unpaired males and later as a greeting ceremony (Glutz and Bauer 1982). Birkhead 1976, 1978) stated that a "bowing" display occurs in razorbills that is comparable to the site ownership display of the same name in the common murre, as well as similar postlanding and ritualized walking postures, although all of these were observed at a higher frequency in the murre, probably because of the greater needs for effective socialization in the latter species. Furthermore, two other displays (fish presentation and alarm bowing) that are present in the common murre appear to be lacking in razorbills, apparently also as a consequence of differences in sociality between these two species.

Copulation may be performed without any specific preliminary displays or may be preceded by billing and rubbing behavior (Bedard 1969c; Glutz and Bauer 1982). It is apparently always performed on a solid substrate. The male mounts from the side, and during treading he balances himself on the female's back by flapping his wings (fig. 41E). During treading the male may remain silent or utter high growling sounds, while the female tilts her head upward, opens her beak, and utters a deep growling call (fig. 41F). The posture assumed by both sexes during copulation is very much like that of the common murre (fig. 41G).

41. Social behavior of the razorbill (mainly after Glutz and Bauer 1982): A, ecstatic and bill down postures; B, foot staring; C, preen solicitation and allopreening; D; nibbling of mate's feathers; E, F, copulation. Also G, copulation in common murre, for comparison.

Reproductive Biology

BREEDING SEASON AND NESTING SUBSTRATE. In North America the breeding season is fairly long, with egg records in the Gulf of Saint Lawrence extending from June 10 to July 25 (half between June 21 and July 4), and Ungava records (2 only) are from June 13 to July

1 (Bent 1919). In Greenland nesting also occurs in mid-June (Salomonsen 1967). On the Saint Mary Islands of the Gulf of Saint Lawrence there is an approximate 40–50 day prereproductive period between the time of arrival (mid-April) and the start of egg laying, and this appears to be similar to the comparable period in Greenland (Bedard 1969c).

Nest substrate characteristics observed by Bedard (1969c) in the Gulf of Saint Lawrence have been described earlier. Hudson (1982) stated that on Skomer Island, Wales, nest sites were either on small, exposed ledges on cliffs having one or two walls but no roof or were burrow or boulder sites having a roof and tended to be either in excavated holes or gaps between rocks. Of 1,688 pairs, 77.3 percent used ledge sites and 22.7 percent used burrow or boulder sites. In the Kandalaksha Bay area of the USSR the birds nested mainly (65 percent) in hollows under boulders or rock fragments, less often (30 percent) in vertical or slightly sloping rock crevices or gaps open to the top, and rarely (5 percent) in rock cavities. The substrate was of bare rock in most cases, rather than of fine-grained materials as is typical of nesting guillemots in the same area (Bianki 1977). It is thus apparent that the razorbill is quite flexible in its nest site requirements, depending upon what is available in the area, but it favors sites better protected than the open ledges usually used by murres and more rocky and crevicelike than the burrows preferred by puffins and guillemots.

NEST BUILDING AND EGG LAYING. Razorbills do little nest building as such. Bianki (1977) noted that nests typically had a poor lining of stems, twigs, and similar materials that seemed to have gotten there by accident but that played a useful role in preventing the egg from rolling about. Nearly all nests contain a single egg. Bianki (1977) reported two eggs present in 0.5 percent of the nests examined at Kandalaksha Bay, USSR, but in at least one of these cases the second egg was addled. Bent (1919) also noted that where two eggs have been found together they were probably laid by two birds. Certainly egg replacement after loss of an egg is quite regular. Lloyd (1979) reported that 25 percent of the eggs lost in her study area were replaced, compared with an estimate of 35 percent by Belopolskii (1957). Lloyd reported an average 14 day interval between loss and replacement of eggs, while Kartashev (1960) noted a similar 12–18 day interval.

INCUBATION AND BROODING. Both sexes incubate, with the male probably slightly less involved (Bedard 1969c). Exchanges of incubation are variable but usually occur at intervals ranging from 30 minutes to more than 6 hours. When the female approaches her mate to take over, she tickles his throat while uttering a cawing sound and pushes him aside. After getting off the egg, the relieved bird throws substrate materials to the side and under itself (Perry 1975). Incubation lasted an average of 35.1 days in a sample of 239 eggs studied by Lloyd (1979); Plumb (1965) reported a range of 34–39 days in a sample of 29 eggs and a mean of 36.2. Hatching success is generally fairly high in razorbills. Plumb (1965) summarized his own and earlier data for Skokholm Island, indicating a hatching success of 36–69 percent for generally small samples of from 31 to 86 eggs. Bianki (1977) observed that 84 percent of 170 eggs laid during two different years hatched, while Lloyd (1979) reported an overall hatching success of about 70 percent (range 60–78 percent, depending on laying interval) for 785 initial and replacement eggs. Hatching success was slightly lower for replacement eggs than for initial eggs and gradually declined as the season progressed. Birds that laid late in the season were mainly younger, so age and previous breeding experience were apparently important influences. The highest rate of breeding success occurred in the 6 to 9 year age-classes, partly because older birds laid relatively large eggs that produced heavy chicks at hatching.

GROWTH AND SURVIVAL OF YOUNG. The prefledging period of chick life has been studied by Tschanz and Hirsbrunner-Scharf (1975), who compared the adaptations of razorbill and common murre chicks to their usual environments (murres typically being reared on ledges in social situations, razorbills typically in burrows under more solitary conditions). Using cross-fostering experiments, they found that razorbill chicks reared by common murres suffered higher mortality rates from chilling, lack of food, and falling from ledges, whereas murre chicks raised by razorbills exhibited minimal chick losses. This was attributed to behavioral differences among the chicks that are associated with the two species' relative probabilities that chicks will get dirty, be taken by predators, fall off the cliff, and have traumatic interactions with strange adults. The chicks remain in the nest for a rather variable period; Plumb (1965) reported three estimates averaging from 15.7 to 18.5 days, with extreme limits of 12 and 24 days. Brun (1958) noted an average chick weight of 165.4 grams immediately before fledging and an average wing length of 70 millimeters at 16.5 days. Most chicks still had traces of down around the neck and under the wings when they left the nest. Hudson (1982) reported a somewhat higher mean fledging weight, with no consistent differences in weight or fledging time for chicks produced in ledge versus burrow sites. However, burrow sites did produce a higher overall breeding success (0.7

young per pair vs. 0.55 for ledges), perhaps in part as a result of the higher egg predation rate on ledges than in burrows. Lloyd (1979) observed a very high (93 percent) overall fledging success for three years, with fledging success of chicks from replacement eggs averaging even slightly higher (97 vs. 94 percent) than for those from initial eggs. Most chick losses occurred during the first week of nestling life, and the incidence of chick mortality was inversely related to chick weight at hatching, suggesting that egg size is critically related to chick survival. Bianki (1977) also reported a very high (96 percent) fledging rate among 143 hatched eggs at Kandalaksha Bay, USSR, and described the actual fledging process. He noted that one adult typically stood by the nest, calling and luring the chick to leave it and enter the water. As soon as the chick did so, the adults led it away from shore. Once the young are on the water, the birds immediately begin to migrate away from their nesting sites.

Perry (1975) has graphically described the nighttime "epic journey" of the chicks as they leave their nests for the first time, fluttering, hopping, and tumbling down cliffsides to the sea, stimulated by the calls of their waiting parents. Ingold (1973) determined that adults can find their own chicks on the water among a group of chicks, apparently by individual recognition of calls.

BREEDING SUCCESS AND RECRUITMENT RATES. An overall breeding success of 81 percent (138 young fledged from 170 eggs) was reported by Bianki (1977), while Lloyd (1979) observed an overall "nesting" (breeding) success of 0.71 fledged young per pair (including renesting efforts) for 735 initial and 54 replacement eggs studied over a three-year period. Bedard (1969c) estimated a 66 percent overall breeding success rate (64 fledged young from 96 eggs) during two years of study. There is probably a fairly high mortality rate of young birds between their precocial fledging and their attainment of adult mortality rates; Lloyd (1974) noted that of 353 recoveries of razorbills banded as chicks 55 percent occurred during the first year after banding, whereas only 18 percent of the birds banded as adults were recovered in their first year after banding, suggesting that juveniles are roughly three times as vulnerable as adults. Bianki (1977) judged that 89.3 to 95.5 percent of the birds banded as chicks die before they begin "mass nesting." Lloyd (1974) estimated that there may be an approximate 25 percent annual mortality rate of young birds to the end of their fourth year of life and an annual mortality rate of 11 percent thereafter. A slightly lower (8 percent) annual adult mortality rate was estimated by Steventon (1979). Lloyd concluded that only 20 chicks per year need to survive to breeding age from every 100 pairs to maintain a stable population. Estimates of recruitment rates in this species, as in other alcids, are impossible to make on the basis of breeding grounds data, and there do not appear to be any estimates of the incidence of first-year birds in wintering flocks.

Evolutionary History and Relationships

There seems little doubt that this is a very close relative of the extinct great auk. Presumably both evolved in the North Atlantic, with the great auk perhaps adapting to colder waters and more arctic climates and probably concentrating on larger fish. Possibly continued generic distinction between them is warranted, based on the secondary flightlessness of the great auk, but even that point might be argued. (After this manuscript was submitted for publication, Strauch 1985 recommended the generic merger of *Alca* and *Pinguinus*.)

Population Status and Conservation

Lloyd (1976b) has surveyed the world distribution of the razorbill, concluding that approximately 208,000 breeding pairs might exist, with 9 percent in Canada and New England and 70 percent in Britain and Ireland. Nettleship (1977) judged that only the eastern Canadian colonies in Newfoundland and Labrador are stable or increasing; elsewhere colonies appear to be declining. A small but slowly increasing number of birds are now nesting in two areas of Maine (Mantinicus Rock and Old Man Island) (Korschgen 1979). In Greenland the numbers appear to be very small (under 500 breeding pairs), but the counts are probably very incomplete. The colonies in Britain and Ireland may be relatively stable, but those in the Gulf of Saint Lawrence have been declining since 1960, perhaps because of oil pollution and toxic chemicals in the birds' diet. Danger from these two sources is considerable to razorbills, which often winter in or migrate through major oil shipping lanes or areas of offshore drilling. Losses from fishing nets, hunting, human disturbance of nesting colonies, and other natural factors such as adverse weather have also been cited as problems for this species.

Great Auk

Pinguinus impennis (Linnaeus)

OTHER VERNACULAR NAMES: Garefowl; gerjrfugi (Danish); grand pingouin (French); Riesenalk (German); beskrypaya gargarka (Russian); isarukitsoq (Greenland).

Distribution

Extinct. Last certain records of living birds were two taken in early June 1844 on Eldey, Iceland. Previously bred on Funk Island (off Newfoundland), on Iceland, probably on Saint Kilda, and possibly on Shetland and the Faeroe Islands.

Description

BREEDING PLUMAGE (sexes alike). Chin, throat, foreneck, and sides of head and neck uniform velvety dark snuff brown or soft blackish brown, passing gradually into brownish black on pileum and hindneck; a large oval patch of white covering greater part of space between bill and eye; upperparts uniform black, the secondaries tipped with white; underparts, including chest, immaculate white, this ending anteriorly in an angle on median portion of upper chest or lower foreneck; bill black, its grooves whitish; iris dark brown; legs and feet black. Tail of 14 rectrices (Ridgway 1919).

WINTER PLUMAGE. Similar to the breeding plumage, but with chin and throat white. The white lore area is reduced, and there is a gray stripe from the eye to the ear region and an extension of the white cheek area forward above this stripe to the eye (Luther 1972).

JUVENILES AND FIRST-WINTER BIRDS. Not described, but probably similar to those of the razorbill.

DOWNY YOUNG. Covered with dark gray down (Luther 1972).

Measurements and Weights

MEASUREMENTS. Wing: approximately 146 mm. Exposed culmen: 80–89 mm. Total body length 736–62 mm (Ridgway 1919). Eggs: average of 40, 123.7 x 75.5 mm (Bent 1919).

WEIGHTS. No weights available. Adults estimated by Bedard (1969d) and Kartashev (1960) to weigh approximately 5,000 g. Bengtson (1984) estimated the egg weight as 325 g, or 6–7 percent of adult weight. Schönwetter (1967) estimated an average egg weight of 372 g, which he judged to represent 11.7 percent of adult weight. The range of fresh egg weights was estimated at 327–466 g.

Epilogue

The great auk has been extinct for almost 150 years, and all that remain as evidence of its existence are about 80 mounts or skins, a moderate number (79) of eggs, and a very few skeletons, scattered around the museums of the world like so many dusty artifacts of an ancient civilization. Indeed, all that biologists can now do to learn any more about this fascinating species is sift these relics like scavengers searching through a refuse pit in hopes of finding anything that might glitter and attract attention. The last record of any living great auks dates from June 1844, when a group of Icelanders captured and killed what was probably the last pair of nesting birds on Eldey, an islet off the southwest tip of Iceland that was the great auk's best-known breeding site. This was an ingnominious end for a species that represents the ultimate in alcid evolution, a bird that had so fully perfected its wings for underwater swimming that they had lost the capability of flight. The great auk thereby abandoned its fate to the avaricious and predatory nature of humans and effectively signed its own death warrant, as had the dodo and various other flightless and defenseless birds before it. As Bent (1919) has aptly stated, the birds initially were hunted for food, later were used for bait or killed for their fat or feathers, and finally and most ironically were exterminated because the rarity of their skins or eggs made them valuable to collectors.

The frustration and sadness associated with this loss are like seeing the scraps of notes for a projected novel by Hemingway or da Vinci's preliminary sketches of a planned monumental sculpture. Yet we have so little evidence about the biology and behavior of the great auk that it is perhaps better to write a simple eulogy than to try to reconstruct a posthumous biography. A fairly complete summary of information on the species was published by Grieve (1885) within a few decades after its extinction, and more recently Bengtson (1984) has assembled as much information as possible on the biology and behavior of the great auk.

Black Guillemot

Cepphus grylle (Linnaeus)

OTHER VERNACULAR NAMES: Sea pigeon; tystie (British); white-winged guillemot; guillemot à miroir (French); Gryllteiste (German); serfag (Greenland); teista (Icelandic); chistik (Russian); tobisgrila (Swedish).

Distribution of North American Subspecies (See Map 15)

Cepphus grylle atlantis Salomonsen
BREEDS from southeastern Quebec, Newfoundland, and southern Labrador south to Maine; and from the Shet-

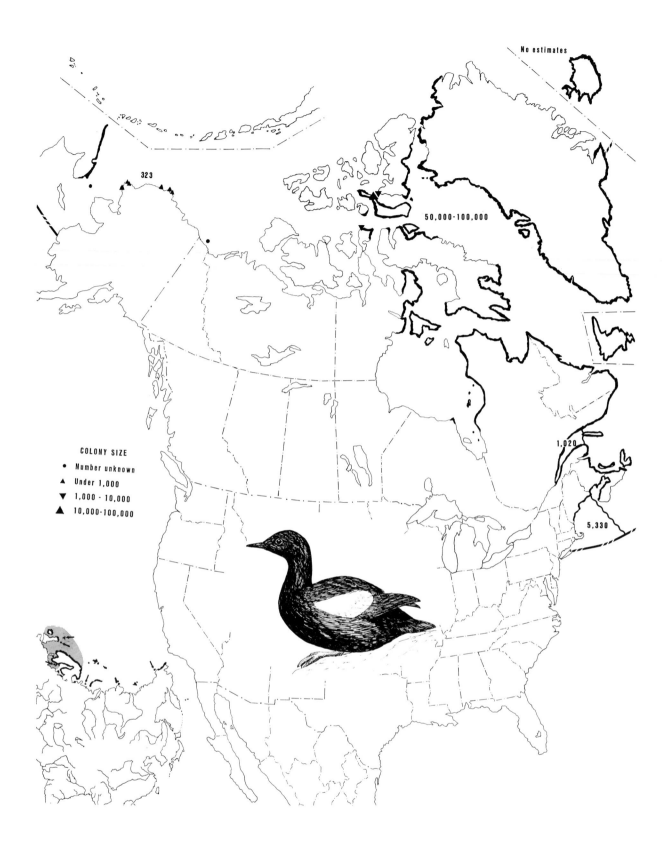

COLONY SIZE
• Number unknown
▲ Under 1,000
▼ 1,000 - 10,000
▲ 10,000-100,000

No estimates

323

50,000-100,000

1,020

5,330

15. Current North American distribution of the black guillemot; symbols as in map 11.

land Islands, northern Norway, northern Finland, and northwestern Russia south to Ireland, Isle of Man, northern England, and islands in the Kattegat.

WINTERS in open waters off the breeding places south to Massachusetts and Rhode Island, rarely to Long Island and New Jersey, northern France, Belgium, Netherlands, northwestern Germany, and southern Norway.

Cepphus grylle ultimus Salomonsen
BREEDS from Melville and Ellesmere islands south to Melville Peninsula, Southampton Island, the eastern shore of Hudson Bay and James Bay, and northern Labrador; western Greenland from Hall Land south to Disko Bay. Has also bred on west shore of James Bay (Cape Henrietta Maria).

WINTERS off the breeding grounds, wherever there is open water, north to northern Greenland, moving south in Hudson Bay and James Bay.

Cepphus grylle arcticus (Brehm)
BREEDS in Greenland, from Disko Bay in the west and Blosseville coast in the east, south to Cape Farewell, intergrading with *C. g. ultimus* near Disko Bay and with *C. g. mandtii* south of Scoresby Sound.

WINTERS in open waters off the breeding range.

Cepphus grylle mandtii (Mandt)
BREEDS from northeastern Greenland, Jan Mayen, Spitsbergen, Bear Island, Franz Josef Land, Novaya Zemlya, Vaigach Island, New Siberian Islands, Bennet Island, Wrangel Island, and Herald Island south to the arctic coast of Siberia. Also local in northern Alaska (Cape Thompson to Barter Island) and Herschel Island, Yukon.

WINTERS in open waters throughout the breeding range, and south in Bering Sea from Bering Strait to Saint Lawrence Island, northwestern Alaska, in the Kara Sea, Barents Sea, and along the arctic coast of Russia and the northern Alaskan coast.

Description (Modified from Ridgway 1919)

ADULTS IN BREEDING PLUMAGE (sexes alike). Plain fuscous black or very dark fuscous, faintly glossed with greenish, especially on back, scapulars, and rump; posterior lesser, middle, and distal half of greater wing coverts immaculate white, forming a large patch on the wing, sometimes superficially uninterrupted but usually broken by exposure of the black of basal portion of greater coverts, which also have the greater part of inner webs black; axillaries and under wing coverts, except along edge of wing, immaculate white; bill black; inside

of mouth vermilion; iris dark brown; legs and feet vermilion, the claws blackish.

WINTER PLUMAGE. Wings and tail only as in summer; rest of plumage pure white, the pileum, back, scapulars, and upper part of rump variegated with black, the whole of concealed and part of exposed portions of the feathers being of the latter color; legs and feet paler red. First-winter birds differ from adults in having transverse brown markings on the underparts and wings.

JUVENILES. Similar to the winter plumage but white wing patch broken by blackish tips to all the feathers, the secondaries and primary coverts with terminal spots of white, rump and underparts indistinctly barred with dusky, and pileum showing little of concealed dusky.

DOWNY YOUNG. Plain deep sooty brown, darker on head, paler and more grayish on abdomen. Bill black, legs and feet dark brown, mouth pink (Harrison 1978).

Measurements and Weights

MEASUREMENTS (of *atlantis*). Wing: males 154.0–174.5 mm (average of 2, 164.2); females 156–71 mm (average of 2, 163.5). Exposed culmen: males 31–34 mm (average of 2, 32.5); females 29.0–33.5 mm (average of 2, 31.2). Eggs: average of 54, 59.5 x 40 mm (Bent 1919).

WEIGHTS. Six breeding-season males from Cape Dorset ranged 333–410 g, averaging 386 g, and four females ranged 325–434 g, averaging 372 g (Macpherson and McLaren 1959). Substantial racial variation occurs. Estimated egg weight of nominate *grylle,* 50 g (Schönwetter 1967). Newly hatched young weigh 29–40 g, averaging 33.9 g (Dementiev and Gladkov 1968).

Identification

IN THE FIELD: This fairly large (teal-sized) alcid is easily recognized in breeding plumage by its entirely black color save for white upper and lower wing coverts and by its red feet. In the water the white upper wing coverts show up as oval patches. In winter the birds are generally white, becoming somewhat darker on the upperparts, but the wings remain unchanged in appearance. First-winter birds have their white wing patches mottled with black-tipped feathers. The species is fairly silent but sometimes utters faint, shrill piping whistles.

IN THE HAND. The brilliant white upper wing coverts of all birds older than yearlings (which have dusky-tipped white coverts) identify this as a guillemot, and the normal presence of only 12 rectrices and of white rather than gray under wing coverts separates it from the pigeon guillemot.

Ecology and Habitats

BREEDING AND NONBREEDING HABITATS. The breeding range of this species extends broadly along the Atlantic coast from arctic to temperate seas having August temperatures ranging from 0 to 16°C. Rocky seacoasts are the primary breeding habitat, although nesting also occurs in a few areas of sandy coastlines where sheltered nesting sites are provided by driftwood, abandoned structures and similar human artifacts, and the like. Steep seacoasts are avoided, and instead the birds tend to occupy areas near the high tide line that contain accumulated rocks and boulders at the foot of rock faces, rock piles along the shoreline, or in areas where caverns and crevices of rocky coastlines are abundant. During the nonbreeding period the birds are highly pelagic, generally remaining close to the limits of pack ice in winter and then often feeding on crustaceans and planktonic materials associated with the edges of pack ice (Voous 1960).

SOCIALITY AND DENSITIES. Black guillemots are relatively noncolonial, with their distributions determined primarily by the occurrence of suitable nesting sites. Asbirk (1979) found that 37 percent (154 pairs) of the population he studied bred as solitary pairs (nests at least 10 meters apart), while the remaining 63 percent nested in colonies of from 2 to 28 pairs. The densest colonies were found where preferred boulder nest sites were numerous and close together. Similarly, Preston (1968) reported that at Kent Island, New Brunswick, the distribution of breeding birds was primarily a function of substrate conditions, and the size and density of the breeding population were determined by the availability and density of nest sites. The colonial or "aggregate" group of birds he studied was associated with the tops of two neighboring ledges on one shoreline, while the "nonaggregate" group was widely scattered along storm beaches and in gaps between the island's cliffs. In the "aggregate" group nearly all the nests were no more than 60 feet from their nearest neighbor, while in the "nonaggregate" group only about 40 percent of the nests were this close together, and some were as far as 600 feet apart. The total estimated breeding population of Kent Island ranged from 35 to 43 pairs during the years of Preston's study. Cairns (1980, 1981) reported populations of 145 and about 200 breeding pairs at two study sites in southern Quebec, with densities up to about 55 nests per 50 meters of coastline, and Asbirk (1979) found about 410 pairs nesting on a 10 hectare island in

Denmark (41 pairs per hectare). Probably the largest and densest concentrations in North America occur in a few areas of arctic Canada along the north coast of Devon Island and its vicinity, where three colonies with from about 2,000 to 10,000 pairs are believed to be present (Brown et al. 1975).

PREDATORS AND COMPETITORS. Asbirk (1979) reported that the most important predators of eggs and chicks in his study area consisted of herring gulls (*Larus argentatus*) and probably also greater black-backed gulls (*L. marinus*) and hooded crows (*Corvus corone*). On Kent Island the primary predators of eggs and young are probably herring gulls, though American crows (*Corvus brachyrhynchos*) have also been implicated as egg predators (Winn 1950; Preston 1968). A number of other possible predators of adults or young have been suggested (Kartashev 1960; Bianki 1977), but there seems to be little direct evidence of their importance.

Competitors include other fish-eating species and species that compete for nesting sites. Preston (1968) found no evidence for significant competition for either foods or nesting sites on his Kent Island study area, and Asbirk (1979) suggested that competition for nest sites with common shelducks (*Tadorna tadorna*) may sometimes force guillemot pairs to move to other locations. Bedard (1969c) compared nest site choice of six species of seabirds in his study area and found a limited overlap in site utilization among the black guillemot, razorbill, and Atlantic puffin, with 2 percent of the puffin nests and 13 percent of the razorbill nests occurring in the nesting sites most favored by black guillemots (shallow vertical rock fissures), while 29 percent of the guillemot nests were in habitats more frequently used by puffins and razorbills. Besides being predators of eggs and chicks, the larger gulls are sometimes significant kleptoparasites of guillemots, stealing the food carried by adults as they try to approach the nest to feed their chicks (Winn 1950).

General Biology

FOOD AND FORAGING BEHAVIOR. A sample of 26 birds collected in coastal waters off Denmark during fall and winter included mostly fish remains. Fish was a more or less important part of the sample in 89 percent of the birds, and 63 percent had also eaten crustaceans. Three of the birds had fed exclusively on fish, 3 exclusively on crustaceans, and 4 had eaten polychaetes. Among the fish, gobies (*Gobius* spp.) were the most important single component, with 73 percent of the sample having eaten this species, while 3 or more birds contained remains of butterfish gunnels (*Pholis gunnellus*), sticklebacks (*Spinachia*), or viviparous blennies (*Zoarces*). More than half of the birds had eaten only one kind of fish, while 8 had eaten two kinds, and 2 had eaten three kinds. Crustaceans were found in 73 percent of the stomachs, with crabs (*Brachyura*), shrimps (Crangonidae), prawns (Palaemonidae), isopods, and lobsters (*Galathea*) all represented. In general, fish constituted about two-thirds of the food and crustaceans one-third, the fishes including both free-swimming and bottom-dwelling forms. Other studies from European waters indicate that *Pholis* is one of the principal foods taken there, and it is likewise one of the most important chick-rearing foods of the species in the Kent Island area (Winn 1950; Preston 1968). Preston observed that *Pholis* made up 68 percent of more than 500 identified prey items brought to chicks, while a sculpin (*Myoxocephalus*) and a shanny (*Ulvaria*) composed 18 and 9 percent respectively. Nearly all of the observed prey items were benthic forms occurring in the littoral zone.

Like the pigeon guillemot, this species forages in fairly shallow waters during the breeding season, mainly diving in depths of 1–8 meters and staying underwater for up to a minute. Probably the birds have a maximum diving depth of 40–50 meters, and the maximum observed diving time is 112 seconds (Piatt and Nettleship 1985). On Kent Island the birds Preston observed fed mainly around emergent ledges and over offshore shoals; the types of food taken varied somewhat by year and by time of day as the birds shifted their foraging areas in accordance with the tides. He did not find any evidence for sexual differences in foods taken or in the kinds of foods eaten by adults and those brought to the young. Slater and Slater (1972) reported that guillemots at Fair Isle, Scotland, fed their young entirely on fish ranging from 8 to 20 centimeters, primarily butterfish, sand launce (Ammodytidae), codlike fish (Gadidae), sea scorpions (*Taurulus*), flatfish (Pleuronectidae), and other undetermined types. Different pairs fed their young on differing proportions of these prey types, which typically were caught near shore, with each pair apparently foraging in similar places during each excursion. Bradstreet (1980) found that foraging during spring by pelagic birds in the Barrow Strait region typically occurred at the ice/water interface, with some feeding done under the ice edges, mostly on arctic cod (*Boreogadus*) and amphipods. At coastal ice edges arctic cod made up almost 100 percent of the food samples, while at offshore ice edges a substantial dietary component was amphipods (*Onisimus* and *Apherusa*).

MOVEMENTS AND MIGRATIONS. Because of the species' tendency to winter in northerly areas, migrations are probably poorly developed or essentially absent for many breeding populations. Salomonsen (1967) reported that Greenland birds begin to disperse in September and October, not seldom northward, with a definite southward movement beginning in October and continuing slowly through November. Many of these birds winter in leads in the ice associated with tides or riptides, or in leads in fast ice, occasionally as far north as northern Thule district and even in Hall Land. Banding results suggest that a uniform displacement of the various northern breeding populations occurs in winter, although the birds in the more southerly open water areas are essentially stationary. Birds of the year especially tend to move north along the coast after fledging in August, in part moving passively with the current. Likewise birds of the year tend to begin their spring migration later than older birds, with most of them moving back to their (presumably) natal breeding areas or spending the summer immediately to the south of them. During spring in the Barrow Strait area guillemots tend to occupy interface areas of land or ice and marine waters, especially ice edges, feeding on such cryophilic foods as fish and crustaceans that are associated with these habitats, particularly the undersurface of the ice (Bradstreet 1979).

Social Behavior

MATING SYSTEM AND TERRITORIALITY. The black guillemot is a monogamous bird that typically retains the same mate from year to year and utilizes the same nest site in successive years. Of 42 pairs Asbirk (1979) marked in one year, all but 14 remained intact the following year. In 6 of the remaining pairs one of the members had probably died, while in 3 pairs both members had found new mates. In the 4 remaining cases of mate change it was not possible to determine if the original mates had died. Five pairs remained intact for at least 3 years, and one pair persisted at least 4 years. The estimated overall "divorce" rate was 7 percent annually. Of 10 pairs that retained their mates for the following season, the breeding success in the second year was 45 percent, while 8 pairs that changed their mates had a breeding success of 36 percent the second year. Similarly, Petersen (1981) reported that of 16 pairs, 4 remained intact for at least 4 years, and that there was a "divorce" rate of only 5 percent annually. Territorial defense in this species appears to be limited to the nest site itself, a perching site (often a stone) in the "assembly area" of the shoreline, and another perching site above the nest, the last being most strongly defended. Occasionally only a single perching site is present (Asbirk 1979). Preston (1968) similarly found that terrestrial activity was concentrated at the nest site, a nearby perch rock, and a communal roosting area. The perch and nest site were both defended, but there was little indication of mate defense.

VOICE AND DISPLAY. Vocalizations of breeding birds include the "scream," a high-pitched and prolonged piping call uttered with a wide-open bill during alarm situations (fig. 42F). A more shrieking sound is uttered during direct lunging attack on another bird and may continue in the air during overt chase as a "duet flight." At lower threat intensities a twittering call is uttered in conjunction with the "twitter waggle" display (fig. 42H), and a whistled note is used as an appeasement gesture during "hunch whistling" (fig. 42C). When sitting on the tops of their nest sites the birds may utter a quick staccato piping call or "nest song" at the approach of other guillemots, and a similar rapid staccato piping is uttered during billing behavior between members of a pair (fig. 42I) (Asbirk 1979). A few other calls are associated with particular postures or situations, as noted below.

Agonistic posturing includes an alert, neck stretching posture associated with the alarm scream (fig. 42F) and several aggressive postures. High-intensity aggression is marked by open-bill lunges, while at lower levels of threat the birds walk toward an intruder in a hunched posture (fig. 42G), with the wings partially opened and the wingtips sometimes dragging. At still lower intensities the bird stands in an oblique posture with bill pointed downward and utters a twittering whistle while waggling the head from side to side in a "twitter waggle" posture (fig. 42H). At times the wings may be raised stiffly over the back while the body is lowered, exposing their white undersides (fig. 42H) and apparently enhancing the appeasement aspect of the display. Birds on established perch sites respond to intruders by "hunch whistles" (fig. 42C), during which the head is tossed upward and backward several times while the bird maintains a crouched posture and alternately opens and closes its bill as it utters rising and falling piping notes. Hunch whistling may also occur between members of a pair, probably as an appeasement signal. During water chases two birds sometimes perform "leapfrogging," when the trailing individual takes off, flies over the lead bird with whirring wings (fig. 42D), and alights ahead of the former lead bird with upraised wings and opened bill (fig. 42E). Another display on water is swimming in line, performed by as few as 2 or as many as 20 birds (fig. 42A), during which the neck is stretched and the bill is opened as a loud peeping call is uttered. This is a variably formalized chase during

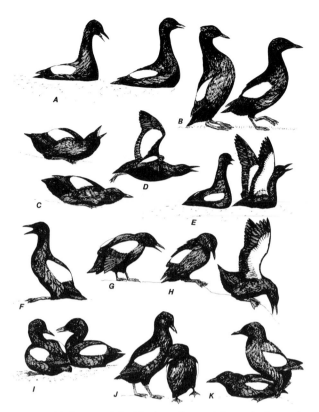

42. Social behavior of the black guillemot (after Asbirk 1979): A, in-line swimming; B, in-line walking; C, hunch whistling; D, leapfrog flight; E, landing after leapfrog flight; F, alarm scream; G, H, aggressive walking, followed by twitter waggling and wing raising; I, billing; J, precopulatory circling; K, copulation.

which actual surface, aerial, or underwater chases may also develop. On land a similar ritualized chasing may also occur (fig. 42B), with the displaying bird assuming a very upright posture and strutting about on its toes. Displays that are limited to mates include billing and copulatory behavior. Billing is often done by birds standing or sitting near each other, with their bills alternately bobbing from side to side as a piping call is uttered. When billing on water, the two birds move in circles around themselves and around one another. A similar circling behavior also occurs on land before copulation. Here the male assumes an erect posture, utters a series of staccato calls, and struts around the female, who walks around in a more hunched posture (fig. 42J). When ready for copulation the female squats and the male mounts, trampling with his feet. The female typically remains motionless during treading (fig. 42K), then she rises up and throws the male off her back.

Copulation evidently never occurs in the water (Asbirk 1979; Preston 1968; Winn 1950).

Reproductive Biology

BREEDING SEASON AND NESTING SUBSTRATE. Egg records from Maine extend from June 12 to July 16, those from the Bay of Fundy and Nova Scotia from June 11 to July 6, and those from the Gulf of Saint Lawrence from June 8 to July 15. A few records from Hudson Bay and Cumberland Gulf are from June 10 to July 24 (Bent 1919). Preston (1968) found that on Kent Island the first eggs were laid in late May or early June, with observed yearly variations in the first egg ranging from May 28 to June 8. Laying continues about a month, with nearly all clutches being started before July and the first chicks hatching in late June or early July. Cairns (1981) found nearly a month's difference in breeding phenology in two study areas of southern Quebec, corresponding roughly to differences in springtime air temperatures at the two locations. Mean hatching dates on the two areas were June 26 (Saint Mary's Islands) and July 15 (Brandypot Island), with approximately a month-long spread in each area. Finally, in northern Alaska there are a few egg records from July 6 to August 1, and nestlings have been seen as late as September 10 (Divoky, Watson, and Bartonek 1974).

The nesting substrate of black guillemots is highly variable, depending on local conditions. Asbirk (1979) found that most (57 percent) of 411 nests in his study area were under and between stones, while 43 percent were under fish boxes or driftwood. In Greenland the birds variously nest among boulders, in talus, under rocks, or in narrow cliff crevices (Salomonsen 1967). On sandy beaches in northern Alaska the birds have been found using crevices in driftwood piles, in depressions of sand dunes, in natural sand burrows, and among or under various kinds of man-made debris, such as pieces of plywood, boxes, or collapsed huts (Divoky, Watson, and Bartonek 1974).

NEST BUILDING EGG LAYING. Evidently guillemots modify their nesting sites little if at all. They have apparently never been seen carrying materials into their nests, and at most they scratch out a weak deepening of the nest cavity to receive the eggs (Asbirk 1979). The egg-laying period is probably dependent upon the age of the birds, with the older birds laying eggs statistically earlier than the younger age-classes and the younger birds also producing statistically smaller clutches. Asbirk found that the average clutch size of 386 Danish nests was 1.85 eggs, with single-egg clutches laid later than average and apparently by young and inex-

perienced birds or possibly by individuals that had changed nest sites. Among known-age birds there was a progressive increase in average clutch size with age, so that birds breeding for at least their third season had an average clutch of 2.0 eggs. Preston (1968) also observed a relation between clutch size and the breeding experience of the female, with first-year breeders averaging 1.44 eggs and older age-classes averaging 1.77 eggs. He also noted that clutch sizes tended to decrease over the laying season (from 2.0 eggs during the first week to 1.0 during the fifth week) and that clutch sizes were larger (1.82 vs. 1.34 eggs) in the nonaggregated component of the population. The average interval between successive eggs of a clutch is about 3 days, with extremes of 2 to 6 days. Replacement clutches are sometimes produced by birds that lose their clutches early in the season, often after intervals of 12–23 days following egg loss (Asbirk 1979). In Iceland egg replacement occurs only if incubation of the initial clutch is not well advanced (Petersen 1981).

INCUBATION AND BROODING. Incubation begins only after the second egg is laid, and often not until 4 or 5 days after laying in single-egg clutches (Preston 1968). Both sexes participate more or less equally, with fairly frequent shifts of incubation typical. Asbirk (1979) observed that minimum incubation shifts ranged from less than an hour to more than 4 hours and judged that these short periods were related to the species' relatively abundant source of nearby food, which requires little searching effort or flying time. The incubation period of the first-laid egg was found by Asbirk to average 31.9 days (38 nests), while the second egg required an average of 28.5 days to hatch (41 nests). The two eggs thus usually hatch within a day of one another. If one of the two eggs should fail to hatch it is removed from the nest cavity, as are the shells of hatched eggs (Preston 1968).

GROWTH AND SURVIVAL OF YOUNG. The newly hatched chicks are brooded for 3 or 4 days, after which they have attained sufficient thermoregulatory ability so the parents can leave for prolonged periods. Feeding typically begins on the day after hatching, and the prey fish are carried to the chicks one at a time by both adults. Feeding trips are nearly continuous through the day, starting at about sunrise and lasting through late afternoon, with up to 4 trips per hour during peak feeding periods but averaging 1–2 trips per hour, with morning and afternoon peaks. Feeding rates are higher in two-chick nests than those with a single chick but not quite twice as high. By experimentally adding a third nestling to two-chick broods, Asbirk (1979) found that in 5 of 9 cases the parents were able to raise all three chicks to fledging, whereas in similar experiments with razorbills, Atlantic puffins, and rhinoceros auklets the adults have generally been unable to raise extra chicks. Asbirk found that no correlation existed between the number of feedings per hour and the age of the chicks, though the length of the fish brought to the nest increased with chick age, and the number of feedings per hour also increased with the number of chicks in the nest. The fledging period was reported as 35–40 days by Preston, 31–51 days (averaging 39.5) by Asbirk, and 39–40 days by Winn (1950). Before fledging there is typically a slight weight loss, in Asbirk's study averaging 6 percent of the chick's peak weight, which in some years averaged slightly higher than average adult weight. At fledging time the adults may lure the chicks to water with a fish, after which they are apparently taken to feeding areas offshore (Winn 1950). Postfledging contacts between adults and young are still uncertain, but the young are believed to be essentially independent of parental care after fledging.

BREEDING SUCCESS AND RECRUITMENT RATES. Preston (1968) found that egg losses constituted 88 percent of total prefledging mortality on Kent Island, mostly as a result of destruction by gulls, chilling, and flooding. Losses were much higher (78 vs. 42 percent) in one-egg clutches than in two-egg clutches, perhaps because the former were probably mostly produced by inexperienced birds breeding in suboptimal nesting sites. However, fledging success for one-chick and two-chick broods was nearly identical (91 and 88 percent respectively), with most chick losses occurring during the first week after hatching, perhaps from chilling. Overall, an average of 0.73 young per nesting pair was fledged, but the productivity in the "aggregate" area was substantially lower than in the "nonaggregate" area (0.58 vs. 0.93 young per pair), which Preston attributed to possible unknown crowding effects on clutch size and hatching success. Cairns (1981) found a similarly reduced hatching success for one-egg versus two-egg clutches (32 vs. 58 percent) and a lower fledging success (38 percent) for chicks hatched from one-egg clutches than two-egg clutches (63–67 percent). The overall breeding success was 12 percent for one-egg clutches and 39 percent for two-egg clutches, with most chick losses occurring during the first few days after hatching. Contrary to Preston's findings, breeding success was evidently not related to habitat structure or nesting density (Cairns 1978, 1980). Asbirk (1979) reported an overall hatching success during three years of 59 percent and a rearing success of 54.8 percent, resulting in a breeding success of 32.3 percent and 0.59 young fledged per nest. Pe-

tersen (1981) determined that 935 Icelandic eggs had a 79.5 percent hatching success, and 89.2 percent of the young were raised to fledging, representing an overall breeding success of 70.9 percent. Asbirk reported significant differences in hatching and rearing success associated with nesting habitat; nearly twice as many eggs and chicks disappeared from nests under stones as from driftwood nests. However, overall hatching success was identical in the two habitats, whereas fledging success and overall breeding success were higher under stones than under driftwood. Hatching success was similarly identical in solitary versus colonial nesters, whereas fledging success and overall breeding success were higher in colonial nesters. Additionally, all earlier nests had a higher overall breeding success than late nests, and older known-age nesting birds had a higher breeding success than younger birds. Asbirk (1979) determined that nesting is sometimes attempted by birds at 2 years, although some 2-year-olds in the colony probably did not breed. He estimated an annual adult survival rate of 82–89 percent, compared with estimates of 77–85 percent by Preston (1968). Few 1-year-olds were seen at the colony studied by Asbirk, and an uncertain number of 2-year-olds bred, so that the actual recruitment rate is impossible to judge at present. Salomonsen (1967) noted that 71 percent of 730 band recoveries of this species were of birds in their first year, and an additional 10 percent were of year-old birds. Thus if 75–80 percent of the postfledging population consists of prereproductive birds, the annual recruitment rate of colonies producing 0.6 young per nesting pair would be about 12–15 percent, or close to available estimates of adult survival rates. The oldest age-group in the band recoveries summarized by Salomonsen was 13-year-olds.

Evolutionary History and Relationships

The close relationship and probable evolutionary history of the black, spectacled, and pigeon guillemots have already been discussed by Storer (1952), who considered them three allopatric species having no known hybrids. He believed that the separation of the black and pigeon guillemots probably occurred when the Bering Strait was closed during the late Pliocene or early Pleistocene, with *carbo* presumably separating from *columba*like stock at the same time in the nearly isolated Sea of Okhotsk. The central position and relatively generalized morphology of the genus *Cepphus* in the family Alcidae have also been noted earlier. Hudson et al. (1969) considered the nearest relative of *Cepphus* to be *Uria*, and Storer (1952) has also discussed the evolutionary history of these two genera. An adaptive radiation of alcid types from a generalized *Cepphus*like ancestor, extending in one direction toward more efficient underwater divers of the murre and razorbill groups and in the other direction toward semiterrestrial birds such as the puffins that are well adapted for burrowing and walking, can be readily visualized.

Population Status and Conservation

Nettleship (1977) did not attempt to estimate the numbers of black guillemots breeding in eastern Canada but judged that in most areas their population trends were uncertain, while in the Scotian Shelf, Gulf of Maine, and Bay of Fundy areas the birds were probably increasing. As noted earlier, numbers are probably highest in arctic Canada, where colonies numbering in the thousands have been reported (Brown et al. 1975). The total population of breeders and nonbreeders in central and eastern Canada may be from 50,000 to 100,000 birds (Renaud and Bradstreet 1980). About 3,400 pairs were nesting along the coast of Maine in the early 1970s, and the birds have been slowly increasing there since 1900 and have expanded their range to the Maine–New Hampshire border (Korschgen 1979). The birds are probably increasing along the north coast of Alaska, where numbers are still only in the hundreds but where they appear to be invading previously unoccupied habitats (Sowls, Hatch, and Lensink 1978). An oil vulnerability index of 70 has been assigned to the species (King and Sanger 1979).

Pigeon Guillemot

Cepphus columba Pallas

OTHER VERNACULAR NAMES: Sea pigeon; guillemot du pacifique (French); Taubenteiste (German); chistik (Russian).

Distribution of North American Subspecies (See Map 16)

Cepphus columba columba Pallas
BREEDS from the Chukotski Peninsula and Diomede Islands to southern Kamchatka, and from Saint Lawrence and Saint Matthew islands and the Aleutians west to Attu, Bogoslof, and Shumagin islands, Kodiak, and southeastern Alaska south to Santa Barbara Island and San Luis Obispo County, California.

WINTERS from the Pribilof and the Aleutian islands to Kamchatka and the Kurile Islands (casually to Sakhalin and Hokkaido) and to southern California. (Birds breeding on the Kuriles represent a different subspecies.)

Description (Modified from Ridgway 1919)

ADULTS IN BREEDING PLUMAGE (sexes alike). General color plain fuscous black, the tips of scapulars and interscapulars more slaty, the rump and upper tail coverts uniformly more slaty (between dark mouse gray and iron gray), the underparts and anterior portion of head more sooty (between clove brown and bone brown); posterior lesser wing coverts, middle coverts, and tips of greater coverts white, the white tips to the last becoming gradually broader on proximal coverts until from about the middle of the series the black is quite concealed and the white blends with that of the middle and lesser coverts, which also are dusky beneath the surface; axillaries and under wing coverts brownish gray, paler toward edge of the wing, the innermost postcarpal coverts mostly dull white, the third series from secondaries broadly tipped with dull white; bill black; interior of mouth vermilion; iris dark brown; legs and feet vermilion, claws black.

WINTER PLUMAGE. Wings and tail as in summer; rest of plumage pure white, the pileum, back, scapulars, and upper part of rump with feathers blackish on concealed and part of exposed portions; legs and feet paler red. Second-winter birds resemble adults, except for dark crossbars generally marking the abdomen and breast, and some mottling on the wing patch (Kozlova 1961).

JUVENILES. Similar to the winter plumage, but white of wing coverts tipped with dusky, secondaries and primaries with terminal spots of white, rump and underparts indistinctly barred with dusky, and pileum with less black. During the first fall and winter the underparts become whiter, with fewer brownish crossbars present (Kozlova 1961).

DOWNY YOUNG. Plain dark sooty grayish brown, darker on anterior portion of head, paler and more grayish on abdomen.

Measurements and Weights

MEASUREMENTS. Wing: males 161–80 mm (average of 11, 173.6); females 166–81 mm (average of 14, 177.5). Exposed culmen: males 30.0–33.5 mm (average of 11, 31.7); females 29.5–34.5 mm (average of 14, 32.8) (Ridgway 1919). Eggs: average of 51, 60.5 x 41 mm (Bent 1919).

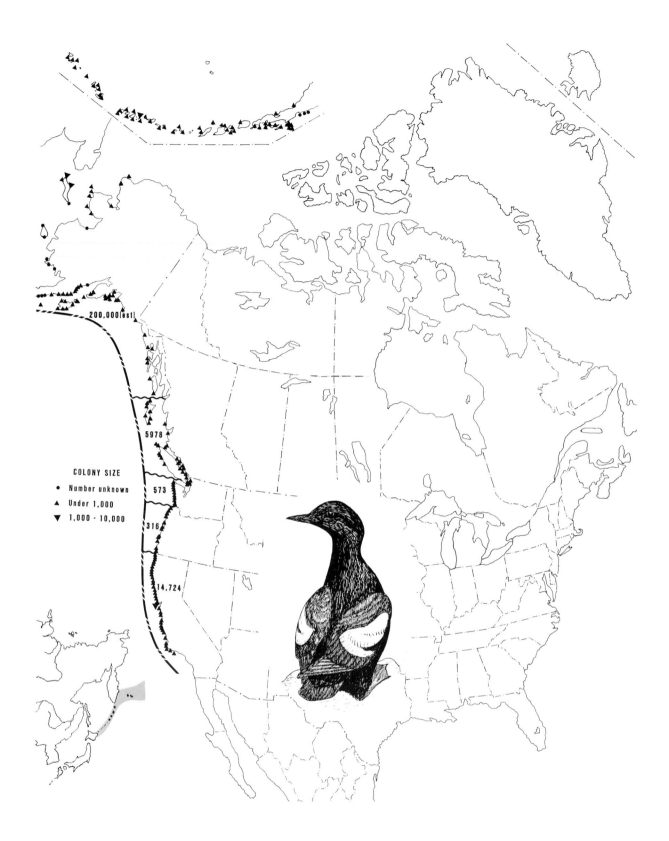

COLONY SIZE

- Number unknown
▲ Under 1,000
▼ 1,000 - 10,000

200,000(est)

5978

573

316

14,724

16. Current North American distribution of the pigeon guillemot; symbols as in map 11.

WEIGHTS. A sample of 53 adults averaged 450 g (Drent 1965). The estimated egg weight is 55 g (Schönwetter 1967). Average hatching weight, 43.7 g (Drent 1965).

Identification

IN THE FIELD. This Pacific-coast counterpart of the black guillemot closely resembles the latter, but in breeding plumage its white upper wing coverts are partially crossed by black bars, so that no clear oval patch is produced in standing or swimming birds. Both species have bright red legs and feet. In winter plumage the pigeon guillemot is perhaps somewhat browner throughout, especially first-winter birds, which are blackish above and mottled with dusky on the throat and breast. Older birds in winter are almost wholly white below, and their scapulars are broadly edged with white. Both species utter faint, shrill whistling notes.

IN THE HAND. This is the only Pacific-coast alcid with white upper wing coverts and brownish gray under wing coverts; the latter feature also separates it from the black guillemot. Furthermore, the black guillemot has 12 rectrices while this species normally has 14.

Ecology and Habitats

BREEDING AND NONBREEDING HABITATS. Pigeon guillemots have much the same general breeding habitat requirements as described for the black guillemot. The southern end of its North American breeding range corresponds to the 16°C isotherm of surface waters in August, and its northern limits to the 5°C isotherm, or slightly more temperate than that of the black guillemot. Not as much information on nest site requirements is available as for the black guillemot, but what is present suggests a similarly plastic adaptation to a variety of possible sites, the usual common criterion being a concealed and sheltered location for the eggs (Thoreson and Booth 1958; Lehnhausen 1980). Like black guillemots, the birds seem disinclined to excavate their own burrows but may enlarge burrows in clay banks previously dug by swallows. At least in rocky habitats the nests are usually very close to water, often near the high-tide line, and since the birds can take off from nearly level surfaces they need not place their nests on strongly sloping substrates as do, for example, puffins (Lehnhausen 1980). During the nonbreeding season the birds are nonpelagic and probably fairly sedentary, although no banding data are available to confirm that supposition. The birds rarely move into water more than 50 meters deep (Gould, Forsell, and Lensink 1982) and instead tend to spread out thinly along coastlines in winter (Forsell and Gould 1981).

SOCIALITY AND DENSITIES. These birds nest singly or in small colonies; as in the black guillemot, nesting distribution is probably dictated by the location of available sites rather than by any colonial tendency of the species. Drent (1965) worked on a colony of 100–110 pairs on Mandarte Island, British Columbia, and Thoreson and Booth (1958) stated that most colonies on Deception Island, Washington, numbered 5–18 pairs. Density estimates seem unavailable, but on Triangle Island, British Columbia, an estimated 100 pairs breed (Vermeer, Summers, and Bingham 1976). This island is only about 1 square kilometer in area; thus a breeding density of 200 birds per square kilometer exists there. Searing (1977) estimated a breeding density of 28 birds per linear kilometer of beach at Kongkok Bay, Saint Lawrence Island, or about one bird per 1,000 square meters of cliffs.

PREDATORS AND COMPETITORS. Probably the usual crow and gull predators affect the egg survival of this species; of these the northwestern crow (*Corvus caurinus*) has been mentioned specifically as a serious egg predator (Bent 1919). In British Columbia the species' breeding distribution is apparently affected by land predators such as the mink (*Mustela vison*) (Drent and Guiguet 1961).

Possible competitive interactions of the pigeon guillemot with other seabirds have been investigated by Scott (1973), who compared breeding biologies and foraging strategies of pigeon guillemots with those of common murres and two species of cormorants. He observed overlapping breeding phenologies but major differences in nesting habitats among the four species. He also found substantial differences in relative offshore foraging distributions, with the cormorants foraging closest to shore, the pigeon guillemot feeding somewhat farther out on bottom-dwelling species, and the common murre primarily concentrated on midwater fish species well away from shore. Lehnhausen (1980) made a somewhat similar comparison of the pigeon guillemot and three other species of alcids that breed at Fish Island, in the Wooded Islands group of the Gulf of Alaska. He found no evidence of asynchronous breeding cycles among these species and also found a high degree of daily foraging synchrony. Thus he determined that all four species had peaks in foraging activity during morning hours and that all but one (the parakeet auklet) had a second evening period of foraging. Pigeon guillemots and parakeet auklets interacted agonistically more often than did any other combination of species, and the guillemots invariably displaced the auklets. Interactions between tufted puffins and pigeon guillemots were also frequent, and in this case the puffins always displaced the guillemots. Lehnhausen suggested that these interactions were related to similarities in nest site requirements for these species and may reflect competitive interactions associated with this limited environmental resource.

General Biology

FOOD AND FORAGING BEHAVIOR. Most of the available information on foods of the pigeon guillemot derives from identification of prey species brought to nestlings. However, since it has been determined in the black guillemot that these are not significantly different from the foods eaten by the adults themselves, they may probably be taken as representative of the species. Drent (1965) observed that of 662 items brought to chicks, 508 were fish, 6 were shrimps, and the remainder could not be positively identified. The fish most frequently tallied were blennies (Xiphisteridae, Pholidae, Stichaeidae, Lumpenidae), while sculpins (Cottidae) were second most numerous. Small numbers of fish representing other families were also observed. A sample of 78 fishes collected from guillemot nests in California included a total of 24 species, of which sculpins composed over 50 percent. This concentration on cottids indicates a strong tendency for the birds to forage in sandy portions of the benthic zone, sometimes to considerable depths, since some of the species represented are known to occur in depths of 40–150 meters. However, most of them are to be found on or over rocky bottoms within the subtidal zone. Eldridge and Kuletz (1979) examined 1,229 food samples brought to chicks in Prince William Sound, Alaska, and established that their relative frequencies were sand launce (Ammodytidae) 53.1 percent, blennies 19.2 percent, sculpins 14.2 percent, codfish (*Gadus*) 2.0 percent, flatfish (Pleuronectidae, Bothidae) 1.5 percent, invertebrates 9.5 percent, and the remainder unidentified. Thoreson and Booth (1958) noted that the major types of foods brought to chicks in their study area in Washington were sand launce, smelt (*Hypomesus*), blennies (*Epigeichthys* and *Xiphister*), snake eels (*Lumpenus*), sole (*Lepidosetta*), and lamprey eels (*Entosphenus*).

According to Storer (1952), pigeon guillemots are less prone to forage close to shore than are black guillemots, and he related their larger body size to the fact that they largely inhabit windward coasts, where they may be forced to forage more pelagically. Drent (1965) observed that foraging by adults near the nesting colony was distinctly rare; they primarily foraged in shoal waters about 4–5 kilometers away. Lehnhausen (1980) saw some birds foraging near shore or tide rocks but judged that they might be inexperienced breeders, nonbreeders,

or off-duty incubating birds. My own observations of diving alcids at Sea World, in San Diego, indicated that, unlike murres and puffins (which engaged in nearly continuous swimming while foraging), pigeon guillemots went directly to the bottom and "hovered" there by paddling their feet as they probed nooks and crannies for food.

MOVEMENTS AND MIGRATIONS. There is little specific information on this, but the general evidence is that pigeon guillemots are highly sedentary. At least as far north as British Columbia and Washington, the birds are probably almost wholly resident, moving in the winter from exposed coastlines to sheltered bays and inlets and rarely occurring more than a mile offshore (Martin and Myres 1969). They are also resident in Alaska, at least throughout the Gulf of Alaska and the Aleutians, where they are abundant in neritic habitats and are generally dispersed as single birds or at most small groups (Gould, Forsell, and Lensink 1982; F. Zeillemaker, pers. comm.).

Social Behavior

MATING SYSTEM AND TERRITORIALITY. Mate retention in this monogamous species was proved by Drent (1965), who observed that 4 marked pairs remained together as long as four successive seasons, and determined only one definite case of "divorce." Similarly, he established that there were 6 cases of birds' using the same nest site for a minimum of four seasons, 7 cases of use for no fewer than three seasons, and 6 cases of at least two seasons. There were 6 cases of nest site changes, in 5 of which the original nest site was unusable. Nesting territories are established before egg laying, and site defense extends to nearby perch sites on beach boulders. Perch sites are used to promote contact between pairs, as places to spend leisure hours, and as copulation sites. Drent observed that these defended beach sites were about 1 meter in diameter, and that pairs that had changed nest sites sometimes retained their perch sites of the previous year. The perch site was often very close to the pair's nest site, but in one case it was about 30 meters away.

VOICE AND DISPLAY. Vocalizations have been described by Drent (1965) and to a more limited degree by Thoreson and Booth (1958). A flight intention call, like a continuously repeated *tsip*, sometimes is uttered, as when flying to the feeding grounds from the colony or when about to bring fish to the young after a delay. Like the black guillemot, alarmed birds stretch their necks, gape, and utter a very long, drawn-out *wheeeoo* note. It may be uttered at the sight of approaching gulls, crows, humans, or other alarming stimuli, and sometimes also during flight, when it has a peculiar vibratory quality. Drent noted two variations of the alarm scream. One included an introductory staccato ticking and was given in response to the sight of tufted puffins; the other was a hissing and reedy variant uttered by a pair surprised on their nest. Other vocalizations occur in association with specific display postures and will be mentioned in that context.

Display postures, as described by Drent, appear to be virtually identical to those of the black guillemot. In agonistic encounters, one bird may make a sudden rush at its opponent, with its bill oriented toward the object of the attack and the wings often slightly spread. There is no call associated with this lunging posture. If the lunge is made on the water the antagonist might dive, and it might also take flight in an attempt to escape. If so it may be followed by the pursuer in the air in a "duet flight," which is probably purely agonistic in nature rather than having anything to do with courtship. Likewise, bill dipping is a signal of mild alarm that is performed at all times of the year under varying stimulation and probably has nothing to do with courtship. As in the black guillemot, the hunch whistle display is performed both in the water and on land, and a very similar or identical posture is assumed (see fig. 42C). The hunched posture is accompanied by a loud piping and repeated *weep* note, and at high intensity when on land the wings are raised and the neck is stretched. This posture discourages the intruding bird or may grade into overt aggression or appeasement behavior. The "twitter whistle" is a common appeasement signal; as in the black guillemot, it is performed either on land or in the water, and by both sexes. The tail is cocked, the wings are slightly spread, and the outstretched head and neck are waggled sideways (see fig. 42H). Twitter whistling is mostly limited to encounters with territorial intruders but may rarely be performed in the context of a triumph ceremony toward the mate after a fight. Paired birds perform mutual twitter billing either on water or on land. As in the black guillemot (fig. 42I), their bills rarely actually touch; instead they simply pass and repass one another as the pair utters soft twittering notes, punctuated at intervals with a more melodious "trilled song" that is uttered with wider bill gaping. Mutual billing and twittering presumably are important pair-forming and pair-maintaining mechanisms and are used as a greeting ceremony by pairs throughout the breeding season. Copulation is performed on land, generally on the pair's perch site. It is generally preceded by mutual billing and twittering, followed by circling behavior apparently identical to the corresponding display of black guillemots (see fig. 42J). During treading the female is usually passive, but rarely she may utter the alarm scream or throw her head back and gape at the male in a murrelike manner. Treading normally lasts 30–75 seconds, and copulatory behavior is most intense during the 12 days immediately before egg laying. No copulation was observed in pairs after the laying of the second egg, and none was observed earlier than 28 days before the laying of the first egg.

Reproductive Biology

BREEDING SEASON AND NESTING SUBSTRATE. Egg records from California are from early May to mid-July, with a peak in mid-June. Washington and British Columbia records are from May 9 to July 13, also peaking in mid-June. A small number of records from southern Alaska are from June 15 to July 5 (Bent 1919). In Drent's (1965) studies on Mandarte Island, he observed an extreme range of 42 days in clutch commencement over four years and an average initiation date of June 11. He saw three types of nest substrates used there, including boulder cavities or cracks in boulders, chambers in the soil capped by boulders, and abandoned rabbit (*Oryctolagus cuniculus*) burrows. Nearly all cavities of suitable size were used. Lehnhausen (1980) analyzed nest sites at Fish Island and found that rocky slopes and cliff face sites were used about equally. In rocky slope habitats nearly all the sites were within 13 meters of the high-water line, and all the nest chambers examined were in dead-end passages. Most of the cliff face sites were in vertical cracks, with highly variable entrance dimensions. The mean entrance height and width were 15.9 and 18.9 centimeters respectively, and similar dimensions were typical of the nest chamber. The birds apparently preferred sites with relatively small entrances and tended to favor those that were generally rectangular. This species has also been seen nesting on bridges in Oregon and beneath wooden piers in the Aleutians (F. Zeillemaker, pers. comm.).

NEST BUILDING AND EGG LAYING. Except when burrows are dug or enlarged in clay banks, nest building as such probably does not normally occur in pigeon guillemots. Apparently no materials are ever brought into the cavities, and the eggs are typically deposited on the bare substrate. Drent (1965) reported a median egg laying interval of 3 days (range 1–4), with the first egg typically hatching 32 days after laying and the second one in 29.8 days. Thus the first-laid egg typically hatches first, though there are occasional exceptions. The incidence of renesting after loss of the initial clutch is still undetermined but probably is similar to that of the black guillemot. Thoreson and Booth (1958) believed that the single-egg clutches they observed toward the end of the season were probably the result of destruction of the initial clutch. Drent determined that in the colony he watched intensively 30 percent of the birds were nonbreeders; these included yearlings and a substantial number of 2-year-olds. Some 2-year-old males were seen trying to copulate, but the age of initial breeding was not established.

INCUBATION AND BROODING. Both sexes incubate, probably at least intermittently from the laying of the first egg and continuously from the day or so after the laying of the second egg. Most incubation bouts last less than an hour, but overnight incubation periods without shifts seem to be the rule. The roles of the sexes are similar, but in one nest studied by Drent the male was present 62 percent of the observed time. Losses of eggs before hatching are sometimes fairly high. At one colony studied by Thoreson and Booth 62.6 percent of the eggs failed to hatch, apparently in part because of human disturbance. Taking all colonies into account, there was a 46.1 percent failure rate for 78 eggs. In one area most egg losses resulted from heavy rainfall that caused chilling or nest desertion.

GROWTH AND SURVIVAL OF YOUNG. The young are initially fed within 24 hours of hatching and continue to be fed until fledging, which occurred an average of 35 days later (range 29–39) in Drent's (1965) study area. The adults carry in prey, one item at a time; fish are grasped by the head, and the tail dangles out one side of the bill. Drent saw 662 items brought in during 553 hours of observation, or 1.2 prey items per hour. Both sexes play active and approximately equal roles in this parental feeding. Feeding occurs throughout the daylight hours, and the young are brought greater numbers of prey as they grow older. The initial hatching weight of approximately 44 grams is increased to an average fledging weight of 411 grams, 91 percent of average adult weight (Drent). At fledging time the chicks are coaxed from the nest by their parents, after which they may simply waddle to the water or, if necessary, fly or glide down from higher sites. The adults then reportedly cease to tend their chicks and instead join the colony of nonbreeders while the young birds feed in the nearby kelp beds (Thoreson and Booth). Alternatively, the chicks may be convoyed out to deeper waters and tended by the adults for about a month after leaving the nest (Storer). Thoreson and Booth saw young in the area for at least 10 days after all the adults had left, probably for the open sea. In their study area there was a high prefledgling survival rate (86 percent of 58 chicks), but no other comparable data seem to be available.

BREEDING SUCCESS AND RECRUITMENT RATES. The limited data at hand suggest that breeding success in this species is probably rather similar to that of the black guillemot, and it is likely that both have essentially the same strategies in terms of average clutch size, period to reproductive maturity, mate retention and nest site tenacity, and patterns of nestling growth and development.

Evolutionary History and Relationships

Some comments on these points have been made in the black guillemot account. Kozlova (1961) has noted that the genus *Cepphus* approaches the gulls in its cranial structure and considers that the guillemots are nearest *Uria, Synthliboramphus,* and *Brachyramphus* within the family Alcidae. She believed that the Pacific Ocean forms of *Cepphus* had their origins in Pliocene times from an Atlantic Ocean ancestral type similar to *grylle.* She accepted three Pacific Ocean species of *Cepphus,* regarding the Kurile Islands population (*snowi*) as a distinct species rather than a subspecies of *columba.* Only in a few northern Alaskan locations do the ranges of the pigeon guillemot and black guillemot overlap (Sowls, Hatch, and Lensink 1978), and no hybrids between them are known.

Population Status and Conservation

No detailed estimates of the North American population size of the pigeon guillemot are available, but tallies of state and provincial breeding colonies suggest that there are at least 20,000 birds south of Alaska, with another much larger group, perhaps totaling 200,000 birds (Sowls, Hatch, and Lensink 1978), in Alaskan waters. In Alaska there are a few relatively large colonies of 3,000 to 4,000 birds at localities such as Anagaksik Island, Mitrofania Island, and Jude Island, all along the Alaska Peninsula, and others in the Aleutians. Manuwal and Campbell (1979) judged that the species might be increasing in British Columbia, but the population trends in southeastern Alaska and Washington were unknown to them. In California the population on the Farallon Islands has slowly recovered from an all-time low in 1911, but the state's population is highly vulnerable to coastal oil spills because of its essentially inshore habitat distribution (Sowls et al. 1980). King and Sanger (1979) assigned this species an oil vulnerability index of 84, based on a variety of distributional, behavioral, and other characteristics, one of the highest given to any of the alcids and substantially higher than the index (70) they calculated for the black guillemot.

Marbled Murrelet

Brachyramphus marmoratum (Gmelin)

OTHER VERNACULAR NAMES: Long-billed murrelet; guillemot marbre (French); Marmelalk (German); madara-umisuzume (Japanese); dlinnaklyuvy pyzhik (Russian).

Distribution of North American Subspecies (See Map 17)

Brachyramphus marmoratum marmoratum (Gmelin) BREEDS apparently on islands and near the coast from southeastern Alaska to northwestern California. In Alaska, probably a common to abundant breeder in southeastern and south-coastal areas, a resident and probable local breeder in the Alaska Peninsula and also the Aleutians, and a casual summer visitor in western areas (Kessel and Gibson 1976).

WINTERS from southern Alaska to southern California, casually north to the western Aleutians and the Pribilof Islands.

Description

ADULTS IN BREEDING PLUMAGE (sexes alike). Upperparts dark sooty brown (dark fuscous), the interscapulars, scapulars, feathers of rump, and upper tail coverts tipped with deep rusty (sayal brown to verona brown), producing broad bars; underparts with feathers mostly white but broadly margined terminally with fuscous, sometimes so broadly as to reduce the white to an irregular spotting; the chest, sides, and flanks sometimes nearly uniform fuscous; axillaries and under wing coverts uniform fuscous; bill black; iris dark brown; legs and feet flesh color (Ridgway 1919).

WINTER PLUMAGE. Upperparts dusky (dark fuscous), interrupted by a nuchal band of white, the interscapulars, rump feathers, and upper tail coverts tipped with gray; scapulars mostly white, especially the inner ones; entire underparts, including malar, auricular, and suborbital regions and lower half of loral region, immaculate white, the sides and flanks more or less striped with dusky grayish; axillaries and under wing coverts uniform fuscous, as in summer (Ridgway 1919). First-winter birds are more brownish above than adults, and the underparts are mottled with brown (Kozlova 1961).

JUVENILES. Above uniform dusky (fuscous blackish), the nape somewhat intermixed with white; white scapular patch less distinct than in winter adults; lores almost wholly dusky; underparts white, transversely mottled with dusky or fuscous nearly everywhere, but more especially on chest, breast, and sides; bill smaller and weaker than in adults.

DOWNY YOUNG. The down is long, soft, and thick, but absent below the eyes and around the bill. The back and head are yellowish buff, the back mottled with black, and the head and neck have distinct black spots. The undersides are light buffy gray, becoming darker on the flanks. The bill is black, and the legs and feet are pinkish white to gray in front and black behind. The iris is brown (Binford, Elliott, and Singer 1975; Harrison 1978).

Measurements and Weights

MEASUREMENTS (of *marmoratum*). Wing: males 121.5–129.0 mm (average of 10, 126.1); females 112–27 mm (average of 6, 121.7). Exposed culmen: males 13.5–16.5 mm (average of 10, 15.1); females 15–16 mm (average of 6, 15.4) (Ridgway 1919). Eggs: average of 11 listed by Day et al. (1983) was 59.8 x 37.6 mm (extremes 57–63 x 35.0–39.5).

WEIGHTS. Sealy (1975a) reported that 37 adult males averaged 217 g (range 196.2–252.5 g) and 37 females averaged 222.7 g (range 188.1–269.1 g). Two eggs weighed 38.5 and 41 g (Kiff 1981). A newly hatched chick weighed 34.5 g (Simons 1980).

Identification

IN THE FIELD. This small auklet-sized bird is the only alcid south of Alaska that is a mottled brownish during the breeding season, and in winter it is the only one south of Alaska that has a white scapular stripe. In both plumages it closely resembles the Kittlitz murrelet, and in southern Alaskan waters these two species can perhaps be separated by the shorter exposed bill of the Kittlitz, its more uniformly brownish color in the breeding season (back mottled with white and not distinctly darker than the flanks and breast), and by its greater amount of white on the face in winter (separating the eyes from the black crown). The call is a hoarse, drawn-out squawk.

IN THE HAND. The combination of small size (wing under 130 mm) and a very short tarsus (15–17 mm) that has an entirely reticulated scale pattern provides separation from all species except the Kittlitz murrelet, which has a much shorter exposed culmen (less than 15 mm, compared with at least 25 mm in the marbled murrelet).

17. Current North American distribution of the marbled murrelet (*lightly shaded*), and inclusive distribution of the Xantus murrelet (*hatched*). The Asian range of the marbled murrelet is shown on the inset map.

Ecology and Habitats

BREEDING AND NONBREEDING HABITATS. The total breeding distribution of this species is poorly understood, but it apparently is limited to fairly warm waters of the west coast of North America and the east coast of Asia, approximately between the August surface water isotherms of 9°C and 15°C. It is most closely associated with the humid coastal areas supporting wet-temperate coniferous forests with redwood, Douglas fir, and other ecologically similar species, but it also inhabits coastlines along tundra-covered uplands along the Alaska Peninsula and in the Aleutian Islands. In winter the birds move farther south, sometimes as far as southern California, but some wintering occurs on protected waters as far north as the Kodiak area of Alaska (Forsell and Gould 1981) and as far west as the Aleutians (F. Zeillemaker, pers. comm.). For most of the year the birds seem to prefer semiprotected waters of bays and inlets, making only limited use of rocky coastlines (Hatler, Campbell, and Dorst 1978), but in California they sometimes occur well offshore in the open ocean during winter months.

SOCIALITY AND DENSITIES. Very little information is available. Certainly the birds are usually found in low densities and appear to be rather nonsocial, although local concentrations may occur in rich foraging areas. Counts during June and July in British Columbia indicate that only rarely are more than 10 birds seen together as a group, and that groups of singles and pairs constitute about 70–90 percent of the total numbers seen (Hatler, Campbell, and Dorst 1978). Sealy (1975a) noted that during late June and early July he observed an average of 7 birds per flock, and a maximum of 11, among 75 flocks in this same general area. As many as 268 birds, including at least 43 pairs, have been seen on May surveys in Barkley Sound in Pacific Rim National Park. Surveys of Barkley Sound in early June have revealed as many as 100 pairs in that area (Guiguet 1971). There may also be a considerable flocking of birds immediately before the appearance of the first young at sea, for which an adequate explanation is still lacking (Sealy 1975a).

PREDATORS AND COMPETITORS. Nothing is known of possible predators on eggs or young, but presumably some of the larger owls might be predators of nesting adults and chicks. Peregrines (*Falco peregrinus*) have been reported as possible predators of adults and juveniles as well (Sealy 1975a). Likewise, little can be said of potential competitors, but the Kittlitz murrelet is probably the most significant one in areas where both species occur. The ancient and marbled murrelets are broadly sympatric in their breeding distributions, but the ancient murrelet forages predominantly on two species of euphausiid crustaceans during the early part of the breeding season and begins to eat fish only toward the end, whereas the marbled murrelet concentrates on fish during most of the breeding season, taking a euphausiid crustacean only during the very early part of this period. The two species also differ considerably in their foraging areas, with the marbled murrelet an inshore forager (within 500 meters of shore and in water usually less than 30 meters deep), while the ancient murrelet is an offshore forager, usually found over 2 kilometers from land (Sealy 1975c).

General Biology

FOOD AND FORAGING BEHAVIOR. Sealy's (1975c) studies on this species are by far the most complete of any available. He examined 75 adult and subadult birds taken between March and August as well as 6 newly fledged juveniles. He found that among invertebrates the most numerous food items were euphausiid crustaceans (*Thysanoessa*), mostly longer than 24 millimeters, and these were eaten mainly before June 1. During the rest of the sampling period fish predominated, primarily sand launce (*Ammodytes*) in the size classes up to 60 mm fork length, but with smaller numbers of viviparous sea perch (*Cymatogaster*) of comparable length. Smaller numbers of fish representing several families (Scorpaenidae, Osmeridae, and Stichaeidae) and of similar lengths were present in small quantities. The major fish prey, sand launce, belongs to a group of fish in which the young of the previous fall and winter tend to migrate to surface waters and move inshore in late spring, when they would become available to the murrelets. The fall and winter diet of the species is essentially unknown, but samples from a few birds suggest that *Cymatogaster* may remain an important food item, and possibly also mysid and schizopod crustaceans (Sealy 1975c).

Foraging in spring is done mainly by pairs or by single subadults, and later in early July mixed flocks of adults and subadults begin to form. Nearly all foraging is done in fairly shallow water, close to shorelines. The breeding season of the species may be influenced by the seasonal availability of inshore fish foods, especially *Ammodytes*, at least in the area of the Queen Charlotte Islands (Sealy 1975c). The abundance of various fish that Sealy found to be murrelet foods more or less corresponded to their apparent relative seasonal abundance in the study area, but evidently the birds select only a small spectrum of the available zooplankton.

MOVEMENTS AND MIGRATIONS. There is no evidence for substantive migration in this species; instead, the birds often seem to move to more protected inlets and bays during the winter season. Perhaps as many as 13,000 marbled and Kittlitz murrelets winter in the Kodiak Island area of Alaska (Forsell and Gould 1981), and almost certainly the great majority of these would be marbled murrelets. There is an extremely large wintering population of marbled murrelets in the northern Gulf of Alaska, with no evidence of migratory movements; instead, fall dispersal occurs through inshore and offshore waters, with a few birds wintering at the heads of bays and fjords (Islieb and Kessel 1973). There are marked seasonal variations in abundance of marbled murrelets in the Queen Charlotte Islands (Sealy 1975a) and also along the west coast of Vancouver Island (Hatler, Campbell, and Dorst 1978), but at least in part these may reflect ecological shifts rather than major geographic changes in location. Certainly marbled murrelets do exhibit occasional marked population movements southward into California, and likewise there have been a few cases when the Asiatic race *perdix* has strayed into North America (Sealy, Carter, and Alison 1982).

Social Behavior

MATING SYSTEM AND TERRITORIALITY. Sealy (1975a) presented evidence that marbled murrelets may remain paired throughout the year, inasmuch as they arrived at his study area already in pairs, and they have also been observed in pairs while at sea, during both summer and winter. Nothing is known for certain of the incidence of mate retention and nest site tenacity in this species. Binford, Elliott, and Singer (1975) suspected that the nest site they found had been used traditionally, based on its well-constructed and seemingly highly tended condition. Hirsch, Woodby, and Astheimer (1981) found a nest only 10 meters away from a previous year's nest site, further suggesting that nest site fidelity does occur. However, the solitary nesting typical of the species suggests that there is no special need for territorial defense behavior.

VOICE AND DISPLAY. Almost nothing is known about the vocalizations or social behavior of this species. Byrd, Gibson, and Johnson (1974) observed courtship displays in May at Adak Island, Alaska, in which both members of a pair would extend their bills upward, utter shrill calls, and paddle around furiously in unison along a seemingly random path. These displays lasted several minutes, and then the birds dived repeatedly. The birds are apparently highly vocal at other times, with flocks of adult or subadult birds reportedly producing a continuous din throughout the day at about the

time the young are fledging (Guiguet 1950) and with individual birds said to utter high twittering notes. Adults are very quiet during incubation or brooding, and chicks are likewise apparently very quiet, producing only muted sounds even when handled (Simons 1980). Binford, Elliott, and Singer (1975) heard no sounds from the fledgling they observed, from the time it was found until it died.

Reproductive Biology

BREEDING SEASON AND NESTING SUBSTRATE. Few actual egg records are available, but Sealy (1974) determined the egg laying period in the Queen Charlotte Islands area from the reproductive condition of females. There the egg laying dates during a two-year period spanned 6 to 7 weeks, between about May 15 and late June or early July. Day, Oakley, and Barnard (1983) summarized data on 8 known and 1 probable marbled murrelet nests; all but one contained unhatched eggs, and the remaining one a hatched chick. Dates of the nests with eggs ranged from June 3 (Kodiak Island) to August 1 (East Amatuli Island, Alaska). They ranged in elevation from 68 to 690 meters above sea level and from less than 1 to 24 kilometers from the coastline (4 less than 1, 3 between 1 and 10, and 1 more than 10). The nest sites varied considerably in slope and directional aspect, though a possible preference for shady north-facing slopes has been suggested. The mean elevation of nests where trees were present was 570 meters; for nests beyond tree limits it was 100 meters, and for all nests it was 304 meters. Of 6 nests, 4 were in 100 percent vegetative cover, ranging from lichens to lush grasses, shrubs, and coniferous trees. Although a few nests have been located on ground sites (e.g., Kiff 1981; Simons 1980; Hirsch, Woodby, and Astheimer 1981; Hoeman 1965), it seems likely that tree nesting is usual (Binford, Elliott, and Singer 1975; Harris 1971; Kuzyakin 1963). The nest found in California's Big Basin Redwoods State Park was 45 meters above the ground, a short distance out on a nearly flat-topped limb of a very tall Douglas fir (*Pseudotsuga menziesii*) in a virgin forest habitat (Binford, Elliott, and Singer 1975). By comparison, a nest found in the USSR was 6.8 meters above the ground in an only moderately large larch (*Larix dahurica*) and was placed on a cushion of lichens growing on a horizontal branch (Kuzyakin 1963). In both cases the nest was between 6 and 10 kilometers air distance from the sea. In areas north of the tree line marbled murrelets apparently nest at low to medium elevations, in generally heavily vegetated areas, thus overlapping to some degree with the nesting habitat of the Kittlitz murrelet. However, the available evidence suggests that they may nest at generally lower elevations and in heavier cover immediately around the nest. When nesting on the ground, both species seem to select crevices or sites at the bases of large rocks, perhaps to improve protection from predators, provide a more stable microclimate, and also give protection from falling rocks or snow (Day, Oakley, and Barnard 1983).

NEST BUILDING AND EGG LAYING. Although nest building behavior is still undescribed, the structure of at least one tree nest has been well described (Binford, Elliott, and Singer 1975). Apparently it was built of materials at hand, with no additional loose materials brought in. The rim was originally composed of naturally growing mosses and later was built up of feces that formed a ring held together by the meshwork of mosses. These droppings were presumably made by the chick but possibly also came from the adults. The bowl of the nest was a depression in the bark of the trunk, with no apparent bill or claw marks. The bark color closely matched that of the breeding adults, and the brownish moss and weathered droppings were similar in color to the paler parts of the chick's natal down. By comparison, one ground nest was simply a shallow depression with no nest materials, and apparently no effort was made to hide the egg. Nonetheless, the downy chick proved to be extremely well concealed in this environment (Simons 1980). Another ground nest was in a grottolike rock cavity, where the egg was placed on bare rock (Johnston and Carter 1985).

All the evidence available (Sealy 1974) suggests that the clutch consists of a single egg, and thus the egg laying period for an individual female is one day. Harris (1971) reported a case of two nestlings' falling out of a tree that had been cut down by loggers, but Sealy believes the tree may have contained two separate nests. A single brood patch is present in each adult, supporting the contention that single-egg clutches are at least the normal situation.

INCUBATION AND BROODING. Sealy (1974) estimated the incubation period of this species at approximately 30 days, based on indirect information. Simons (1980) discovered an incubating murrelet on a ground nest on East Amatuli Island, of the Barren Islands group, Alaska, on July 8, which hatched on August 1. He followed the incubation behavior carefully and found that, except for a brief period following a severe storm, the nest was attended every day. Both adults incubated and were almost entirely motionless and quiet during this time. The incubating birds faced the sea (75 meters away) and flattened themselves at the sight or sound of ravens (*Corvus corax*) or glaucous-winged gulls (*Larus glaucescens*). The adults performed 24-hour incubation shifts, exchanging places during the evening. Hatching occurred 25 days after the nest was discovered, and on the apparent day of hatching the chick was already alert, with well-developed vision and hearing, and exhibited self-defensive reactions.

GROWTH AND SURVIVAL OF YOUNG. Simons's (1980) observations were the first prolonged ones for this species. He judged that initial brooding following hatching lasted 12–24 hours, and the chick may have also been brooded later on during the night. He found that the chick weighed 39 grams on the probable day of hatching, doubling this weight by the 5th day and tripling it by the 9th. The primaries began to emerge on the 5th day, and by the 21st day the chick had well-advanced growth of juvenal feathers in spite of still being covered by most of its down. Feather development was nearly complete by day 25, and during the next 2 days nearly all the down feather tips had fallen or been preened away, revealing the distinctive juvenal plumage. Evidently much of the down was actually swallowed by the chick after periods of self-preening. The chick apparently was fed daily by the adults, which in two observed cases appeared shortly after sunset, fed the chick, and remained at the nest for an average of only 7 minutes. Up to three small fish were brought simultaneously by a parent. On one occasion the adult was seen to present the chick with a single fish, probably a capelin (*Mallotus*), about 8 centimeters long. Indirect evidence suggested that the chick may have been fed several times a night. A feeding frequency of twice a day (once by each parent) was indicated by observations in the same area the following year (Hirsch, Woodby, and Astheimer 1981). By weighing the chick after feeding, Simons was able to estimate that the average load size was 14.3 grams. Chick growth was relatively rapid, and the upper asymtote of total weight was estimated at 144 grams, together with a fledging weight of about 150 grams. This was estimated to occur about 27–28 days after hatching, although the exact time of fledging was not determined. When last observed at that age, the chick was easily capable of walking the short distance from the nest to the sea. Its wings were then fully developed, and it appeared almost ready for independent flight. Observations of a chick at nearly the same location the following year indicated a 28 day fledging period. In that case the chick's weight at fledging was 140 grams, and its upper asymtote was reached at 166 grams. In this case fledging evidently occurred when both parents came to the nest at twilight and called to the chick. It then left its nest and walked about on the rock ledge above it. By the next morning it and its parents had left the site, but a few days later a pair with a chick were observed in a cove only 0.5 kilometer from the nest (Hirsch, Woodby, and Astheimer 1981). It seems likely that chicks in elevated tree nests fly to the coast, rather than walk or swim, in spite of the substantial distances that are sometimes involved (Binford, Elliott, and Singer 1975). Sealy (1975a) judged that the newly fledged chicks probably fly out to sea at night. Of the first 12 fledglings he saw at sea in early July, 11 flew at his approach, some for at least a kilometer. He believed the young birds assume independent lives once they reach the sea and noted that although adults sometimes are seen with fledged chicks the chicks are more often seen alone. The young birds feed by diving among kelp beds, feeding largely on sand launce. Over half (62.2 percent) of his observations of young were of lone chicks; the next most commonly observed groupings were of two chicks (14.1 percent) and of a chick with two adults (9.1 percent) (Sealy 1974).

BREEDING SUCCESS AND RECRUITMENT RATES. No direct information is available on this. Carter and Sealy (1984) estimated that some 1,445 breeding pairs in the Barkley Sound area produced about 1,200 fledged young, or 0.83 young per pair. The one-egg clutch size and seemingly hazardous nest sites used by this species suggest that a rather low reproductive rate might be typical. No information on longevity is available.

Evolutionary History and Relationships

Certainly the marbled and Kittlitz murrelet represent a typical superspecies, and both presumably had their origins in the northern Pacific area. The Kittlitz murrelet is distinctly more arctic-oriented and thus may have speciated in the Bering Sea, while the marbled murrelet probably had its origins farther south, presumably along North American or Asian coastlines of the northern Pacific. The only significant distinguishing structural difference that they have evolved is in their bill lengths, which possibly relates to undocumented minor foraging differences. The breeding plumage differences between them are apparently adapted for maximum protective coloration advantages in heavily shaded (marbled) versus open alpine tundra (Kittlitz) habitats. Possible behavioral isolating mechanisms between them are still unknown; there are obviously both temporal and ecological overlaps in nesting. Structurally the genus *Brachyramphus* is closely related to *Synthliboramphus*, and both forms were probably derived from an original *Cepphus*like ancestor in the Pacific area (Kozlova 1961), perhaps in late Miocene or Pliocene times. These murrelets have a pelvis structure that is unusually broad and short and that presumably makes them less efficient divers than the other murrelets and perhaps is related in part to at least the marbled murrelet's unusual nesting behavior (Storer 1945; Kuroda 1954).

Population Status and Conservation

Islieb and Kessel (1973) estimated a total marbled murrelet population of several hundred thousands, possibly in the millions, in the North Gulf Coast and Prince William Sound region of Alaska. An additional total of about 15,000–20,000 *Brachyramphus* murrelets is believed to winter in the Kodiak archipelago region of Alaska (Forsell and Gould 1981), and collectively these two regions constitute the prime wintering areas of seabirds in south-coastal Alaska. There are no collective estimates for British Columbia, Washington, or Oregon, and there is no clear indication of the species' possible population trends there (Varoujean 1979; Manuwal and Campbell 1979). A speculative state total estimate of 2,000 birds has been made for California (Sowls et al 1980), with the assumption that the population is declining because of loss of redwood forest breeding habitat. Another marbled murrelet population of unknown size occurs on the Asian side of the Pacific Ocean. The species has been assigned an oil vulnerability index of 84, somewhat above the average calculated for the entire group of Alcidae (King and Sanger 1979). The birds evidently are highly scattered, as well as extremely elusive, on their breeding areas. Thus the only foreseeable major threats to them seem to be possible oiling losses in major wintering areas (such as Prince William Sound) and, to a much lesser extent, breeding habitat destruction. Carter and Sealy (1984) documented a fairly substantial local mortality resulting from gill net fishing in Barkley Sound, British Columbia, representing about 7.8 percent of the potential fall population and about 32 percent of the estimated fledged young of the year.

Kittlitz Murrelet

Brachyramphus brevirostre (Vigors)

OTHER VERNACULAR NAMES: Short-billed murrelet; guillemot de Kittlitz (French); Kurtschnabelalk (German); korotkoklyuvy pyzhik (Russian).

Distribution of Species (See Map 18)

BREEDS locally along the Alaskan coast from Point Hope and the Wales area southward in coastal and adjacent montane tundra to the tip of the Alaska Peninsula and west on the Aleutians at least to Atka; also east to Glacier Bay. Also breeds locally in Siberia, probably from Wrangel Island to the Okhotsk basin.

WINTERS in the southern parts of its breeding range, from the Aleutians east to Glacier Bay, and off the coast of Asia south to northern Japan.

Description

ADULTS IN BREEDING PLUMAGE (sexes alike). Predominant color of upperparts dusky (varying from nearly black to nearly dark gull gray according to angle of view), the surface glossy and this dusky color broken everywhere (except on wings and tail) by irregular streaks or longitudinal spots of light buff, these broadest on scapulars, rump, and upper tail coverts, the nape with buff predominating; wings grayish dusky, the middle and greater coverts and secondaries narrowly margined terminally with pale gray or grayish white (the distal coverts also narrowly edged with the same), the inner webs of secondaries broadly tipped with white; middle rectrices narrowly tipped with white, the outermost rectrix white with a dusky shaft streak on distal portion, the next two (on each side) similar but with the dusky distal streak broader, the fourth (from outside) with inner web white, the fifth and sixth with inner web mostly white; loral, suborbital, auricular, and malar regions, chin, throat, and upper foreneck light buff, narrowly and sparsely streaked with blackish; rest of underparts white, the lower foreneck, upper chest, and sides of lower neck thickly marked with U-shaped bars of blackish, the sides and flanks similarly but more irregularly marked (the markings on outer portion assuming the form of irregular spots), the lower chest, breast, and abdomen with much fewer and narrower irregular bars of dusky, the vent region and under tail coverts nearly immaculate white; axillaries and under wing coverts uniform deep brownish gray (nearly hair brown); bill black; iris dark brown; legs and feet blackish (Ridgway 1919).

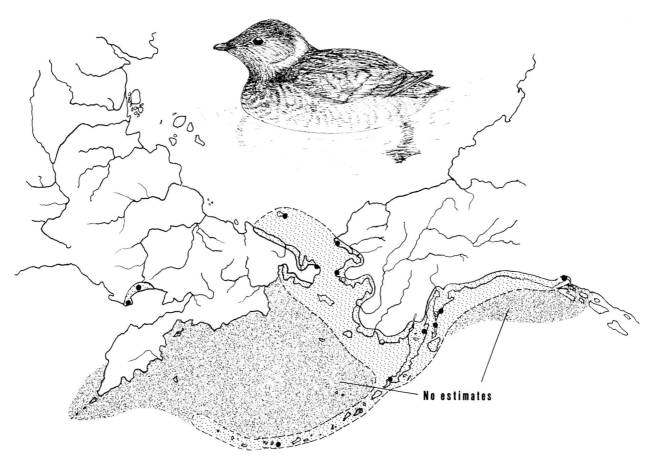

18. Inclusive known breeding distribution of the Kittlitz murrelet (*lightly shaded*), with the probable wintering range (*darkly shaded*) and reported nesting sites (*dots*) also indicated.

WINTER PLUMAGE. Pileum, crescentic bar immediately in front of eye, a broad bar across sides of upper chest (the two nearly meeting in front), and upperparts generally, deep slate gray, with a silky gloss, the feathers of back and rump narrowly tipped with white, many of them showing a darker slate color beneath surface; scapulars mostly white, with slate gray predominating on outer webs; entire underparts, and all of head and neck except as described (including a collar across nape), immaculate white; wings and tail as in summer (Ridgway 1919).

JUVENILES. Very similar to the winter plumage, but with barring or vermiculations on the face, nape, and underparts and with dark barring on the tail (Devillers 1972).

DOWNY YOUNG. The down of the body is mostly medium gray, with blackish bases showing in places and the back suffused with buffish yellow. The head and throat are buffish yellow with black spotting, becoming gray on the breast and pale gray on the belly. The bill is black, the legs and feet are pink in front, becoming brownish black behind. The iris is dark brown (Thompson, Hines, and Williamson 1966; Harrison 1978).

Measurements and Weights

MEASUREMENTS. Wing: males 129–41 mm (average of 2, 135); female 127 mm. Exposed culmen: males 9.5–10.5 mm (average of 2, 10); female 10.5 (Ridgway 1919). Eggs: the average of 9 listed by Day et al. (1983) was 60.0 x 37.3 mm (extremes 57.8–62.5 x 35.6–39.0).

WEIGHTS. The average of 14 adults was 224 g, and a single egg weighed 34 g (J. Bedard, quoted in Sealy 1972). Estimated fresh egg weight, 41 g (Schönwetter 1967). A newly hatched chick weighed 35.7 g (Thompson, Hines, and Williamson 1966).

Identification

IN THE FIELD. This murrelet is likely to be observed from the Aleutian Islands to the southern coast of Alaska, where it might be confused with the very similar marbled murrelet. However, in breeding plumage the Kittlitz murrelet is more uniformly patterned with mottled brown above and below and lacks the dark brownish back and contrasting white scapular pattern typical of marbled murrelets. Its bill is also shorter, which might be a helpful field mark with immature birds. In nonbreeding plumage the Kittlitz murrelet is the only murrelet in which the white of the face is so

extensive as to extend above the eye, separating it from the dark crown.

IN THE HAND. This species can be identified by its unique combination of fairly small size (wing less than 142 mm), a very short tarsus (15–17 mm) that is entirely reticulated in its scale pattern, and a bill with an exposed culmen length of only 9–11 mm. In plumage it is often very similar to the marbled murrelet, but the lateral rectrices tend to be white or mostly white rather than brownish or only narrowly edged with white on the outermost pair. In young birds the outer rectrices are mostly white with brown spots or bars, while in adults a variable number of outer rectrices are entirely white except for dusky shaft streaks.

NEST IDENTIFICATION. There is a complete overlap of size and color of the eggs of Kittlitz and marbled murrelets, although most Kittlitz eggs are olive green as opposed to greenish yellow in ground color. Nests of marbled murrelets are usually at low to medium elevations in heavily vegetated habitats, while those of Kittlitz murrelets are usually at higher elevations in rocky, unvegetated areas. Close observation or photography of the adult is thus necessary for certain nest identification.

Ecology and Habitats

BREEDING AND NONBREEDING HABITATS. This species has a breeding range that is still largely uncertain. It seemingly is limited in North America to rocky glacial moraines in alpine or arctic tundra communities near the coastlines of Alaska from about the vicinity of Le Conte Bay west through the Aleutians and north to Point Barrow, along a narrow gradient of August surface ocean temperatures of approximately 6–12°C. Scree and talus accumulations along fairly steep slopes may be an important component of breeding habitat, based on the very limited number of nests found so far (Day, Oakley, and Barnard 1983). Very little is known of the habitats used in the USSR, but the description of a single nest (Kishchinski 1968) in a rocky alpine setting 700 meters above sea level corresponds closely to the conditions that have been described for North American nests. Nonbreeding birds or off-duty breeders spend the summer months in inshore areas, especially along glaciated coastlines. During winter the birds tend to disperse through both inshore and offshore waters but probably are most abundant in bays and nearshore habitats, becoming rare at the heads of fiords (Islieb and Kessel 1973; Gould, Forsell, and Lensink 1982). Their winter distribution patterns are difficult to discern because of similarities to marbled murrelets in appearance and ecology, but surveys in the Gulf of Alaska indicate a ratio of marbled to Kittlitz murrelets of about 16 to 1 (Gould, Forsell, and Lensink 1982).

SOCIALITY AND DENSITIES. Apparently nesting is done in a solitary, dispersed manner, since two nests have never been found together. There is no way of judging either breeding or wintering densities with any accuracy.

PREDATORS AND COMPETITORS. These presumably include any of the mammalian or avian predators likely to be active on tundra environments, including gyrfalcons (*Falco rusticolus*), snowy owls (*Nyctea scandiaca*), and foxes (*Alopex lagopus,* possibly also *Vulpes fulva*). It is likely that the Kittlitz murrelet's major competitor is the marbled murrelet, as noted in the account of that species. However, they exhibit only limited overlap in breeding habitats, and their comparative foraging behavior is still unstudied.

General Biology

FOOD AND FORAGING BEHAVIOR. Practically nothing is known of the foods of this species. Presumably they are very similar to those of the marbled murrelet, namely small fish and planktonic crustaceans. Kishchinski (1968) believed that because of its small and weak bill the Kittlitz murrelet may feed only on invertebrates, in contrast to the stronger- and longer-billed marbled murrelet, which is largely fish eating. However, it seems likely that at least during the chick-raising period the birds depend on fish for feeding their young.

MOVEMENTS AND MIGRATIONS. What little is known of this species suggests that migratory movements are probably very limited, with the birds shifting ecological locations in winter but probably not migrating very far, at least at the southern limits of their range. In the waters of upper Unakwik Inlet and upper College Fiord and in those adjoining the Malaspina-Bering ice fields the birds winter in great numbers, outnumbering all other alcids in this area. A substantial number of nonbreeders or postbreeders move to Prince William Sound in late July and August; about 57,000 birds were observed there once, mainly in the fiords and bays at the northern and western edges of the sound (Islieb and Kessel 1973). The birds have also been seen in very small numbers in the Aleutian Islands during winter, but probably those nesting on the west coast of Alaska migrate south into the Gulf of Alaska. The birds that presumably breed on Saint Lawrence Island also no doubt move out of that area in winter.

Social Behavior

MATING SYSTEM AND TERRITORIALITY. Nothing is known of these, but presumably the birds follow the usual alcid pattern of mate retention and possibly also nest site tenacity. So far there seem to be no cases of nests found in the same area in successive years, and for a solitary species precise nest site tenacity would be of questionable value in any case.

VOICE AND DISPLAY. Nothing has been written on these subjects.

Reproductive Biology

BREEDING SEASON AND NESTING SUBSTRATE. Much of the available information on this species' breeding season is indirect. Sealy (1977) analyzed patterns of wing molt from museum specimens and reported that the prealternate (prenuptial) molt occurs rapidly between mid-April and mid-May, bringing the birds into reproductive plumage. The prebasic (postnuptial) molt is marked by a simultaneous loss of the flight feathers, which does not begin until late August. All told, the breeding season from initial egg laying to fledging probably spans the period from early June to mid-August. Day, Oakley, and Barnard (1983) analyzed the information available from both published and unpublished sources and listed a total of 14 definite and 3 probable nest records. Counted collectively, the records include a span of egg records from June 10 to July 26, with one record for the first ten days of June, four records for the second, and five for the third, none for the first ten days of July, five for the second, and two for the third. Females with shelled eggs in their oviducts have been collected as early in the season as May 29 (Thayer 1914). One egg from a nest near Cape Thompson probably hatched on July 27 or 28 (Thompson, Hines, and Williamson 1966). Likewise the pipping egg in the nest found near Cold Bay by Bailey (1973) probably hatched shortly after its discovery on July 22. All of this suggests that the eggs are typically laid in June and hatch by late July. This would mean that most or all fledging should occur by mid- to late August, assuming a fledging period of between 13 and 24 days (Bailey 1973).

The nest substrate can now be fairly readily characterized, thanks in large measure to the efforts of Day, Oakley, and Barnard (1983). Elevations of the known nests have ranged from 230 to 1,070 meters, averaging 570 meters, and all but one of the three additional probable nests also fell within this range. Where coastal forests are present the birds tend to nest relatively high. Such nests averaged 800 meters elevation, while those found where no trees occur averaged only 520 meters. In the arctic tundra at the northern edge of the nesting range the average elevation of nests was 340 meters, and the mean elevation of all nests in areas beyond the tree limits was approximately 420 meters. Of 9 nests, 7 faced in northerly directions, and the average slope of 7 nests was 40°. The average straight-line distance to the coastline was 16.4 kilometers for 11 nests but was 75 kilometers for one nest and 23 for another, suggesting a great energy drain for foraging adults and considerable stress for fledging young in reaching the coast. In 6 nests there was a stream large enough to support fledged young at an average mean distance of 600 yards. Only 2 of 10 nests had a vegetative cover of more than 5 percent, and even in these the vegetation was mostly lichens, mosses, and short herbs. Eight of the nests were on the lower side of a rock at least as large as the incubating bird, and in another case the nest was in a natural depression made by a frost heave. One nest was at the top of a mountain, but the average vertical distance below the peak was about 145 meters for 6 others. All told, it appears that the birds are largely limited to rocky sites in tree-free habitats, in particular talus and scree areas with an abundance of lichen-covered rocks that closely match the dorsal color of the nuptial plumage. Nesting below such rocks not only may provide some protection from visual predators but might also offer shelter from winds and falling rocks. Such rocks might also offer visual clues for adults to find their nests when returning in twilight. Particularly in the northern parts of their breeding range, the birds may favor nesting sites with streams nearby to help transport the chick to the coast. This is in contrast to marbled murrelet chicks, which evidently fledge at a later state of development and probably regularly fly to the coast (Day, Oakley, and Barnard 1983).

NEST BUILDING AND EGG LAYING. It is doubtful that any nest building as such occurs in this species; there is no indication that materials are brought to the nest site, and the sites are otherwise seemingly unmodified by the birds. The eggs distinctly are colored for concealment (mostly olive green with darker brownish and blackish spotting, rather than mostly greenish yellow with darker spotting as in marbled murrelets). Since all the evidence indicates that a single egg constitutes the clutch, the egg-laying period may be considered as lasting only a day. Nothing is known of possible renesting after egg loss. As in the marbled murrelet, the egg is very large relative to the weight of the female and presumably is formed over a period of several days. This fact, plus the short breeding season, suggests that renesting following clutch loss would be infrequent at best.

INCUBATION AND BROODING. An incubation patch is present in both sexes, and there is no reason not to believe that both participate fully in both incubation and brooding. The incubation period is not known with certainty but is probably very close to the approximately 30 days in the marbled murrelet.

GROWTH AND SURVIVAL OF YOUNG. Nothing has been reported on the feeding of the chick, the rate of chick growth, or other details of nest life. Bailey (1973) observed a nest that had contained a pipping egg on July 22 and held a downy chick when next visited on July 28. The nest was again observed on August 4, when the chick had wing quills but otherwise was still downy. When the nest was visited again on August 15 the chick was gone, presumably having fledged. If this is the case, the chick would have left when 13–24 days old, depending on the hatching date and the actual date of fledging. A chick that was found making its way to sea in early August was only 40 percent of adult weight and had a wing-chord length 79 percent that of an adult (Day, Oakley, and Barnard 1983). By comparison, Sealy (1975a) judged that marbled murrelet chicks fledge at about 21 days, at about 70 percent of adult weight and with a wing length 86 percent that of adults. The fledging period is now known to be about 28 days (Hirsh, Woodby, and Astheimer 1981), with 40 percent of adult weight attained by about 20 days after hatching. This suggests that a 20–21 day fledging period may be close to the situation typical of Kittlitz murrelets, or about a week shorter than the fledging period of the marbled murrelet. This seemingly shorter fledging period and appreciably lower body weight by 3 weeks after hatching may reflect the fact that the birds tend to nest a good deal farther inland than do marbled murrelets, and adequate food carrying by adults for their chick may simply become more stressful earlier in the posthatching period. Day, Oakley, and Barnard (1983) suggested that Kittlitz chicks probably fledge primarily by fluttering down hillsides into nearby stream drainages and thus eventually reach the sea. This is presumably done without any parental guidance or postfledging care, since no adults were seen with the chick just mentioned, and adults have not been observed in company with chicks in coastal areas.

BREEDING SUCCESS AND RECRUITMENT RATES. No information is available on these subjects.

Evolutionary History and Relationships

As noted in the account of the marbled murrelets, these two forms constitute a superspecies and differ mainly in their relative ecological adaptations for breeding in high arctic habitats. They may also differ in their foraging behavior and foods taken, but that remains to be proved.

Population Status and Conservation

So little is known of this species that almost nothing can be said with any certainty. The largest number of birds have been observed in the area of Prince William Sound and the North Gulf Coast of Alaska, where the midsummer population may number a few hundred thousand birds. The highest counts (57,000 birds) made in Prince William Sound during this period (late July, early August) presumably include a substantial number of immature nonbreeders and possibly some failed breeders but would exclude all unfledged juveniles and probably most breeders. Nothing is known about the size of the Asian component of this species' population. It has been assigned (King and Sanger 1979) an oil vulnerability index of 88, the same index given the whiskered auklet, which is the highest of any alcid and indeed the highest of any bird. Certainly the apparently high concentration of these birds into a fairly small area in summer renders them highly vulnerable to possible oil losses. On the other hand, the birds seem able to nest widely over the high arctic tundra of coastal Alaska, and it seems unlikely that breeding habitat losses are likely to influence its status.

Xantus Murrelet

Synthliboramphus hypoleucus (Xantus de Vesey)

OTHER VERNACULAR NAMES: Scripps' murrelet; guillemot de Xantus (French); Lummenalk (German); pato nocturno (Spanish).

Distribution of North American Subspecies (See Map 17)

Synthliboramphus h. scrippsi Green and Arnold BREEDS on Anacapa and Santa Barbara islands, southern California, and on Los Coronados, Todos Santos, San Benito, and Natividad islands, western Baja California.

WINTERS north to Monterey Bay, casually to Point Arena, California.

Synthliboramphus h. hypoleucus (Xantus) BREEDS on Guadalupe Island, western Baja California.

WINTERS coastally, north casually to the vicinity of Catalina Island, southern California, and south to Cape San Lucas.

Description (Modified from Ridgway 1919)

ADULTS IN BREEDING PLUMAGE (sexes alike). Upperparts, including whole of loral region and upper half of auricular region, slate color or deep slate gray, the scapulars, interscapulars, and wing coverts slate blackish centrally, the primaries and primary coverts dull blackish slate narrowly margined with slate color; a narrow white crescent mark beneath lower eyelid; entire underparts, except outer portion of sides and flanks, immaculate white; outer portion of sides and flanks slate gray, some of the feathers tipped with white; under wing coverts immaculate white; inner webs of primaries grayish white passing into gray distally and toward shafts; bill black, basal portion of mandible pale bluish; iris dark brown; inner side of tarsi and upper surface of toes and webs pale blue, the outer side of tarsi and underside of feet dusky.

WINTER PLUMAGE. Similar to the summer plumage, but with white on sides of head involving most of the loral, suborbital, and auricular regions.

JUVENILES. Not yet described but probably comparable to those of *craveri*.

DOWNY YOUNG. Upperparts uniform black; underparts, including malar and suborbital regions, immaculate white; a flank patch of grayish black or dusky gray, confluent with black of rump; thighs blackish. The downy plumage of this species is considerably lighter underneath than in *craveri*, the light area extending in *scrippsi* up the face to the auricular area (Jehl and Bond 1975).

Measurements and Weights

MEASUREMENTS. Wing: 111–28 mm (average of 33 from Los Coronados, 119.1); females 115–27 mm (average of 37 from Los Coronados, 120). Exposed culmen: males 15.6–21.4 mm (average of 41 from Los Coronados, 18.0); females 16.0–21.3 mm (average of 36 from Los Coronados, 18.2) (Jehl and Bond 1975). Eggs: average of 152, 53.5 x 36 mm (Bent 1919).

WEIGHTS. Males 138–85 g (average of 25, 155.4); females 130–84 g (average of 21, 161.8) (Jehl and Bond 1975). The estimated egg weight is 37 g (Schönwetter 1967). The weight at hatching is 17.3 percent of average adult weight, or about 27 g (Thoreson, in press).

Identification

IN THE FIELD. This species may be seen from southern California southward and resembles a miniature murre in its black-and-white plumage pattern. Its bill is narrow, black, and pointed, separating it from similar auklets, and the eye is dark rather than white. It differs from the extremely similar Craveri murrelet in that it has entirely white under wing coverts. It also has less of a semicollar in front of the wing and a more sinuous junction of dark and white on the side of the head, and the dark cheek markings do not extend below the base of the lower mandible. Its call is a series of high, thin whistles.

IN THE HAND. The combination of small size (wing under 130 mm), a moderately long tarsus (22–25 mm), and a moderately long tail (30–33 mm) separates this species from all other murrelets except the Craveri. Measurements apparently do not adequately separate these species; instead the wing covert pattern and other minor plumage differences mentioned above must be utilized. Minor differences in bill ratios between the two species are noted in the account of the Craveri murrelet.

Ecology and Habitats

BREEDING AND NONBREEDING HABITATS. This species breeds primarily in rocky habitats on islands off the California coast, with more limited nesting on sandy beaches. Apparently it is essentially residential, with its breeding limits lying approximately between the August surface temperature isotherms of 15–20°C or the February isotherms of 12–18°C. The largest colony is on Santa Barbara Island, where the climate is mild and the sparsely vegetated coastline consists mostly of sheer cliffs and a few narrow, rocky beaches. Slopes with rock outcrops and abundant crevices are probably the preferred nesting habitat (Murray et al. 1983). During the nonbreeding season the birds are pelagic, often occurring well away from shoreline.

SOCIALITY AND DENSITIES. At least on Santa Barbara Island, nest sites are clumped, probably because of the patchy distribution of suitable nest sites. There the nearest-neighbor distances were estimated by Murray et al. (1983) as averaging 5 meters, with 172 such internest measurements ranging from 0.15 to 40.0 meters. The Santa Barbara nesting population has been estimated as 6,000–10,000 birds (Murray et al. 1983), and the island has an area of 2.6 square kilometers; thus the average estimated breeding density is about 2,000–3,800 birds per square kilometer.

PREDATORS AND COMPETITORS. In areas where feral cats (Felis cattus) occur they are probably important predators, but they no longer occur on Santa Barbara Island. There the major predator of adults is the barn owl (Tyto alba), which seems to concentrate on this species and on the Cassin auklet. Although western gulls (Larus occidentalis) may take some chicks, the major cause of mortality to unattended chicks and eggs was found by Murray (1980) to be deer mice (Peromyscus maniculatus). The island fox (Urocyon littoralis) may also be a significant predator in some areas, although it is absent from Santa Barbara Island. One remarkable observation of a western gull's capturing an apparent healthy adult murrelet on the water and swallowing it whole is of interest (Oades 1974). Peregrines no longer occur on Santa Barbara Island, but they probably were once important predators too.

Probably the Cassin auklet and storm petrels such as the ashy storm petrel (Oceanodroma homochroa) compete with this species for nest sites, though no specific information on this seems to be available.

General Biology

FOOD AND FORAGING BEHAVIOR. Little specific information has been available on the foods and foraging adaptations of this species. Recent studies by Hunt et al. (1979) indicate that larval stages of the northern anchovy (Engraulis mordax), sauries (Scomberesocidae), and rockfish (Scorpaenidae) are the most important food resources. Of these the anchovy is the most important, and its relative availability may have important impact on the breeding success of the birds.

MOVEMENTS AND MIGRATIONS. Probably few if any real migratory movements occur in this species, but certainly dispersal (primarily northward) by nonbreeding or postbreeding birds does occur. Thus, vagrants (presumably of scrippsi) have been observed off British Columbia, Washington, and Oregon, and, more remarkably, a possibly mated pair of hypoleuca was collected off Cape Flattery, Washington (Jehl and Bond 1975). This location is well over a thousand miles from the species' northernmost regular breeding area, although extralimital nesting of this subspecies has recently been documented on Santa Barbara Island (Winnett, Murray, and Wingfield 1979). This form apparently regularly wanders north to the coast of California and perhaps occasionally strays substantially beyond.

Social Behavior

MATING SYSTEM AND TERRITORIALITY. Both mate retention and nest site tenacity have been established in this species. Of 5 pairs of birds banded at their nests in 1977 and observed in 1978, 3 retained both their mates and their nest sites. Of 20 banded individuals, 13 maintained the same nest sites for three years and 4 for four consecutive years, while a 5th bird remained under the same bush but moved its nest location slightly (Murray et al. 1983). Any territorial behavior is certainly limited to the nest cavity itself, inasmuch as nests have been found as close as 0.15 meter apart. Visits to prospective nest sites may begin as early as 2 months before egg laying, although most visits began 2–3 weeks before egg laying according to Murray et al. (1983).

VOICE AND DISPLAY. Vocalizations have not been described in detail but consist in part of a trilled whistling "song" produced by congregations of birds on the water near their breeding colonies (Jehl and Bond 1975). Similarly, there are no descriptions of social behavior patterns in this species, which like the other murrelets is nocturnal on the breeding grounds.

Reproductive Biology

BREEDING SEASON AND NESTING SUBSTRATE. Egg records from the Coronados Islands are from March 30 to July 6, with a probable peak between May 27 and June 17 (Bent 1919). Jehl and Bond (1975) determined a similar spread of dates but calculated a slightly earlier peak of May 15–20. They also indicated a peak laying period of between mid-April and early May for the Guadalupe Islands. The egg records from the Santa Barbara Islands extend from March 8 to early July, with a usual peak in April but with substantial year-to-year variations in laying patterns. An estimate of laying synchrony (the spread of laying dates for the 80 percent of the clutches falling closest to the mean laying date) indicated that in three different years this varied from 24 to 47 days, with an average of 35.6 days (Murray et al. 1983). Nest locations are quite variable, but on the Santa Barbara Islands the birds most commonly use rock crevices. Murray et al. (1983) reported that 70 percent of 224 nest sites they observed were in rock crevices, while a shrub (*Eriophyllum nevinii*) accounted for 21 percent, other plants 6 percent, the burrows of rabbits and burrowing owls (*Athene cunicularia*) 3 percent, and man-made structures 2 percent.

NEST BUILDING AND EGG LAYING. Murray et al. (1983) found no indication of active burrowing or other nest construction, including lining nests with vegetation. Visiting of nesting burrows begins about 2 or 3 weeks before the onset of egg laying, with the birds typically initially grouping offshore in staging areas at dawn and dusk. Each night the birds begin flying to the nesting grounds immediately after dark, with a peak movement 2 or 3 hours after dark and a second activity peak just before dawn as the birds return to the sea. At least dur-

ing daylight the first egg is unattended by the adult birds after it is laid and before the laying of the second egg, which on average is 8 days after the first (range 5–12 days). This unusually long interval between the two successive eggs is related to the very large relative egg size, which Murray et al. (1983) estimated as 22.2 percent of adult weight and Sealy (1975b) judged to be 23.7 percent. Normally the clutch is of two eggs; of 296 nests examined 25 percent had single-egg clutches and 17 contained three or four eggs, the latter almost certainly the efforts of two females. There is no evidence that the birds ever rear more than a single brood per season. However, laying of replacement clutches was confirmed by Murray et al. (1983), who found that a pair that abandoned their first clutch in late May after it failed to hatch began a second clutch 20 days later. They found a high incidence of egg loss as a result of predation by deer mice. Thus, of 470 eggs laid, 28 percent were lost to mice before clutch completion, and another 16 percent were lost after the laying of the second egg, at least in part because of human disturbance. Other causes of egg loss included abandonment (10 percent before clutch completion, 4 percent afterward) and accidents (3 percent). Another 5 percent of the incubated eggs failed to hatch. All of these losses resulted in a hatching success of only 39 percent.

INCUBATION AND BROODING. Incubation typically does not begin immediately after the laying of the second egg; instead there is an average 2 day interval before incubation commences. Even after incubation begins the eggs may remain unattended for up to 4 days. On average there was a cumulative total period of 2.9 days during the incubation period during which the eggs were left unattended in 45 nests (range 0–19 days), and over 60 percent of the nests were unattended for at least one day during the incubation period. Embryos have been found to survive as long as 4 days of continuous neglect, though the chicks hatched from such neglected eggs weighed less on hatching than those from more fully attended nests. Egg neglect also extends the length of the incubation period, which in continuously attended nests averages about 32 days. The birds exhibit unusually long incubation shifts, which in one year of study averaged 2.77 days (66.5 hours) and in a second year 3.14 days (75.4 hours), virtually identical collectively to the 72 hour average that has been reported as typical of the marbled murrelet. Murray, Winnett-Murry and Hunt (1979) hypothesized that egg neglect in this species allows the adults more time to forage for patchily distributed food resources that are difficult and time consuming to locate. During the period of most prevalent egg neglect the adult birds managed to gain more weight foraging than they had lost in incubation, and by the late stages of incubation they had regained an average weight only slightly below that typical of the prelaying stage. Hatching in this species is fairly prolonged, with initial pipping 2–5 days before hatching. Where both eggs are present the two eggs usually hatch almost simultaneously, but sometimes up to 24 hours may elapse between the emergence of the two chicks.

GROWTH AND SURVIVAL OF YOUNG. Newly hatched young are highly precocial and average about 15 percent of the adult weight. However, the length of their tarsi is 98 percent of the average adult length. They are not fed at the nest before leaving it, and Murray et al. (1983) reported that they lost an average of 8 percent of their hatching weight before departure. Chicks 1 and 2 days old have body temperatures within 3°C of adult temperatures and thus probably can withstand substantial chilling. They typically left their nests 1 or 2 nights after hatching and rarely remained in the nest as long as 5 days. Departure was exclusively at night and followed a period of intense vocalization, after which both adults and chicks emerged from the nesting cavity. Typically the parents took off and flew to the sea almost immediately, leaving their chicks behind. These then made their way to the cliff edges, where they were blown off or jumped into the surf 75 meters or more below. Although no subsequent observations of families were made, it is likely that there are rapid reunions between parents and chicks and a gradual movement offshore during the first night at sea. The young are able to dive easily at this stage, apparently using their still downy wings and feet for propulsion under water. In subsequent sightings of 5 broods both parents were in attendance in all cases, and in 3 of these the families were more than 18.5 kilometers offshore (Murray et al. 1983).

BREEDING SUCCESS AND RECRUITMENT RATES. There is no direct information on these topics. Hatching success was very low (39 percent) in the only available study (Murray et al. 1983), but apparently there was low chick mortality on the island. Losses of chicks after they reach the sea are undocumented and probably impossible to judge.

Evolutionary History and Relationships

The relationships of the Xantus and Craveri murrelets, which often have been placed in a separate genus (*Endomychura*) from the other murrelets, are clearly with *Synthliboramphus* rather than with *Brachyramphus* (Jehl and Bond 1975). As noted by Jehl and Bond, this group of species forms an interesting morphological north-south cline in decreasing wing length and leg length and progressively longer and weaker bill structure, although the possible ecological reasons for these relationships are still obscure.

Population Status and Conservation

The Santa Barbara Islands population of this species is certainly the largest single population component, and at least in the 1970s it probably numbered about 6,000–10,000 birds. It is possible that the deer mouse population, which represents the single greatest source of egg mortality, may have increased in this century with the increase in introduced grasses. All the other known colonies are small, no more than 150 birds. Because of their restricted breeding range, the species is particularly vulnerable to possible losses from oil drilling off the southern coast of California.

Craveri Murrelet

Synthliboramphus craveri (Salvadori)

OTHER VERNACULAR NAMES: None in general English use; guillemot de Craveri (French); Craveri-lummchen (German); pato nocturno de Craveri (Spanish).

Distribution of Species (See Map 19)

BREEDS on islands in the Gulf of California, north to Consag Rock, and probably on the west coast of Baja California north at least to the San Benitos–Cedros area (Jehl and Bond 1975). Occurs in autumn north to Monterey, California.

WINTERS within the breeding range, south to the Sonoran coast.

Description

ADULTS IN SUMMER AND WINTER. Upperparts, including whole of loral and orbital regions and upper portion of auricular region, plain blackish slate; a whitish bar or crescent mark beneath lower eyelid and a less distinct one immediately above upper eyelid; underparts, except sides and flanks, immaculate white; sides and flanks plain dull slate color or brownish slate; under wing coverts brownish slate gray, some of the larger ones and tips of some of the smaller coverts white or grayish white; inner webs of primaries grayish white only to-

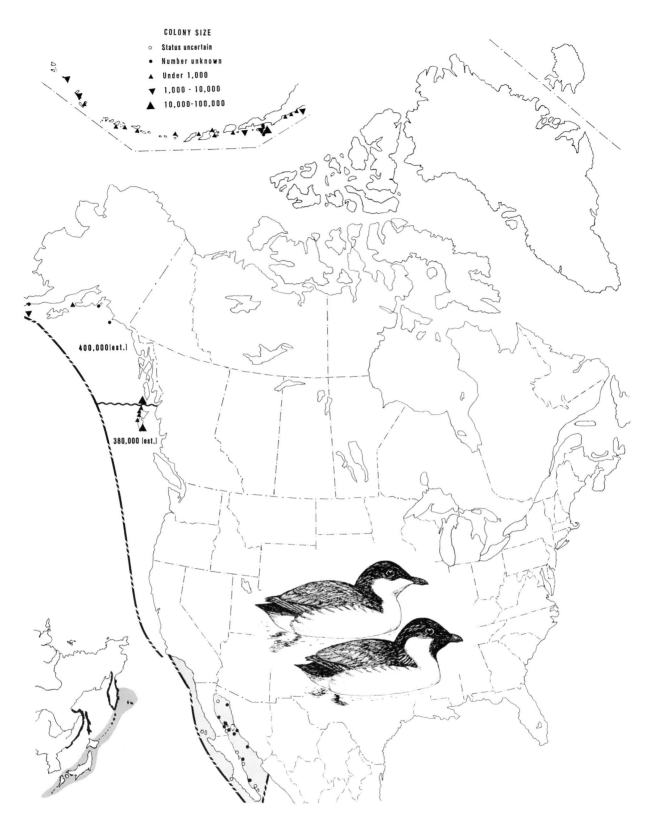

COLONY SIZE

- ○ Status uncertain
- ● Number unknown
- ▲ Under 1,000
- ▼ 1,000 - 10,000
- ▲ 10,000-100,000

400,000(est.)

380,000 (est.)

19. Current North American distribution of the ancient murre-let, (*unshaded*), and the inclusive distribution of the Craveri murrelet (*shaded*), with broken lines indicating limits of non-breeding ranges. The Asian range of the ancient murrelet is shown on the inset map.

ward base; bill black, iris dark brown; legs and feet bluish (Ridgway 1919).

JUVENILES. Very similar to adults, but with darker (nearly dead black) upperparts and numerous fine but rather conspicuous blackish spots on the tips of the feathers of the sides of the breast and body (Bent 1919).

DOWNY YOUNG. Seal brown down covers the upperparts, which are slightly redder and paler than in adults, and fine transverse markings of whitish besprinkle the back and rump, but not the crown or wings. The throat is grayish, the abdomen is white, and the sides of the body and wing surfaces are nearly the same shade of brown as the crown and back (Bent 1919).

Measurements and Weights

MEASUREMENTS. Wing: males 107–23 mm (average of 41, 116.1); females 111–24 mm (average of 30, 117.8). Exposed culmen: males 18.0–22.5 mm (average of 42, 19.8); females 18.2–22.4 mm (average of 29, 19.9) (Jehl and Bond 1975). Eggs: average of 34, 52.3 x 34.9 mm (Bent 1919).

WEIGHTS. Males, 128–49 g (average of 6, 137.1); females 131–37 g (average of 5, 134.8) (Jehl and Bond 1975). The estimated egg weight is 35 g (Schönwetter 1967).

Identification

IN THE FIELD. This species is largely limited to the Mexican coastline and closely resembles the Xantus murrelet, but it has grayish rather than white under wing coverts, it lacks white tips to the feathers of the flanks, and its black cheeks extend below the base of the lower mandible. It also has more of a blackish semicollar in front of the wing, producing a rather sinuous border below the black and white. Its call is a trilling whistle.

IN THE HAND. The combination of small size (wing under 130 mm), a moderately long tarsus (21–25 mm), and a fairly long tail (over 30 mm) separates this species from all other murrelets except the Xantus murrelet. This species has a slightly shorter average wing length than the Xantus and a very slightly longer average culmen length, although there is almost total overlap. However, bill shape is slightly different in the two species, with the Craveri murrelet having a relatively less robust bill. The resulting bill ratio (bill depth relative to culmen length) is thus 0.27 for the Craveri murrelet and 0.31–0.35 for the Xantus murrelet. Additionally, the

plumage differences noted above provide for in-hand separation (Jehl and Bond 1975).

Ecology and Habitats

BREEDING AND NONBREEDING HABITATS. This species has the southernmost range, and one of the most restricted, of all the Alcidae, being limited to the islands and perhaps the adjoining coastlines of the Gulf of California and probably also to a few islands off the west coast of the Baja Peninsula, north apparently to the San Benitos Islands, where it would encounter the Xantus murrelet during the breeding season. Surface ocean temperatures during August in the Gulf of California average about 25°C, while those on the adjoining lower west coast of the Baja Peninsula average 20–25°C. February surface water temperatures within the Gulf of California are not much colder, averaging up to about 20°C. Within that region, certainly most and perhaps nearly all breeding is limited to island situations, but coastline nesting might occur on areas of shoreline that are isolated and protected from terrestrial predators by precipitous and extensive cliffs (DeWeese and Anderson 1976). In general the birds are known to breed only near high-tide limits in rocky habitats having crevices large and deep enough to provide both access and relative safety. Otherwise they are essentially pelagic, often being found well offshore in open ocean.

SOCIALITY AND DENSITIES. Density limits on this species are probably set by the abundance and distribution of rock crevices suitable for nesting, but there do not appear to be any estimates of breeding densities available.

PREDATORS AND COMPETITORS. Probably terrestrial predators, such as feral cats and introduced rats (*Rattus* spp.), are important at least locally, but DeWeese and Anderson (1976) did not find any positive evidence of their influence in the nesting areas they observed. Both occur on some of the islands of the Gulf of California and are likely to be locally significant. DeWeese and Anderson did find evidence of predation on adults by peregrines (*Falco peregrinus*) and also by barn owls (*Tyto alba*), leading them to conclude that these two species are the most important of the naturally occurring predators for nesting birds. If the Craveri and Xantus murrelets do exhibit any sympatry in their breeding ranges, as has been suggested by Jehl and Bond (1975), the two must compete for nesting sites and possibly for other resources such as food. The two forms may even hybridize on the San Benito Islands, although the extent and genetic significance of such hybridization is unknown.

General Biology

FOOD AND FORAGING BEHAVIOR. Very few specimens have been analyzed as to their food consumption, but DeWeese and Anderson (1976) examined the stomach contents of 5 birds. Three groups of fishes predominated in this sample, and all were present in 4 of the 5 birds analyzed. These include rockfish (*Sebastes,* Scorpaenidae), thread herring (*Opisthonema?* Clupeidae), and a lantern fish (*Benthosema,* Myctophidae). Of these, the rockfish was found in the largest numbers and the lantern fish in the smallest. The largest fish were 40–70 millimeters long, and similar-sized fish were seen being fed to dependent chicks. Fish found in small numbers and in only a single stomach included jacks (*Caranx?* Carangidae), mackerels (*Scomber,* Scombridae), and an unidentified flatfish as well as two other unidentified fish. Except for the lantern fish, all the specimens found were larval fishes. Additionally, many unidentified shrimps and a squid were found in a single specimen.

The composition of these stomach contents indicates that the birds had been foraging at or near the surface, over deep waters. DeWeese and Anderson (1976) noted that they never saw the murrelets feeding in close association with other seabirds; in 99 of 104 observations the birds were foraging alone. Additionally, the groups of foraging murrelets were typically small, only rarely consisting of large numbers of pairs or multiple parent-chick groups. The maximum number of pairs seen in any group of paired birds observed between mid-February and early March was 17, but only 12.6 percent of 142 paired birds observed were seen as solitary pairs. Thus it seems that foraging is mainly done in very small groups that tend to be rather widely scattered.

MOVEMENTS AND MIGRATIONS. After the breeding season there evidently is a northward movement of birds during fall (August to October) toward California, when they are regular from northwestern Baja California to Monterey and have rarely strayed as far north as Oregon. Confusion in the field with the Xantus murrelet has obscured the actual migratory status and nonbreeding distribution of the species, especially at the northern end of its range. The birds also may range an uncertain distance south beyond the Gulf of California along Mexico's west coast during the fall and winter, at least to the vicinity of Mazatlán and Sonora, and they have been reported offshore in the Pacific as far west as Guadalupe Island. By late December the numbers in the Gulf of California begin to build, with the birds evidently by then already in pairs. Occupation of breeding colonies may begin as early as February, and the northward postbreeding dispersal begins during June and July (DeWeese and Anderson 1976).

Social Behavior

MATING SYSTEM AND TERRITORIALITY. No studies of marked birds are available to test the degree of monogamy, mate retention, and nest site fidelity typical of this species. However, the occurrence of paired birds as early as February might suggest either that mates are retained through the winter or that pairs are formed while still on the ocean, at least a month before initial nesting activity. Indeed, a high frequency of male/female duos among birds collected at sea during the nonbreeding period suggests a year-round monogamous mating system (DeWeese and Anderson 1976). These authors suggested that most birds arrive at the nesting areas between January and April. There is one case of a nest site that was known to be occupied yearly from 1972 through 1975 (when it was destroyed), although the identities of the individual birds using the site were not established.

VOICE AND DISPLAY. Nothing is known of the social behavior patterns of this species. The calls of both the Xantus and Craveri murrelets have been described as trilling whistles, which might be slightly less musical and "drier" in the Craveri, but the calls are so similar that vocalizations are probably unlikely to play any significant role as an isolating mechanism between the two (Jehl and Bond 1975). An early observer described the species as having three distinct vocalizations, of which the one associated with "displeasure" is very harsh (Bent 1919).

Reproductive Biology

BREEDING SEASON AND NESTING SUBSTRATE. Egg records from the Cape Region of Baja California range from February 6 to April 11, with a probable peak between February 14 and 24 (Bent 1919). DeWeese and Anderson (1976) assembled information suggesting that the nesting period extends from February to early April and the period of hatching and nest departure probably occurs primarily in April and May. Bancroft (1927) noted that during April chicks could be seen that ranged from only a few days old to almost full-grown, suggesting that there is a rather unsynchronized breeding season. The usual nesting substrate is evidently a rock cavity or crevice, but there are also records of the birds' nesting in ground burrows, under large rocks, and among bushes. Of 9 sites found by DeWeese and Anderson (1976), 3 were in narrow crevices, 2 were in shallow holes in rocky slopes, 2 were under rocks, 1 was in a shallow hole in a cave, and 1 was under a low cliff base. These sites ranged from 0.3 to 5.5 meters above the high-tide level, averaging 3.0 meters. The historical records cited by Bent (1919) suggest that the birds most often nest in rock crevices, the nest site being a depression in the earth at the end of the crevice. According to Bancroft (1927) the birds nest only in entirely dark sites and may slightly work the sand in the nest itself to provide a suitable substrate.

NEST BUILDING AND EGG LAYING. The nest sites described above suggest that little actual nest building is normally done, though it is possible there is some burrowing or digging. There is no information on the interval between eggs, but with an average adult female body weight of about 135 grams and an estimated egg weight of 35 grams, the relative egg weight would be 25 percent, or higher than any of the ratios of the other murrelets reported by Sealy (1975c). This suggests that there is probably a substantial interval between the laying of the two eggs, as in the ancient murrelet. Two-egg clutches are probably normal; DeWeese and Anderson (1976) noted that 6 of 8 nests they observed had 2 eggs, and both of the single-egg clutches were unattended and the eggs damaged or old. Of 63 museum sets that they examined, 53 were two-egg clutches, 3 were three-egg clutches, and 7 were single eggs, for an overall average of 1.94 eggs per clutch. Excluding the three-egg clutches and including their own field data, DeWeese and Anderson calculated an average clutch of 1.88 eggs for 67 clutches.

INCUBATION AND BROODING. Certainly both sexes incubate, but nothing is known of their relative roles, the lengths of incubation bouts, or related information. An early collector (W. W. Brown) reported that many males were collected on the nests during daylight hours and estimated the incubation period as 22 days (Bent 1919). This would be a very short incubation period, and it seems very probable that the incubation duration might be closer to 31–33 days and thus comparable to those of the Xantus and ancient murrelets. This seems especially likely considering the species' very large relative egg size and the fact that chicks are highly precocial.

GROWTH AND SURVIVAL OF YOUNG. From 2 days of age the chicks take to the sea and become pelagic (Bent 1919; Bancroft 1927). This probably occurs under cover of darkness, though the actual departure from the nest sites does not seem to have been described. Once the chicks have departed they are usually attended by both parents (Bancroft 1927). The young evidently remain with their parents at least until late June, after which they cannot be distinguished from them.

BREEDING SUCCESS AND RECRUITMENT RATES. DeWeese and Anderson (1976) reported an average brood size of 0.84 young per adult during brood counts in April and May of 1972 and 1974. They tallied a total of 62 adult-chick groups, which collectively had 111 adults present, or 1.8 adults per brood. These 62 groups had 93 chicks present, indicating an average brood size of 1.5 chicks. They noted that during April the number of chicks per adult approached 1.0, while May counts averaged slightly less than 0.5 per adult, suggesting substantial early chick mortality. There are no estimates available of recruitment rates in this species.

Evolutionary History and Relationships

Jehl and Bond (1975) noted that there is some evidence of extremely limited hybridization between the Xantus and Craveri murrelets on the San Benito Islands, where they estimated that the breeding murrelet population might contain 20–30 percent *craveri*, based on observations in the late 1960s and early 1970s. Assortative mating seems to be occurring there between the Xantus and Craveri murrelets, but there some individuals with facial patterns intermediate between the two races of *hypoleuca*. Evidently on the San Benito Islands these two races are undergoing limited gene flow between them. On Guadalupe Island the local birds (subspecies *hypoleuca*) have longer, thinner bills than are typical of the Xantus murrelet, more closely resembling *craveri*, although Jehl and Bond did not suggest hybridization as a reason for this trend. They did suggest that the Craveri murrelet may have very recently extended its breeding range north to Guadalupe and the San Benito Islands, where it has come into contact with both races of the Xantus murrelet. The taxonomic problems associated with this complex situation are severe, but it seems clear that at least the Craveri murrelet should be regarded as a species distinct from *hypoleuca*. It is fairly easy to imagine *craveri* speciating from a disjunct population of an ancestral form similar to *hypoleuca* that became isolated in the Gulf of California and adopted an essentially sedentary existence there. In the process it evolved some slight differences in bill shape and length as it adapted to local foraging conditions, but it has not undergone any great divergence in plumage and probably evolved few if any modifications in its social behavior and vocalizations.

Population Status and Conservation

It is virtually impossible to guess the total population of the Craveri murrelet. Thoreson (in press) made a tentative estimate of 6,000–10,000 birds, but without indicating its basis. In any case, this species is almost certainly the rarest of the North American alcids, and since its breeding grounds are outside the national boundaries few if any direct conservation measures can be applied to it.

Ancient Murrelet

Synthliboramphus antiquum (Gmelin)

OTHER VERNACULAR NAMES: Black-throated murrelet; gray-headed murrelet; guillemot antique (French); Silberalk (German); umisuzume (Japanese); starik (Russian); pato nocturno antiguo (Spanish).

Distribution of Species (See Map 19)

BREEDS from the Commander Islands and Kamchatka to Amurland, Sakhalin, the Kurile Islands, Korea, and Dagelet Island; and from the Aleutian, Sanak, and Kodiak islands to Graham and Langara islands in the Queen Charlotte group, British Columbia, casually to northwestern Washington (Carrol Island).

WINTERS from the Commander Islands south to Fukien, Taiwan, and the Ryukyu Islands; and from the Pribilofs and Aleutians to northern Baja California. Casual inland occurrences during winter are frequent in this species.

Description

ADULTS IN BREEDING PLUMAGE (sexes alike). Pileum, hindneck, and loral region uniform dull black, the malar, suborbital, and auricular regions, chin, throat, and upper foreneck uniform deep fuscous or clove brown, with a rounded or convex outline on foreneck; supra-auricular region and sides of occiput narrowly streaked with white, forming a more or less broken broad stripe, the lower hindneck similarly but more sparsely streaked; back, scapulars, and rump uniform gray (between slate gray and neutral gray); wing coverts duller or more brownish gray (between neutral gray and deep mouse gray), the flight feathers darker; upper tail coverts and tail dull blackish or dusky; sides of neck, lower foreneck, and rest of underparts except sides and flanks immaculate white; sides and flanks uniform sooty black or fuscous black; under wing coverts white; bill grayish lilaceous white or bluish white, darker basally, with a stripe of black along culmen; interior of mouth bluish white; iris dark brown; legs and feet grayish white faintly tinged with violet blue, the outer side of tarsus more bluish, the joints and webs dark bluish gray (Ridgway 1919).

WINTER PLUMAGE. Throat immaculate white, the chin (sometimes upper throat also) slate grayish; white streaks on supra-auricular region, sides of occiput, and lower hindneck wanting; sides and flanks white, the outermost portion striped with slaty or grayish dusky;

otherwise as in summer. First-winter birds have darker chins and many dark flank feathers (Kozlova 1961).

JUVENILES. Birds in their first fall plumage have the throat mostly or wholly white, sometimes with dusky on the chin. Some short, off-white feathers are present on the crown and nape and on both sides of the crop (Kozlova 1961).

DOWNY YOUNG. The upperparts are jet black, including the back, wings, crown, and sides of the head to a point below the eyes; there is a whitish auricular patch behind the ear, and the dorsal region and occiput are clouded with bluish gray. The underparts are pure white, slightly tinged with yellowish (Bent 1919).

Measurements and Weights

MEASUREMENTS. Wing: males 130.5–140.5 mm (average of 3, 134.7); females 132–38 mm (average of 6, 135.5). Exposed culmen: males 13–14 mm (average of 3, 13.3); females 12.5–15.0 mm (average of 6, 13.2) (Ridgway 1919). Eggs: average of 51, 61.1 x 38.6 mm (Bent 1919).

WEIGHTS. The average weight of 75 adult males was 206.3 g, while 79 adult females averaged 205.7 g. The average of 15 fresh eggs was 44.9 g (Sealy 1976). Newly hatched chicks average 30.7 g (Sealy 1976).

Identification

IN THE FIELD. This is one of the commonest murrelets offshore along the western states, and it sometimes occurs well inland after storms. In breeding plumage it has a distinctive black face and throat, with a white stripe above and behind the eye, white "eyelids" and a distinctive white bill, and a unique area of black and white barring along the back and sides of the neck, somewhat like a loon's. In winter it resembles the marbled murrelet with a black "cap" and a paler back. Its nest call is a shrill whistle; piping notes are sometimes uttered while at sea that are quite different from those uttered at the nest.

IN THE HAND. The combination of fairly small size (wings under 141 mm), a moderately long tarsus (25–28 mm) that has a scutellate pattern on the lower front surface, and an outer toe longer than the middle toe serves to separate this species from the other murrelets with which it might be confused. In all the other North American murrelets the tarsus is entirely reticulate and the outer toe is no longer than the middle toe.

Ecology and Habitats

BREEDING AND NONBREEDING HABITATS. In North America the breeding distribution of the ancient murrelet is restricted to an area of coastline in which the August surface water temperatures range from about 9°C to 14°C, while in Asia the breeding range extends south to where August surface water temperatures approach or reach 20°C. The birds are associated with rocky shorelines as well as sandy ones, but areas supporting rank growths of matted grasses are apparently preferred for nesting. Nesting also commonly occurs among rather dense growths of tall coastal forests, such as those of western hemlocks (*Tsuga heterophylla*) and other rain-forest types, in which the underbrush is rather sparse and a thick moss carpet covers the slopes, and nests are usually placed among and under tree roots. Typically nesting is done within a few hundred meters of the shoreline, usually on vegetated slopes and embankments or the slopes and tops of bluffs (Sealy 1976). During winter the birds favor inshore areas of the coastline, often foraging inside the belt of kelp that lies several hundred yards offshore and provides a breakwater for the surf. They also forage to some extent in the surf itself and occasionally may be found several miles out in the open ocean (Bent 1919). However, they apparently only rarely extend out beyond the limits of the continental shelf (Gould, Forsell, and Lensink 1982).

SOCIALITY AND DENSITIES. This is a highly social species; Nelson and Myres (1976) estimated that in the late 1960s and early 1970s perhaps as many as 50,000 birds were breeding along 1.6 kilometers of the Langara Island coastline. Earlier observations suggest that even higher densities occurred in the 1940s and 1950s, when populations may have been five to ten times higher than these more recent numbers. In Alaska an estimated 400,000 birds nest at 40 known sites, an average of 10,000 birds per site. The largest known colony there is of an estimated 60,000 birds at Forrester Island (Sowls, Hatch, and Lensink 1979).

PREDATORS AND COMPETITORS. Certainly the peregrine falcon is a serious predator of this species. On Langara Island the ancient murrelet is the peregrine's principal prey species, at least during the breeding season, and in recent years the sharp decline of the murrelet population has been paralleled by comparable declines in falcons and their productivity (Nelson and Myres 1976). In that area the introduced black rat (*Rattus rattus*) is the only mammalian predator of significance. The rat seems to concentrate on eggs but occasionally might take newly hatched nestlings and even possibly incubating adults. After hatching and during the early fledging of the chicks from their nests they are probably able to avoid most predation by the large gulls and the northwestern crow (*Corvus caurinus*) because of their nocturnal departure (Sealy 1976). Competition

with the marbled murrelet may occur to some degree, but as noted in the account of that species these birds utilize rather distinctly different foods, at least during the breeding season. Likewise, there may be local competition for nest sites with Cassin auklets and possibly also with storm petrels, but since the murrelets seem so highly adaptable to varied nesting sites, this is unlikely to be a serious problem.

General Biology

FOOD AND FORAGING BEHAVIOR. The best information on the foods of this species comes from the work of Sealy (1975a), who collected 61 adults and 30 subadults during the prebreeding and breeding season, as well as 9 newly fledged juveniles. The adult ancient murrelets apparently feed almost exclusively on the planktonic crustacean *Euphausia* early in the season, at least from late March through early to mid-April, after which they suddenly shift to *Thysanoessa*. In the first case the prey species are mostly less than 24 millimeters in length, while in the latter they are mostly larger. *Thysanoessa* continues to be an important food for adults through the summer, but during late May and June there is a significant addition of fish, primarily sand launce (*Ammodytes*) and secondarily viviparous sea perch (*Cymatogaster*) to their diets. Since ancient murrelets are not known to feed their young, this dietary change is not a reflection of parental feeding but instead is evidently an actual dietary shift, possibly brought about by increased availability of sand launce as they migrate to the ocean surface and begin to move toward shore at this time. Subadult ancient murrelets collected during that same period were also foraging on a mixture of *Thysanoessa, Ammodytes,* and *Cymatogaster,* together with small amounts of larval decapods and other minor prey types. The samples from newly fledged chicks indicate that the birds concentrate almost entirely on larval *Ammodytes*.

Sealy observed that both adult and subadult ancient murrelets fed in flocks, and although it occurred throughout the day, feeding seemed to be concentrated during the morning hours from about 6:00 A.M. to noon. Breeding individuals typically would spend 72 hours at sea foraging before gathering at a common staging area and then returning to the breeding colony, while those that had been incubating 72 hours would leave the colony by night and fly to the staging areas. In the morning some individuals would leave such staging areas in small groups of 4–12 individuals and fly directly to the foraging grounds, which were usually from at least 2 kilometers up to 15 kilometers offshore. Subadults were found to have a daily cycle similar to that of

23. Crested auklet, adults in breeding
plumage. Photo by author.

24. Crested and least auklets, adults in breeding plumage. Photo by author.

25. Least auklet, juvenal plumage.
Photo by Frank S. Todd.

26. Whiskered auklet, juvenal plumage.
Photo by C. Fred Zeillemaker.

27. Whiskered, least, and crested auklets, breeding adults. Painting by
Mark C. Marcuson.

28. Rhinoceros auklet, adult in breeding plumage. Photo by Frank S. Todd.

29. Tufted puffin, adults in breeding
plumage. Photo by author.

30. Atlantic puffin, breeding colony.
Photo by Frank S. Todd.

31. Atlantic puffin, adult in breeding
plumage. Photo by Frank S. Todd.

32. Horned puffin adults in breeding
plumage. Photo by author.

adults, but it was not learned whether they too had a 3 day foraging cycle. Although some subadults were observed foraging well away from shore, they also were regularly seen feeding in coves and bays near land. Sealy did not see any downy young feeding with adults on the foraging grounds, and his downy young were all collected in bay areas near shore. Observations of the chicks indicated that they can readily swim and dive, using their feet rather than their wings for propulsion, but their actual methods of feeding are still largely unknown.

MOVEMENTS AND MIGRATIONS. There are certainly some migratory movements in these birds. They seem to be rather rare during winter in the Gulf of Alaska, but some wintering occurs north to the southern Bering Sea and west into the Aleutians. During winter the birds regularly migrate south as far as California's southern coastlines, and stragglers have reached Baja California. There have also been an almost astonishing number of inland records for this species, occurring as far east as Ontario, Michigan, Quebec, and Louisiana. Munyer (1965) has summarized these and concluded that most relate to weather disturbances over the Pacific coast. Most of the records are from late October or November, suggesting that most sightings are made during the major fall migration. In northern California the birds arrive during November, at a time when surface water temperatures are declining, and they depart from California rather abruptly in March, when water temperatures are again increasing (Ainley 1976). When they arrive on their breeding areas at Langara Island the surface water temperatures are about 7°C, and these waters have risen to about 11°C in June as family groups begin to leave the colonies. The birds apparently begin to disperse widely after this departure, in part moving southward into areas of British Columbia where they are not known to nest. By mid-July, when the young are adult sized, they begin moving back into inshore waters, while the adults stay well offshore and undergo their postbreeding molt (Sealy and Campbell 1979).

Social Behavior

MATING SYSTEM AND TERRITORIALITY. Sealy (1976) classified age-groups as yearlings, 2-year-olds and adults and considered the first two groups "subadults." About 20 percent of the birds on the nesting colonies were found to be subadults; although they vocalized and engaged in aggressive behavior, Sealy was unable to learn if any of these birds bred successfully. He did determine that three pairs of birds banded one year were recaptured the next at the same nest site, suggesting that both mate retention and nest site tenacity prevail. Further, single members of two additional pairs used their same nests the following year, although the identities of their mates could not be determined. Territorial defense of nesting burrows is strongly developed, and a good deal of aggressive chasing occurred in conjunction with this, according to Sealy. Courtship probably occurs on the nesting slopes during the night, but details of pair formation are unknown.

VOICE AND DISPLAY. Vocalizations are only poorly described, but they include a whistling call note variously described as "rather shrill," "faint," "piping," and "low and plaintive" (Bent 1919). Sealy (1976) heard a "rasping" call uttered by an adult when it and two flightless young were closely approached by a boat.

Displays likewise are totally undescribed for this essentially nocturnal species. Sealy (1976) did not see a single copulation during some 300 hours of observation during the breeding season and suggested that perhaps it occurs in the nesting burrow at night. Vocalizations and probably also displays are common on the nesting slopes during the night throughout the prelaying and incubation periods, and apparent "play" in the form of short flights, chases, and simultaneous dives also occurs on the staging areas during this period, according to Sealy.

Reproductive Biology

BREEDING SEASON AND NESTING SUBSTRATE. Egg records for the Sanak Island area of southern Alaska extend from June 11 to July 28, with a peak in late June. Other southern Alaskan records are from May 1 to July 16, with a probable peak between May 20 and June 11 (Bent 1919). Studies by Sealy (1976) during two years on Langara Island of the Queen Charlotte Islands group indicated an overall range of clutch initiation from April 22 to May 25, with a peak during the last week of April and the first week of May. By mid-June 90 percent of the adults had left for sea with their chicks, indicating that nearly all hatching had occurred by mid-June, though there are apparently a few records of breeding lasting as late as mid-July. Sealy found no evidence of renesting in murrelets that had lost their clutches and considered that this was probably not a regular part of their breeding biology. He believed that sightings of apparently breeding birds on the nesting slopes in mid-July may have been of nonbreeding subadults. Of 151 nests he observed, most were along the tops and slopes of bluffs and were usually within 500 meters of shore. Burrows were under fallen trees, in the roots of standing trees, and in grass-covered talus slopes. In almost 90 percent of the nests there was an accumulation of salal (*Gaultheria shallon*) leaves and/or the needles of western hemlock as well as a grass lining. In nearly all the nests the birds incubated in total darkness. Typically the burrow is excavated to a length of up to 4 feet, and there is an apparent preference in British Columbia for sites under stones, roots, or fallen logs over nesting in grassy slopes (Drent and Guiguet 1961). However, on Sanak Island they have often been found nesting among rank grasses, under which a shallow cavity only a few inches deep is dug out and lined with dried grasses (Bent 1919).

NEST BUILDING AND EGG LAYING. Adult females producing their first eggs evidently do not visit the colony for about a week, or until the shell is fully formed and the egg is ready to be laid. They next appear about a week later and lay the second egg. Thus all initial burrow defense and nest preparation must be done by the males. These birds typically arrive at their nesting colonies on Langara Island in early April, about 2 weeks before the first eggs are laid. Thus there must be a minimum 2 week period of territorial establishment and burrow digging or renovation of old nesting sites. Arrival on the nesting slopes is closely associated with twilight, just as departure is closely associated with dawn, the first arrivals and last departures averaging about an hour after sunset and before sunrise. Of a total of 151 active nests Sealy observed, 147 had clutches of two eggs, indicating that deviations from two-egg clutches must be quite rare. These exceptions are typically one-egg clutches; the even rarer clutches of more than two eggs probably reflect the efforts of two females. In the case of 19 clutches, the intervals between the laying dates ranged from 6 to 8 days, averaging 7. The relatively large eggs of murrelets, averaging about 22 percent of the adult female weight in this species, probably accounts for the unusually long interval between the deposition of the two eggs (Sealy 1975b).

INCUBATION AND BROODING. Both sexes incubate, beginning after the second egg of the clutch has been laid. Both sexes share equally in the task, with 72-hour shifts, and the changeover invariably occurs at night. This is the longest incubation shift period known for any of the Alcidae. Once incubation has begun, the two birds are evidently never in the burrow together during the day. Since incubation begins with the second egg, hatching of the chicks is essentially synchronous, or at least occurs within about an hour. In Sealy's (1976) study, the second-laid egg hatched 34.9 days after it was laid, and the first-laid one 42.1 days after it was laid. Hatching is relatively prolonged, with the chick typically emerging 4 days after the first signs of pipping.

The weight of 26 newly hatched chicks averaged 30.7 grams, or 15 percent of average adult weight according to Sealy. He did not provide any estimates of hatching success.

GROWTH AND SURVIVAL OF YOUNG. The newly hatched young are highly precocial and are not fed by the adults between the time they hatch and their departure for the sea, which in Sealy's study averaged 2.2 days after hatching. The chicks normally hatch at night, and during their prefledging period they are brooded continuously, usually by the same adults. A small proportion of chicks might spend 3 or 4 days in the nest before departing, but since they are not fed during this time and suffer substantial weight loss of about 10 percent per day, there would be a significant penalty for remaining in the nest any longer than absolutely necessary. Upon leaving the nest the chicks have attained a body temperature nearly as high as that of adults, and so presumably the timing of departure is largely set by the attainment of this level of thermoregulation. There is apparently some variation in the chicks' being accompanied by their parents during nest departure. Sealy observed groups of up to 10 unaccompanied chicks as well as adults with young. On several evenings he saw young birds begin to scramble down to the sea before the nighttime arrival of adults from the staging areas, though presumably they were still in contact with the adults that had been brooding them. Essentially all of this departure is done during the hours of darkness; by dawn adults with young are not found within 10 kilometers of shore, indicating a truly remarkable mobility of the chicks so shortly after hatching. Sealy observed that the tarsi of newly hatched chicks are nearly as long as those of adults; thus the young are able to walk and swim very effectively at an early age as well as to dive using their feet for propulsion. It is unknown how many of these newly hatched chicks are taken by predators, but the combination of their nighttime departure and the probable "swamping" effect of so many prey appearing so suddenly and being available for such a relatively short time (about a month's maximum duration) may reduce the effectiveness of predators. Very little information is available on the postfledging behavior and movements of these birds. Family groups of adults with still-downy young have been observed as far as 30–40 miles offshore, although relatively few sightings have been made. It is thus likely that substantial dispersal of families occurs after the breeding areas are vacated, and during the period of chick growth the adults undergo their postnuptial molt. Sealy and Campbell (1979) summarized the available data on observations of family groups; of 20 such sightings 14 were of no more than 2 adults and 1 or 2 downy young. Of these, 11 cases had 2 adults in attendance and 3 had a single adult. In all, there were 1.3 young per adult in the groups, and 1.8 young present per group. These data support the notion that biparental care is typical following the nest departure and that there is little gregariousness among fledged families of birds.

BREEDING SUCCESS AND RECRUITMENT RATES. Nothing of a substantive nature can be said about nesting success or fledging success in this species, or about possible recruitment rates.

Evolutionary History and Relationships

The nearest relative of the ancient murrelet is the Japanese murrelet, which differs only slightly from it in appearance, primarily by having a distinct head crest. What is known of the biology of the Japanese murrelet also strongly suggests the two species have nearly identical breeding behavior and ecology (Thoreson, in press). Kozlova (1961) judged that these two species of the genus *Synthliboramphus* had their origin in Pacific Ocean waters and, together with *Brachyramphus*, evolved from a *Cepphus*like ancestor. The genus is also obviously a close relative of *Brachyramphus*, in which the legs are even shorter and the pelvis is even broader than in this genus, presumably reflecting differential walking abilities.

Population Status and Conservation

The breeding population of Alaska has been estimated as 400,000 birds (Sowls, Hatch, and Lensink 1978), though this figure reflects the considerable uncertainty inherent in censusing a nocturnal species that is dispersed for most of the year. There may be as many as 190,000 pairs in at least 30 breeding colonies in British Columbia (Sealy and Campbell 1979); in both southeastern Alaska and British Columbia the directional trend of the population is still unknown (Manuwal and Campbell 1979). At least in Alaska the species has been seriously affected by the introduction of arctic foxes into islands where the birds formerly bred in large numbers, and thus it is likely that the general trend has been downward. Likewise in British Columbia some colonies are a tiny fraction of their earlier numbers for reasons that are still unknown but might be related to changes in planktonic density in recent years (Nelson and Myres 1976). The species has been assigned an oil vulnerability index of 74, below the average calculated for the Alcidae (King and Sanger 1979).

Cassin Auklet

Ptychoramphus aleutica (Pallas)

OTHER VERNACULAR NAMES: Aleutian auk; sea quail; starique de cassin (French); Dunkelalk (German); aleutskiy lyzhik (Russian); alcuela nortamericana (Spanish).

Distribution of North American Subspecies (See Map 20)

Ptychoramphus aleutica aleutica (Pallas)
BREEDS on the Sanak Islands, Shumagin Islands, and locally in the Aleutians, south to San Geronimo and San Martin islands, Baja California, and Guadalupe Island.

WINTERS from southern Alaska to northern Baja California.

Ptychoramphus aleutica australe van Rossem
BREEDS off the west coast of Baja California from the San Benito Islands south to Asunción and San Roque islands.

WINTERS within the breeding range.

Description (Modified from Ridgway 1919)

ADULTS (sexes alike). Upperparts plain grayish dusky or dull blackish slate, inclining to dull brownish slate on hindneck and postocular region, the rump and upper tail coverts more decidedly slaty, the back and scapulars tinged with the same; a whitish spot above the upper eyelid and a less obvious one on the lower eyelid; malar and rictal regions, chin, throat, and foreneck plain brownish gray (between quaker drab and deep mouse gray); under wing coverts brownish gray, some of the larger coverts grayish white; rest of underparts immaculate white; bill black, the basal third of mandible yellowish or flesh colored; iris white; legs and feet bluish and dusky. This plumage is initially attained in the third year of life (Manuwal 1972).

JUVENILES. Similar to adults in coloration, but with a whitish throat and a paler breastband and generally less blackish throughout; iris brown.

DOWNY YOUNG. Upperparts deep sooty grayish brown, the sides and chest similar but paler; chin, throat, breast, and abdomen dull grayish white. The iris is brown, and the legs and feet are pink. There is a bare area of skin around the eyes. By 10 days the legs darken to grayish black, with a light blue cast on the tarsus and upper toes (Thoreson, in press).

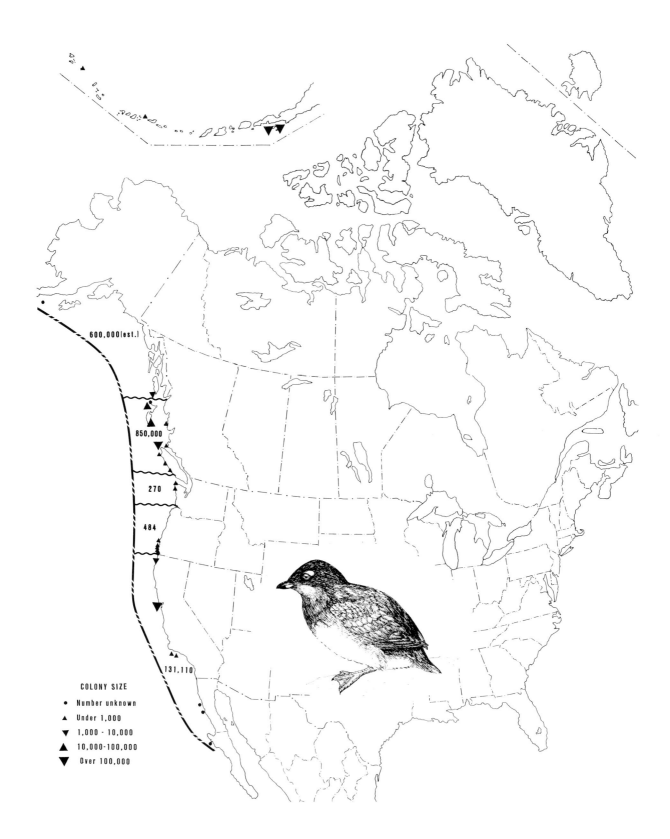

COLONY SIZE

• Number unknown
▲ Under 1,000
▼ 1,000 - 10,000
▲ 10,000-100,000
▼ Over 100,000

600,000(est.)

850,000

270

484

131,110

20. Current inclusive distribution of the Cassin auklet, including colony locations and limits of nonbreeding range.

Measurements and Weights

MEASUREMENTS. Wing: males 109.5–129.0 mm (average of 8, 120.7); females 120–22 mm (average of 3, 121). Exposed culmen: males 18.5–20.0 mm (average of 8, 19.3); females 18.5–19.0 mm (average of 3, 18.8). Eggs: average of 60, 46.9 x 34.3 mm (Bent 1919).

WEIGHTS. In winter and spring, males 146–209 g (average of 15, 173.0 g); females 141–83 g (average of 10, 164.5) (museum specimens). An unsexed breeding sample of 25 birds averaged 188 g (Vermeer and Cullen 1982). The calculated egg weight is 28 g (Schönwetter 1967). Newly hatched chicks average 17.8 g (Manuwal 1972).

Identification

IN THE FIELD. This little auklet is fairly common in offshore areas south of Canada and appears almost uniformly blackish while swimming, except for white eyes and a pale spot at the base of the lower mandible. In young birds the throat is somewhat whitish, and their bills are smaller. There are no significant changes during winter. Their calls resemble the creaking of a rusty gate or a chorus of frogs.

IN THE HAND. This small auklet has a wing length of less than 130 mm and a bill that is tapered, subconical, and wider than high at its base, without any special ornaments or bright coloration at any time. In all plumages the birds are generally slate gray, with whitish spots near each eyelid, though only the upper one is conspicuous. The species is difficult to sex externally, but any bird with a bill depth of 10.3 millimeters or more is likely to be a male (Nelson 1981).

Ecology and Habitats

BREEDING AND NONBREEDING HABITATS. This species breeds on rocky to sandy coastlines of North America, mostly in grassy or other nonforested habitats, on both flat and sloping terrain and sometimes several hundred meters from the coast. Substrates that provide preexisting cavities or those that the birds can easily dig into are used for nesting. The breeding distribution extends from a zone where August surface water temperatures range from about 10°C in the north to 20°C in the south, with peak densities near the colder limits. In the winter the birds are primarily found well offshore in the open ocean, sometimes up to 50 miles from the coastline.

SOCIALITY AND DENSITIES. Triangle Island in British Columbia, which has a total area of little more than a square kilometer, had a breeding population of approximately 359,000 pairs in 1977, representing an almost incredible density of roughly 7,400 birds per hectare or 0.74 breeding bird per square meter. Obviously the breeding birds were distributed nonuniformly, with minimum densities of 0.1–0.19 pair per square meter typical of the shrubby central plateau area that was mostly covered with salmonberry (*Rubus spectabilis*), salal (*Gaultheria shallon*), and Pacific crabapple (*Malus fusca*). Densities were highest (up to 1.1 pairs per square meter) on an open summit area dominated by herbaceous forms such as saxifrage (*Saxifraga newcombeii*), brome grass (*Bromus sitchensis*), licorice fern (*Polypodium vulgara*), and spiny wood fern (*Dryopteris austriaca*). Breeding densities were also high in other vegetationally similar areas where there was short herbaceous vegetation and low, wind-pruned salmonberry interspersed with bare ground. Manuwal (1974b) found the highest breeding densities (1.09 burrows per square meter) in vegetated depressions and similar areas with soft soil substrates, and the lowest densities (0.02 burrow per square meter) on rocky plains with shallow soil. His overall burrow density figures for all habitat types on Southeast Farallon Island averaged about 0.13 burrow per square meter, and the total population of the 37 hectare island had an extrapolated average density of 2,850 breeding birds per hectare.

PREDATORS AND COMPETITORS. At least at the northern limits of the range in the Aleutian Islands the introduced arctic fox (*Alopex lagopus*) is a very serious predator. It can easily dig out Cassin auklets' burrows in their usual soft soil substrates, and apparently many areas that once supported the species have virtually lost them to arctic foxes (Sowls, Hatch, and Lensink 1978). Farther south introduced rats (*Rattus* spp.) are known to be damaging to nesting birds (Sowls et al. 1980), and western gulls are sometimes significant predators on fledging chicks (Thoreson 1964). It is very likely that the larger owls take adult birds, but the auklet's essentially nocturnal behavior probably helps to reduce predation from diurnal aerial predators. Nevertheless, peregrines (*Falco peregrinus*) have been mentioned as serious predators near auklet colonies on the Coronado Islands (Bent 1919), though the species has since been extirpated from that area.

Competitors of the Cassin auklet include other cavity-nesting birds that use similar crevices or burrows. Manuwal (1974b) listed among these the tufted puffin, the pigeon guillemot, and two storm petrels (*Oceanodroma homochroa* and *O. leucorhoa*). Of these, only the storm petrels are small enough to use auklet burrows, and typically they nest in crevices with openings too small for auklets to pass through. Farther north the auklets nest in company with the rhinoceros auklet, but this species is limited to nesting on slopes, and so competition between the two is limited to such areas. Where puffins and rhinoceros auklets nest in company with Cassin auklets, the latter occur in low densities, perhaps as a result of displacement through the extensive burrowing of these larger species (Vermeer et al. 1979). Cassin auklets probably nest in close proximity to some of the other small auklets as well, but these are almost exclusively rock crevice and cavity nesters. The ancient murrelet overlaps substantially with the Cassin auklet in its breeding range, and it seems quite likely that some competition for nest sites occurs between these species.

General Biology

FOOD AND FORAGING BEHAVIOR. This species' major breeding season prey in California waters consists of euphausiid crustaceans (especially *Thysanoessa spinifera*), hyperiid amphipods (*Phromema*), larval squids, and the megalop stage of a decapod crab, all of which are primary components of micronekton (Manuwal 1972). Studies in British Columbia (Vermeer and Cullen 1982) indicate that there the birds rely, at least during the breeding season, on copepods (especially *Calanus*), euphausiids, amphipods, and various small fish; the fish are primarily taken near the end of the breeding season as plankton populations decline. Their planktonic prey ranges in size from 6 to 30 millimeters, and their fish prey from 15 to 45 millimeters. The species' heavy reliance on *Thysanoessa spinifera* in southern California waters may relate to its being the only euphausiid around the Farallon Islands that is abundant in surface waters during daylight hours. The same species is also a major food of Cassin auklets off Vancouver Island (Payne 1965).

This species, together with the crested, least, and parakeet auklets, has a sublingual pouch adapted to food storage in adults that must carry foods back to their developing chicks. Since foraging occurs diurnally and the chicks are fed at night, this pouch allows the birds to store foods for from 24 hours to possibly as long as 36 hours before passing it on to their chicks. The pouch occurs in adults of both sexes and apparently develops at the time of the first breeding attempt, reaching its greatest length at or near the time the young fledge. A bird with a full pouch can carry as much as 35 grams of material (or nearly 20 percent of adult body weight), although the average meal size is about 25 grams (Speich and Manuwal 1974). Vermeer and Cullen (1982) reported a slightly smaller (17.6 grams) average meal size for birds in their study area.

MOVEMENTS AND MIGRATIONS. There is little direct evidence of any major migrations in this species, though the winter distribution of the rather large Alaskan breeding population is still unknown. The birds are common in the Gulf of Alaska from April through November but either are lacking in winter or at least have not yet been found in large numbers (Gould, Forsell, and Lensink 1982; Islieb and Kessel 1973). They may winter in small numbers off the west coast of Vancouver Island, but they are highly pelagic at that time and are easily overlooked. It seems likely that the adult birds disperse well offshore in areas rich in zooplankton and probably move only the minimum distance necessary from their breeding colonies.

Social Behavior

MATING SYSTEM AND TERRITORIALITY. This species is monogamous, with pair bonding lasting at least up to 3 years (4 known instances) and possibly permanently. At least in California, where the birds are resident, a pair may periodically visit their nesting burrows throughout the year, and there is a high level of nest site tenacity. Seven of 16 birds from the apparently excess potentially breeding ("floater") population that attained nesting sites in one year were found back in the same burrow the following year. Additionally, at least some fledged young returned to within 100–150 meters of their hatching site when they matured (Manuwal 1972, 1974a). The birds are highly territorial in spite of their propensity for dense nesting, and territorial defense typically centers on the burrow entrance. If both members of the pair are present near the burrow, the mate is also defended. However, it is still not certain that the male is the primary defender, since there is no sexual dimorphism in color and very little in weight. Territorial defense varies directly with population density and with the reproductive cycle. Much of this defensive behavior is directed toward "floater" birds that are seeking available nest sites. Manuwal judged that slightly over half these "floaters" were adults (3 years old or older), while the rest were yearlings and 2-year-olds. (Manuwal 1972, 1974b).

VOICE AND DISPLAY. Adults on their breeding grounds are highly vocal; Thoreson (1964) recognized at least ten different variants of their calls. During mating, greeting, and other social activities the sounds made resemble the chirring of katydids or crickets. The individual *kreet* notes are sometimes trilled and vary somewhat in duration, pitch, and frequency of utterance. Communal chorusing is common, with sudden changes in rhythm and in intensity of trilling. Paired birds not only perform mutual trilling calls but also utter some twittering notes, while fighting individuals produce more growling sounds. The warning call is a loud *kreer*, and during greeting ceremonies in the nest chamber a series of gutteral notes are uttered. During calls the throat is puffed out considerably (the sublingual pouch possibly serving as a resonating chamber), and the entire body vibrates.

Display postures have also been described by Thoreson (1964), whose sketches provide the basis for figure 43. According to him, paired birds perform various recognition and greeting movements. Probably one of the more important of these is billing (fig. 43A), during which mated birds nibble one another's bills and the feathers at the base of the bill, while uttering repeated *krr* or *chirr* notes. Two other displays of mated pairs are circling and passing. During circling, one member of the pair moves partly around its mate to a facing or mounting position, occasionally raising the wing nearest its mate while doing so (fig. 43D). The wings may also be raised during antagonistic encounters, such as when a bird is being threatened or pecked by a neighbor (fig. 43C). In threatening another bird, the bill is directed toward it and the back feathers may be ruffled (fig. 43F). During passing, one of the pair members gets up from a sitting position and moves ahead of its mate, half-running and half-hopping (fig. 43B). This action may in turn be performed by its mate, producing a kind of leapfrogging extending as far as 5–6 meters. This behavior often terminates with one member of the pair's turning to perform billing with the other, or with both birds' flying back to the original starting place. Head bowing, head bobbing, and head waggling movements also are extremely common. In head bowing (fig. 43G) the head is rather quickly lowered to a nearly vertical position and then returned to the horizontal. In head bobbing (fig. 43H) the vertical movements are less extreme but also rapid. During head waggling (fig. 43I) the head and bill are moved laterally. Head bobbing and head waggling are both commonly performed by paired birds.

One observation of copulation was made by Thoreson (1964). This occurred at night, among a group of more than 50 birds, on a rock he had seen defended by a pair. Treading was preceded by head bowing, head waggling, billing, twittering calls, and intermittant *kreek* notes. Then one member of the pair squatted and the other mounted. Postcopulatory behavior included mutual billing and head waggling as well as probable comfort movements such as wing flapping and feather rustling. This is the only fairly detailed description of copulation yet available for any auklet, and it will be of interest to learn if it or comparable behavior (nocturnal terrestrial copulation outside the nesting chamber on a regular resting site) is typical of the other auklets as

43. Social behavior of the Cassin auklet (after Thoreson 1964): A, billing; B, passing; C, attacking; D, wing raising; E, regurgitation of food; F, threat; G, head bowing; H, head bobbing; I, head waggling.

well. Certainly it is rather similar to the situation in guillemots, where copulation often occurs on a defended resting rock near the nest site.

Reproductive Biology

BREEDING SEASON AND NESTING SUBSTRATE. Egg records for the Farallon Islands extend from April 3 to July 20, with a peak during the first half of June (Bent 1919). According to Thoreson (1964), fresh eggs have been seen on the Farallon Islands until mid-August and possibly even to late November. Lower California records are from March 10 to June 8, those from the Santa Barbara Islands are from May 16 to June 29, and those from the Sanak Islands (Alaska) are from June 6 to July 3 (Bent 1919). A few egg records from British Columbia are from April 18 to June 1, and nestlings have been seen between June and August. Manuwal (1979) observed that during two successive years the duration of laying initial clutches on the Farallon Islands was 47 and 65 days, and there were also overlapping periods of replacement egg laying that lasted 46 to 94 days. Finally, in one of the two years there was a 53 day period during which second clutch (double-brooding) eggs were laid. In that year there was a total period of 134 days when eggs were known to be present. This period generally corresponded to times of high availability of zooplankton in the adjoining waters. During the two years of the study an average of 13 percent of the pairs laid replacement eggs when their first eggs were lost, and 4.5 percent laid second clutches following successful rearing of a chick that same season. Thus this is the only species of alcid for which double-brooding has been proved, although it has been suggested as possible for Xantus murrelets, based on their equally long egg season (Bent 1919).

The nesting substrate is quite variable, but the highest nesting densities were found by Manuwal (1972, 1974a) in vegetated depressions, railroad beds, and grassy plains with deep soil, all of which allow easy burrowing. Some nesting also occurred in rock cavities or crevices, such as in talus slopes, rock walls and piles, and bare rock outcrops. However, nest site temperatures are more stable in sod burrows than in rocky sites, which might help explain the species' preference for soil sites over rock crevices or cavities. In all, about half of the nests Manuwal found in the Farallon Islands were in rock crevices and the rest were in burrows, apparently because of the limited number of preferred burrow nesting sites. Vermeer et al. (1979) reported that slope angle, density of tufted hairgrass (*Deschampia caespitosa*), density of rhinoceros auklet burrows, and altitude were all statistically significant as factors influencing burrow densities in British Columbia. Thus the birds prefer nesting in short vegetation such as hairgrass and away from rhinoceros auklets. Both the slope angle and the altitude factors may be related to the fact that the birds evidently avoid nesting near rhinoceros auklets, which are limited in their nest sites to steep slopes from which they can easily take flight. Excluding this factor, it may be said that the birds preferred to nest in open and short vegetation on all slopes and at all elevations, at least up to 100 meters. On forested islands the birds typically nest at the edges of the forests, burrowing in moss- or grass-covered ground.

NEST BUILDING AND EGG LAYING. This species typically excavates its own burrows in soft soil. Thoreson (1964) reported that excavation of new burrows and repair of old ones may begin as early as December, with both members of the pair actively taking part. Most burrows begin at the base of a solid object, such as a tree root, and they may vary from less than 2 feet to about 4 feet (0.7–1.2 meters) in length, depending on location and the hardness of the soil. Digging is done only at night, and a burrow may require at least 3 months to complete. Tunnels of separate pairs do not intersect,

and blind side branches are only exceptionally present, dug mainly when a replacement egg is to be laid after losing the first. About 24 percent of burrows checked in early February (a month before earliest egg laying in the Farallons) contained apparently paired birds, suggesting a fairly early occupancy of available burrows. Although resident pairs in a colony do not appear to lay their eggs with any clear degree of synchronization, Manuwal (1974a) mentioned that one "floater" bird that had occupied a vacated burrow the day before was already incubating an egg, suggesting that at least some "floaters" would nest if they could find nest sites. Indeed, Manuwal determined that from 37.5 to 70.0 percent of the "floaters" that he provided with nest sites (by removing original territory owners) laid eggs, though their reproductive success rates were quite variable during the two years of his experiment. Eggs are normally laid near the back of the burrow, and incubation typically begins almost immediately. However, Manuwal (1974a) noted that about 8 percent of the eggs laid are not immediately incubated, and indeed most of these are never incubated, for reasons he could not explain. Only one egg is laid, even though the birds have two well-developed incubation patches.

INCUBATION AND BROODING. Both members of the pair incubate, typically shifting incubation duties every 24 hours. Manuwal (1974a) determined an average incubation period of 37.8 days, with a range of 37–42 days, for 86 eggs. One female that lost her mate continued to incubate alone for 10 days, incubating the egg every other day, before she finally deserted it. Egg losses are fairly high during the preincubation and incubation stages; Manuwal (1979) found that 26.2 percent of 664 first-clutch eggs were unsuccessful, and there were higher rates of egg loss in replacement clutches and attempted second nestings. Collectively there was a 28.8 percent mortality of eggs in all nests studied, with clutches laid before the mean laying date more successful than later ones.

GROWTH AND SURVIVAL OF YOUNG. The parents alternate brooding their chick for the first 5–6 days after hatching, and it is then left alone in the burrow while the adults forage, returning each night to feed their chick. The average fledging period for 16 chicks was reported as 41.1 days (range 35–46) by Manuwal (1974a) and 44.7 days (range 41–50 days for 17 birds) by Thoreson (1964). By the time the young are 37–38 days old they reach their maximum rate, after which they begin to lose weight at the rate of 2.5 percent per day until they fledge. The chick is fed once a night by each of its parents, who regurgitate food stored in their sublingual pouches (fig. 43E). Toward the end of the fledging period the intensity of parental feeding declines, but some adults may feed their chicks up to the time of fledging, and others may even continue to return to the nest with food after the chick has left it. Evidently the parents do not always accompany their fledged chicks, since Manuwal found many pairs back in their nest sites several days after their chicks had fledged. The chicks make several short flights before the extended flight that occurs at fledging. At this time the birds are vulnerable to predation by western gulls, though some gulls will also pull auklet chicks from burrows or even take adults that are nesting in shallow burrows. Nevertheless, most chick mortality evidently occurs close to the time of fledging, especially as a result of gull attacks.

BREEDING SUCCESS AND RECRUITMENT RATES. Thoreson (1964) found that 26.6 percent of the pairs in 75 nests succeeded in rearing a chick to fledging. Of 664 first-clutch eggs, Manuwal (1974a) estimated that 60.1 percent resulted in fledged chicks, while the breeding success of 91 replacement eggs was 52.7 percent and that of 97 second-nesting eggs was 15.51 percent. Besides the reproductively active component of this population there is a second "floater" component of surplus potential breeders that was estimated by Manuwal (1974b) to be about 25 percent of the total population. About half of these are adults, and 80 percent have had no previous breeding experience. These "floaters" are able to enter the breeding population as established breeders die or they become otherwise able to obtain nest sites. Manuwal was not able to judge mortality rates of the "floater" component but judged the annual mortality rate of breeders to be about 19 percent per year.

Evolutionary History and Relationships

Judging from an analysis of leg and wing musculature, this genus may not be very distantly removed from the puffin group (Hudson et al. 1969). Storer (1945) detected no major differences between the hind limbs of *Ptychoramphus* (which typically digs its own nest burrow) and those of the other auklets, all of which nest in rock cavities or crevices. He judged it to be the most primitive of the auklets based on its lack of specialization of the bill rhamphotheca and its absence of head plumes. Certainly the auklets and puffin group form a fairly homogeneous assemblage, with a progressively higher development of digging adaptations apparent as one proceeds from the typical auklets through the Cassin auklet and the rhinoceros auklet to the puffins. I believe the Cassin auklet is perhaps best retained as a monotypic genus, which best expresses its somewhat uncertain degree of relationship to the other auklets and the puffins.

Population Status and Conservation

In California the Cassin auklet has a variable conservation status, having disappeared from some historical breeding sites, but it now may be more common on the Farallon Islands than at any other known time in history (Sowls et al. 1980). This habitat is now essentially saturated, judging from the high percentage of "floaters" in the population, and the high philopatric tendencies of the birds probably reduce chances of range extension. The birds are still abundant on the coast of British Columbia, but the population trends there as well as in western Washington are unknown (Manuwal and Campbell 1979). In Alaska the birds have suffered greatly from introduced arctic foxes in some areas, but at present, with the foxes gone from many islands, they may be recovering (Sowls, Hatch, and Lensink 1979). The species was assigned an oil vulnerability index of 84, one of the highest of all the alcids (King and Sanger 1979), and certainly in southern California the danger of oil pollution poses a serious threat to its continued survival (Sowls et al. 1980).

Parakeet Auklet

Cyclorrhynchus psittacula (Pallas)

OTHER VERNACULAR NAMES: Baillie brushkie (Aleutians); starique perroquet (French); Rotschnabelalk (German); umiomu (Japanese); belobryushka (Russian); sukluruk (Saint Lawrence Island).

Distribution of Species (See Map 21)

BREEDS from the Sea of Okhotsk to the vicinity of Kolyuchin Bay, northeastern Siberia, and from the Diomede Islands, Fairway Rock, the Commanders, the Pribilofs and Sledge, Saint Lawrence, and Saint Matthew islands south to the Aleutian Islands and to Prince William Sound, southern Alaska.

WINTERS from the Bering Sea south to Sakhalin Island, the Kuriles, and Honshu, and to the coast of Alaska; much more rarely south to California.

Description (Adapted from Ridgway 1919)

ADULTS IN BREEDING PLUMAGE (sexes alike). Upperparts plain dull slate blackish, gradually passing into dark hair brown or grayish fuscous on chin, throat, and foreneck (sometimes chest also), the sides and flanks uniform grayish fuscous; rest of underparts immaculate

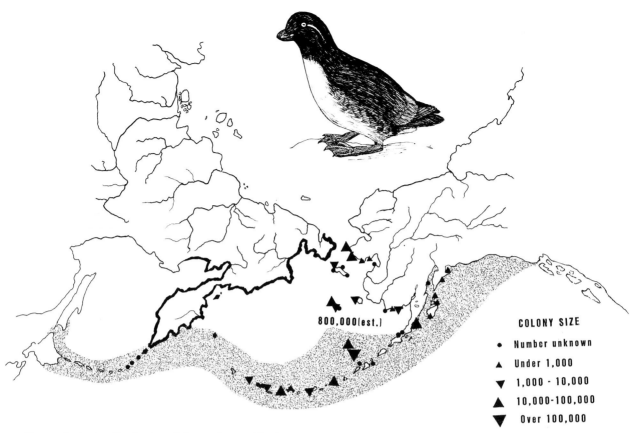

21. Current inclusive distribution of the parakeet auklet, including colony locations (Alaska) and general breeding range (Asia). The wintering areas are shaded.

white, the chest, however, usually more or less clouded with grayish fuscous; whole undersurface of wing plain grayish brown (between hair brown and fuscous), some of the larger coverts (sometimes, at least) with a narrow shaft streak and small terminal spot of grayish white; elongated pointed plumes extending in a line from lower eyelid backward and downward across auricular region, white; bill orangy red or salmon, the nasal shield dark horn color, the tomial tumor pale flesh color; iris white; legs and feet pale bluish gray or bluish white, the side of tarsus and toes blackish, webs blackish centrally, and joints of toes dusky; interior of mouth whitish. The auricular plumes and bill plates are probably initially acquired during the second summer after hatching (Bedard and Sealy 1984), but the definitive breeding plumage is apparently not attained until the third year.

WINTER PLUMAGE. Similar to the summer (breeding) plumage, but throat, foreneck, sides, and flanks white, like rest of underparts, or partly so, and white auricular

plumes reduced. Several accessory bill ornaments are also lost during the nonbreeding period.

JUVENILES. Similar to the winter plumage, but bill smaller and duller red (inclining to brown) and entire underparts, including throat, and foreneck, immaculate white. The iris is bluish gray, the bill pale black, the mouth cavity yellow, and the tarsi are bluish above and gray below (Bedard and Sealy 1984).

DOWNY YOUNG. Above uniform deep smoky gray or sooty grayish brown, the chin, throat, and chest similar but paler; rest of underparts pale brownish gray. The iris is black, the bill gray, and the tarsi and feet are gray, becoming darker below (Bedard and Sealy 1984).

Measurements and Weights

MEASUREMENTS. Wing: males 144.5–152.0 mm (average of 10, 148); females 140.5–152.0 mm (average of 10, 145.8). Exposed culmen: males 14.5–16.5 mm (average

222

of 10, 15.2); females 13.0–15.5 mm (average of 10, 14.2) (Ridgway 1919). Eggs: average of 33, 54.3 x 37.3 mm (Bent 1919).

WEIGHTS. The average of 7 breeding males was 317.6 g (Bedard 1969a), and 17 unsexed adults averaged 280 g (Sealy 1968). The average weight of 3 eggs was 37.5 g (Sealy 1968); calculated fresh weight is 42 g (Schönwetter 1967). Newly hatched young average 28.1 g (Sealy 1968).

Identification

IN THE FIELD. The bill of this species serves to identify it in any plumage. It is short and stubby, with the lower mandible more strongly upcurved than the upper mandible is decurved, producing a nearly circular profile. In the breeding season the bill is orangy red, and the black face has a white stripe extending back from the eye along the side of the neck. In winter the white stripe is lacking, the bill is blackish, and the body is distinctly bicolor, with white flanks and underparts and a uniformly dark head and upperparts. The call is a clear, vibrating whistle, uttered on the breeding grounds.

IN THE HAND. This fairly large auklet (wing 140–52 mm) can be identified by the unique circular bill, which has a depth about equal to its length and an upper mandible that is strongly convex both in the culmen profile and in the profile of the cutting edge. The lower mandible thus has a unique concave cutting edge, and its tip is sharply pointed and strongly recurved.

Ecology and Habitats

BREEDING AND NONBREEDING HABITATS. Breeding in this mostly high-arctic species occurs along rocky coastlines of the Bering Sea, primarily within the area where surface ocean temperatures in August range from 5°C to 10°C. Nesting is mostly confined to the rocky crevices of cliffs and cavities of associated talus and scree accumulations near the sea itself but may possibly extend locally a very short distance inland on talus- or scree-covered slopes. During the nonbreeding season the birds are pelagic, possibly overwintering far from their breeding colonies in subarctic to temperate or possibly even subtropical oceanic environments (Gould, Forsell, and Lensink 1982).

SOCIALITY AND DENSITIES. Little specific information is available on colony densities in this species, but an estimated 150,000 birds breed on Saint George Island in the Pribilofs, which has about 3.5 square kilometers of coastal cliff habitat, representing a density of 43,000 birds per square kilometer. Similarly, about 34,000 nest on Saint Paul Island, which has about 0.5 square kilometer of cliff habitat, or 68,000 birds per square kilometer. Finally, some 20,000 birds nest on Little Diomede Island, a tiny rocky promontory off the west coast of the Seward Peninsula. Under such conditions the breeding density might approach 100,000 birds per square kilometer of nesting habitat. In general, however, the birds are somewhat less colonial than the smaller auklets, since they are more prone to nest in crevices in cliffs and thus perhaps somewhat less restricted to the cavities associated with talus slopes. Searing (1977) estimated an average density of 30 birds per hectare on a cliffside habitat but noted that most were nesting under large boulders near the top of the cliff and thus were excluded from the estimated densities.

PREDATORS AND COMPETITORS. Although little specific information is available, it is likely that many of the same avian and mammalian predators affect the parakeet auklet as mentioned for the *Aethia* species. Sealy (1968) indicated that cliff-nesting individuals are probably less vulnerable to such predators as arctic foxes (*Alopex lagopus*) than are those that nest in talus slopes, but as with the *Aethia* species there seems to be a significant loss of both eggs and nestlings to voles (*Microtus* and *Cleithrionomys*).

General Biology

FOOD AND FORAGING BEHAVIOR. Bedard (1969a) has provided the most complete information on the foods of this species, basing his conclusions on the analysis of 12 samples from the gullets of birds collected before hatching time and from 85 neck-pouch samples obtained during the chick-raising period. The early summer samples, as well as later ones, had a preponderance of carnivorous planktonic forms of crustaceans such as hyperiids and pteropods as well as various larval fish (probably mostly Ammodytidae and Cottidae) and cephalopods. Additionally, the birds took a higher proportion of larger plankton than did either the least or the crested auklet. The appearance of bottom-dwelling flatfishes and Mysidacea in the samples suggest that the upturned mandibles may be used for scooping up foods from the bottom or near-bottom levels, though some surface-dwelling forms also appeared in the samples. Hunt et al. (1980) analyzed 55 throat-pouch samples taken in the Pribilof Islands area and found a high use of euphausiids and polychaetes. Fish remains amounted to 26 percent of the contents by volume and consisted mostly of walleye pollock (*Theragra*). In both areas the species evidently ate a wide array of plankton, invertebrates, and fish larvae and as such was able to utilize both oceanic and neritic waters.

MOVEMENTS AND MIGRATIONS. It is possible that this species undergoes a greater migration than the other auklets, judging from the numbers of birds that have washed ashore or been seen along the coasts of the western states from Washington to California. There are no winter records from the Gulf of Alaska and few from southeastern Alaska or British Columbia, so it is possible that wintering occurs in oceanic areas farther to the south (Gould, Forsell, and Lensink 1982). They have been observed around the Pribilofs as late as December and as early in spring as February and March, suggesting that at least some birds probably overwinter in the vicinity of breeding areas (Preble and McAtee 1923).

Social Behavior

MATING SYSTEM AND TERRITORIALITY. Sealy (1968) judged that parakeet auklets "probably" retain their mates from year to year, but he was unable to follow banded birds for more than one year. He judged that nest site tenacity does occur, since a pair displayed on snow directly over a nest crevice that had been used the previous year and occupied it a few days later when the crevice became snow-free. Manuwal and Manuwal (1979) were unable to identify definite territorial behavior and judged that any territory that is defended is very small. As in the other auklets, it is probably limited to the defense of the nest site itself. Lehnhausen (1980) only rarely observed aggressive interactions among parakeet auklets, and Manuwal and Manuwal observed threats in conjunction with defense of individual distance on both shore and water but rarely observed direct combat.

VOICE AND DISPLAY. Manuwal and Manuwal (1979) have described a flight intention call uttered from land or water at the approach of danger or at other times before flight. Similar calls have been observed in the Cassin auklet and rhinoceros auklet. Thoreson (in press) has also mentioned that the birds utter a loud, trilled *chil, chi, chi, chi-chirrrrip* note and other calls. The usual trilled call is rather musical and tends to rise in pitch. He noted that a mated bird would often land near the nest site and uttered trilled or warbled songs until its mate arrived, after which the two would duet and bill momentarily. If an intruder appeared the birds would raise and open their bills, utter trilled chirring notes, and wave their heads toward the intruder.

Displays have so far been only very poorly described for this species. During limited observations on the Pribilof Islands, I observed repeated billing ceremonies be-

44. Social behavior of the parakeet auklet (after photos by author): A, resting posture; B, C, trilled call postures; D, bill raising threat; E, billing with wing raising; F, billing without wing raising.

tween apparent pairs (fig. 44E,F), during which the birds would face each other, utter duetting calls, and pass their bills back and forth in front of one another, sometimes touching. On some occasions the wings would be partially raised during this ceremony. During probable threats the birds assumed a relatively erect posture, with the bill pointed upward nearly vertically (fig. 44D). Manuwal and Manuwal (1979) recognized two stages of this display, with a low-intensity form that simply involved neck stretching but not bill raising. They also observed water chasing, involving one bird's lunging toward another on the water surface. At times single birds of unknown sex will utter a call from the perch site. During this call the neck is extended forward, the bill is opened, and the head is tilted upward briefly (fig. 44B,C), while the neck is somewhat enlarged and possibly the throat pouch is expanded. No descriptions of copulation are available, but Manuwal and Manuwal (1979) observed two attempted copulations on water following intensive duetting between pairs. They also noted that duetting occurred primarily but not exclusively between paired birds.

Reproductive Biology

BREEDING SEASON AND NESTING SUBSTRATE. Actual egg records for this species are not very numerous. Sealy and Bedard (1973) judged that on Saint Lawrence Island egg laying during one season extended from June 21 to July 7, with a mean egg-laying date of June 23. Most other information also indicates that egg laying occurs during June. Hunt, Burgeson, and Sanger (1981) estimated that on the Pribilof Islands egg laying probably occurred the third week of June and hatching the last week in July. Sealy and Bedard similarly estimated that the hatching period on Saint Lawrence Island extended from July 24 to August 3 and the fledging period from August 29 to September 7. The best description of the nesting substrate is that of Lehnhausen (1980), who examined 15 nest sites. Of these, 13 were in rock and boulder slopes between high-tide line and the base of cliffs, one was in a cliff face crevice, and one was in a crevice between soil-covered rocks below a cliff. No nests occurred in soil burrows or in grassy slopes, and all were enclosed and inaccessible. The nest entrances were quite variable in size but had an average area of 228.3 square centimeters. Entrances were typically rectangular, and there was an average distance of 122.5 centimeters to the nest (4 samples). Slopes into the nest varied equally between positive and negative, and in rocky slopes the nests were always on the lower portions of such areas, usually less than 20 meters above high-tide line. Sealy and Bedard (1973) examined 49 nest sites and found a wide variety of cavity and crevice sites being used. In every case, however, the egg was situated in almost total darkness and was always under a rock or peat (Sealy 1968). Sealy noted that the average nest entrance perimeter for parakeet auklets averaged 38.4 centimeters, only slightly less than is typical for crested auklets, and also apparently slightly less than the average entrance size encountered by Lehnhausen (1980). Both observers reported that in most cases the bird entered the nest from above.

NEST BUILDING AND EGG LAYING. There is probably little, if any, actual nest building. Sealy and Bedard (1973) noted that the birds do not accumulate nesting materials, though the egg is sometimes deposited on a layer of small pebbles. Only a single egg is laid, but one case of replacement laying was documented by Sealy and Bedard, in which a new egg was laid about 16 days after the first had been destroyed. They noted that of 31 eggs laid, 10 were lost during incubation, representing a 67.7 percent hatching success.

INCUBATION AND BROODING. Both sexes incubate, beginning immediately after the egg is laid. Sealy (1968) reported the weights of fresh and pipped eggs as averaging 37.5 and 34.8 grams, respectively, and representing 13.3 and 12.4 percent of average adult weights. He established an average incubation period of 35.2 days for 4 eggs (range 35–36 days) but was unable to determine the relative roles of the two sexes or the durations of the incubation shifts. However, during mid- to late June the birds show a daily activity cycle on Saint Lawrence Island with a single peak early in the morning (Searing 1977), suggesting that changeovers probably occur each day at about that time. During mid-July, about when hatching should be starting, the period of colony attendance still peaked in the morning, but a larger number of birds were present throughout the day. Manuwal and Manuwal (1979) found a similar trend toward later arrival during the latter part of the breeding cycle as well as a longer time spent at the colony. During the late stages of incubation or the early nestling phase the birds spent about 60 percent of their time at sea feeding and about 40 percent at or near the colony. The peak frequency of flights to the nesting rocks occurred in mid-July. Evidently parakeet auklets spend more time feeding each day during the nesting season than do least and crested auklets, which exhibit a double peak of daily colony attendance (Sealy and Bedard 1973).

GROWTH AND SURVIVAL OF YOUNG. Sealy and Bedard (1973) found that of 21 hatched chicks, 16 survived to fledging, representing a 76.2 percent fledging success. The most important predators of nestlings were found to be microtine rodents (*Microtus* and *Clethrionomys*), though wounds made by these animals were not always fatal. They also established an average fledging period of 35.3 days, with extremes of 34–37 days, for 6 chicks. The weight of chicks at hatching averaged 28.1 grams and peaked at 28–29 days, when it averaged about 250 grams. Thereafter it declined slightly to an average fledging weight of 222.6 grams, about 78 percent of adult weight. Thermoregulation is attained by only 3–4 days after hatching. Before fledging the birds spent a good deal of time flapping their wings near the entrance to the nest, and all of 4 birds observed to fledge did so between 3:00 and 5:00 A.M. They flew directly to the sea, with no apparent involvement of their parents in the departure or in subsequent activities on the water.

BREEDING SUCCESS AND RECRUITMENT RATES. The estimate by Sealy and Bedard (1973) of a breeding success of 51.6 percent, based on 16 chicks fledging from 31 eggs, is the only reasonable sample size. Hunt, Burgeson and Sanger (1981) noted that 4 chicks fledged from 6 nests under observation. It is believed that it takes 3 years to attain sexual maturity (Sealy and Bedard 1973), but no estimates of recruitment rates are available.

Evolutionary History and Relationships

Kozlova (1961) pointed out a number of similarities in the skulls of this species and the typical puffins and noted some adaptations in the pelvic area associated with improved terrestrial locomotion. Like the other auklets, this species does have a throat pouch, although Kozlova noted that it is smaller than in the other auklets. It seems probable to me that the parakeet auklet is most closely related to *Aethia*, though an affinity with the rhinoceros auklet and typical puffins is certainly not impossible. (Strauch 1985 recommended the merger of *Cyclorrhynchus* and *Aethia* after the submission of this manuscript.)

Population Status and Conservation

All the North American breeding colonies are found in Alaska, which has been roughly estimated to support about 800,000 birds in 125 known breeding sites (Sowls, Hatch, and Lensink 1978). Nothing is known of possible population trends. The species was assigned an oil vulnerability index of 80 by King and Sanger (1979).

Least Auklet

Aethia pusilla (Pallas)

OTHER VERNACULAR NAMES: Choochkie (Aleutians); knob-billed auklet; starique minuscule (French); Zwergalk (German); ko-umisuzume (Japanese); konyuga-kroshka (Russian).

Distribution of Species (See Map 22)

BREEDS on the north coast of Chukotski Peninsula, the Diomede Islands, and from Cape Prince of Wales south through the islands in the Bering Sea, including the Pribilofs, to the western Aleutian and Shumagin islands, and east to the Semidis.

WINTERS at sea off the coast of eastern Siberia south to Kamchatka, Sakhalin, and the Kurile Islands, off northern Japan, and off the Aleutian Islands.

Description (Modified from Ridgway 1919)

ADULTS IN BREEDING PLUMAGE (sexes alike). Upperparts slate blackish (sometimes inclining to glossy black), passing into dark slate color on suborbital and malar regions and chin, the scapulars intermixed with more or less white, the proximal secondaries (sometimes proximal greater coverts also) more or less distinctly tipped with white; acuminate feathers on forehead and lores, and elongated sharp rictal and auricular plumes white; underparts mostly white, more or less spotted or blotched with blackish or blackish slate, this frequently forming a distinct and uninterrupted band (of variable width) across foreneck, usually in abrupt contrast anteriorly to the immaculate white of throat; axillaries and under wing coverts white and pale gray; bill dusky basally, dark reddish terminally; iris white; legs and feet pale bluish. Adult plumage ("definitive alternate") is probably attained in the third year of life. Second-year birds exhibit head ornaments, a knobbed culmen, a white iris, a speckled throat, and heavily mottled underparts. About 20 percent of birds on the breeding grounds are in this plumage (Bedard and Sealy 1984).

WINTER PLUMAGE. Similar to the summer plumage, but underparts, including sides of neck, continuously white; the chin, however, is slaty, as in summer; white pointed feathers of forehead and such usually less developed, sometimes almost entirely lacking, bill without the knob at base of culmen. Iris white, at least in adults.

JUVENILES. Similar to the winter adult but bill smaller; a few short white feathers on head, and more white on scapulars. Iris gray.

DOWNY YOUNG. Entirely plain dark sooty grayish black, the underparts paler and more grayish. Iris blackish gray, bill medium gray, tarsi and feet gray.

Measurements and Weights

MEASUREMENTS. Wing: males 90.0–97.5 mm (average of 10, 92.9); females 88.5–96.0 mm (average of 10, 93.6). Exposed culmen: males 8–9 mm (average of 10, 8.6); females 7.5–9.5 mm (average of 10, 8.5) (Ridgway 1919). Eggs: average of 57, 39.5 x 28.5 mm (Bent 1919).

WEIGHTS. The average of 26 breeding males was 86.3 g (Bedard 1969a), and 125 unsexed adults averaged 92 g (Sealy 1968). The average weight of 14 eggs was 17.5 g (Sealy 1968); estimated fresh weight is 17 g (Schönwetter 1967). Newly hatched young average 12.3 g (Sealy 1968).

Identification

IN THE FIELD. This is the smallest of the auklets and is likely to be encountered only in arctic waters around Alaska. In summer the small size and white throat patch, bounded by black on the head and sometimes the

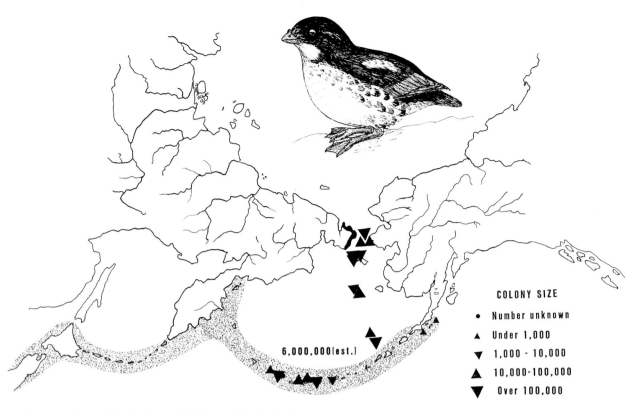

22. Current inclusive distribution of the least auklet, including colony locations (Alaska) and general breeding range (Asia). The wintering areas are shaded.

COLONY SIZE

- Number unknown
- Under 1,000
- 1,000 - 10,000
- 10,000-100,000
- Over 100,000

6,000,000(est.)

breast, are distinctive; the bill is orange with a yellow tip and a small dorsal knob, there are short white cheek plumes resembling a small mustache, and a white stripe extends back from the eye. In winter the combination of a white scapular stripe and a very small size serves to identify the species. First-year birds resemble adult breeders but have smaller bills, lack the knob on the bill, and have smaller or absent cheek plumes. Their voices are a mixture of twittering, cackling, and squealing notes.

IN THE HAND. The tiny size of this species (wing under 100 mm), together with the very short bill (culmen 7–10 mm) that is slightly higher than wide at its base and has a small compressed knob during the breeding season, provides for easy identification.

Ecology and Habitats

BREEDING AND NONBREEDING HABITATS. This is a high-arctic species with a coastal breeding distribution that extends south only to the Alaskan Peninsula and the Aleutians in North America, or approximately be-

tween the coastline limits having adjacent surface waters ranging from 5°C to 10°C during August. Additionally, the birds nest only in talus slopes, where crevices and interstices are provided by rock rubble of appropriate dimensions. Typically these habitats consist of cobbles and boulders up to about 5 meters thick, with a base of parent rock or an accumulation of smaller weathered particles that have gradually been deposited there. The range of rock diameters in the rubble most suitable for least auklets is from less than 0.3 to about 0.5 meter, with a sharp drop-off in usage in areas with average rock sizes above 0.3 meter, at least in part because of competition with crested auklets and probably also the other larger auklet species (Bedard 1969b; Sealy 1968). However, Byrd and Knudtson (1978) found no direct correlation between average boulder size and breeding densities of this and two other species of *Aethia* at their study area; the highest concentration of least auklets was in a boulder field partially covered by a layer of soil and vegetation. Outside the breeding season the birds are pelagic; the most northerly breeding birds must move south in advance of the winter ice limits, and at least in spring they apparently concentrate along the ice edge to the north and east of the Pribilof Islands (Gould, Forsell, and Lensink 1982).

SOCIALITY AND DENSITIES. Bedard (1969b) estimated breeding densities of least auklets on Saint Lawrence Island, finding that in various quadrats the estimated densities ranged from about 1 to 120 birds per 200 square meters, averaging about 46 birds for 30 quadrats. Searing (1977) reanalyzed Bedard's data for the Kongkok basin and added new information of his own that indicated a considerably higher average breeding density of 65 birds per 200 square meters. He thus believed that the colony there had increased about 95 percent in the 10 year period separating the two studies. Densities were found to vary with several physical environmental factors as well as the presence of crested auklets. Contrary to Bedard, Searing found no negative effect from the presence of crested auklets; instead, there was a positive correlation. Densities were highest on very steep slopes, on deeper mantles of scree, and on plots with angular rather than rounded rock rubble. There was also a negative correlation between auklet density and the distance of the plot from the nearest edge of continuous scree.

PREDATORS AND COMPETITORS. Sealy (1969) reviewed various possible predators of auklets on Saint Lawrence Island and suggested that, at least there, such potential predators as peregrines (*Falco peregrinus*) and gyrfalcons (*F. rusticolus*) are rare, while the more common jaegers (*Stercorarius* spp.) and snowy owls (*Nyctea scandiaca*) probably have a negligible effect on the birds. However, herring gulls (*Larus argentatus*) and glaucous-winged gulls (*L. glaucescens*) are probably important predators of chicks and possibly take some eggs as well. Glaucous-winged gulls are the most important predators of adults and fledged young in the Aleutians (F. Zeillemaker, pers. comm.). Certainly the arctic fox (*Alopex lagopus*) is a very serious predator on nesting birds and their chicks wherever it has been introduced, as are free-running dogs where they are present near colonies. Last, voles (*Cleithrionomys rutilus* and *Microtus* spp.) probably cause significant loss of nestling auklets locally by puncturing eggs, killing chicks, and occasionally even wounding adult birds (Sealy 1982). On Buldir Island in the Aleutians the peregrine may be an important predator on least auklets, and a moderate number may also be taken by bald eagles (*Haliaeetus leucocephalus*) (Searing 1977).

Competitors no doubt include the other auklets, with which the least auklet must compete for nest sites. As noted earlier, the least auklet can use rubble of smaller diameter than can the other considerably larger species, but the whiskered auklet is only about 44 per-

cent larger than the least auklet and can thus be expected to compete most strongly with it. Byrd and Knudtson (1978) found no significant differences between the crevice sizes used by least and whiskered auklets, though their small sample sizes precluded strong conclusions on this point. In any case, whiskered auklets made up no more than about 2 percent of the combined species population in their study area, and thus at least in that location they would not have constituted a serious competitive factor.

General Biology

FOOD AND FORAGING BEHAVIOR. Bedard (1969a) provided the first detailed analysis of least auklet foods, examining a total of 269 samples of food taken from gullets or neck pouches from the time of arrival until the end of the chick-rearing period. He found that during that time least auklets forage almost exclusively on planktonic crustaceans, especially the copepod *Calanus finmarchicus*, which has a cycle of summer abundance similar to its relative occurrence in the diet of the least auklet. At times during the summer when this form is scarce, the birds use such prey as small hyperiids, caridean larvae, and some other planktonic forms. In general the birds' diet seems to overlap strongly with that of the crested auklet, but they take substantially smaller prey items. The birds can hold up to 2.5 cubic centimeters of food in their gullets and another 11 in their neck pouches, although the average actual amount carried in the neck pouch was less than 8 grams, or somewhat under 10 percent of the adult body weight. A smaller sample of stomach remains (of 10 birds) was analyzed by Searing (1977), who noted a much larger proportion of decapod larvae eaten in his area, perhaps because his birds were collected only during July. Analysis of foods taken from 12 food pouches indicated that nearly all the foods brought to chicks during August were copepods (mainly *Neocalanus*). Finally, a sample of 258 throat-pouch contents (mainly regurgitated material) from the Pribilof Islands analyzed by Hunt et al. (1980) indicated a similar summer dependence upon calanoid copepods for feeding their young, though the adults themselves may feed more on euphausiids at this time. Recently Springer and Roseneau (1985) concluded that the distribution of copepod biomass in the Bering Sea controls the number and distribution of nesting least auklets in that region.

Foraging during the breeding season is apparently done quite close to the nesting colony; Searing (1977) reported that it occurred mainly within 2 kilometers of shore and along approximately 5 miles of beach near the nesting colony. However, a part of the population apparently flew considerably farther to feeding areas off the northwestern cape of the island. There the birds foraged mostly alone, in pairs, or in small aggregations. Bedard (1969b) noted that foraging occurs in a bimodal daily rhythm, at daybreak or early morning and again in the afternoon. Although Bedard reported that birds tend to be scattered when foraging, he sometimes saw them in very dense groups, foraging in a narrow strip of the littoral zone.

MOVEMENTS AND MIGRATIONS. Very little information is available on migration in these birds, though it is generally believed that they are fairly sedentary, moving only as far as is needed to provide open water for foraging during winter. Evidently no large wintering flocks have ever been found, although the birds' small size and their tendency to forage in a dispersed manner would reduce the probability of finding such groups. In any case it is apparent that few if any winter in the Gulf of Alaska (Forsell and Gould 1981; Gould, Forsell, and Lensink 1982), and more likely that they winter near the limits of sea ice in the Bering Sea, or close to their breeding areas in the case of the Aleutian Island colonies. There are questionable unsubstantiated reports by natives that the birds sometimes continue to use rock crevices in the Aleutians for winter shelter, thus providing food for arctic foxes (Murie 1959). In the Saint Lawrence Island area they are usually seen near shore in mid-May, about a month before initial egg laying (Searing 1977), whereas in the Pribilof Islands they are usually seen by mid- to late April, almost 2 months before egg laying.

Social Behavior

MATING SYSTEM AND TERRITORIALITY. Sealy (1968) has demonstrated that mate retention and nest site tenacity occur in this species. Of 3 pairs of birds banded in one year, 2 were recaptured the following year at the same nest sites, and one member of the third pair was also recaptured. Of 5 nest sites used by marked birds, 4 had at least one adult present from the pair that had used the nest the previous year, and the fifth nest site had been destroyed in the interim. Apparently the arriving birds can recognize the location of their nest sites even when the slope is still covered by snow, and they distribute themselves accordingly. The birds apparently defend only the single nesting interstice or actual nest site, although the hidden nature of these sites made it impossible for Sealy to obtain direct information on this point.

VOICE AND DISPLAY. So far no detailed descriptions of voice and display have been provided for the least auk-let. As with the Cassin auklet, a great deal of chorusing is typical, and it is thus very difficult to discern and describe the calls of single birds. These include rather high-pitched chattering or squeaking notes sometimes described as sounding like tree frogs (*Hyla*), *Calidris* sandpipers, or budgerigars (*Melopsitticus undulatus*). Thoreson (in press) recognized at least three distinct calls among the birds he observed. One is a warning or alert call, in which the bird stands erect and utters short repeated *cheeps* or "squeaky *chirrs*." A second aggressive note, *chee-chee-chee*, is accompanied by pecking, lunging, or other agonistic behavior. Finally, mated pairs perform duets of constantly repeated *chee* notes while billing and head waving.

Probably a good deal of courtship occurs at sea or within the nesting cavity. The only obvious displays that occur on nesting slopes are the extended billings and head wavings done by paired birds, which probably help form and strengthen pair bonds. Copulation has never been observed on the surface of the nesting slopes and was only once seen (by J. Bedard) at sea. Mr. V. Byrd (in litt.) once observed an attempted copulation in water about 0.5 kilometer from shore, and he informed me that a similar attempt was seen a few days later (in mid-May) by another observer. Sealy (1968) happened to interrupt a copulation while searching crevices with a flashlight for active nests. In that case the male was already mounted on the female's back, and the birds stopped all activity upon being disturbed. Sealy believed that copulation probably occurs during the rather short period (at least at Saint Lawrence Island) between snow melt and the start of egg laying.

Reproductive Biology

BREEDING SEASON AND NESTING SUBSTRATE. On Buldir Island egg laying occurs over a short period from late May (estimated earliest date May 24) to early June, with an estimated 80 percent of the eggs laid by June 1 (Byrd and Knudtson 1978; Knudtson and Byrd 1982). Egg records from the Pribilofs extend from May 24 to July 7, with most eggs probably laid in June (Bent 1919; Preble and McAtee 1923). During two different years Sealy (1968) observed a total spread of egg-laying dates from June 12 to July 5, although he noted that J. Bedard had found an extreme date of July 21, which possibly was a re-laying attempt. In each of the two years observed by Sealy there was a distinct concentration of laying into a period of about 10 days. Although his data were limited, he believed that most eggs were laid in early-morning hours.

As noted earlier, all nesting by this species occurs in talus and scree accumulations of cobble and boulders, especially those that provide openings large enough for least auklets but too small for the larger auklets. Of 26 nest sites studied by Knudtson and Byrd (1982), nearly twice as many had soil as rock substrates, and most of the eggs were deposited on flat rather than depressed surfaces. The average crevice volume was 117.6 square centimeters, or about 30 percent smaller than the average crevice used by whiskered auklets and not a great deal larger than the volume of the adult bird. Sealy (1968) noted an average nest-entrance perimeter of only 23.9 centimeters, almost half that typical of crested and parakeet auklets. The average distance from the nest entrance to a landing perch was only 0.5 meter.

NEST BUILDING AND EGG LAYING. No nest is built, and probably the nest chamber is not modified in any way, judging from the fact that most eggs are laid on flat rather than depressed substrates (Knudtson and Byrd 1982). Only a single egg is laid in this species, and there is very little evidence of regular egg replacement in the case of early egg mortality. Thus, Bedard (1967) observed only one apparent case of egg replacement, and Sealy (1968) noted two possible instances (out of a two-year sample of 64 clutches); both of the latter cases occurred when eggs were chilled by flooding during snow melt. He calculated that fresh eggs weighed an average of 19 percent of adult body weight, while pipped eggs averaged 14.6 percent, or among the highest relative weights of any alcids, exceeded only in the murrelets. During two years of study the average egg-laying date varied by 17 days, but the standard deviation of the spread in dates varied only from 3.1 to 4.4 days in those two years.

INCUBATION AND BROODING. Both sexes incubate, probably beginning immediately after the egg is laid. However, there is still no information on average shift lengths or other aspects of the roles the two sexes play in this part of the breeding cycle. Byrd, Day, and Knudtson (1983) pointed out that during the breeding season there is a net movement of birds to the colony, suggesting that both members of many pairs spend the night on land. They judged that during incubation birds are able to feed only every other day, indicating 24-hour incubation shifts. Sealy (1968) established an average incubation period of 31.2 days for the least auklet (range 28–36 days in 15 nests), and Byrd and Knudtson (1978) reported an average period of 35–36 days. Searing (1977) reported a hatching success of 49 percent for 70 nests, though part of this loss was attributed to investigator-induced predation by gulls and foxes attracted to the nest markers. Knudtson and Byrd (1982) found a hatching success of 68 percent for 28 nests, with infertility or embryo death a major source of prehatching mortality.

GROWTH AND SURVIVAL OF YOUNG. After hatching, the chick is brooded continuously for about 5 days, until it is able to maintain a constant body temperature. Thereafter the chick is fed by both parents, who make multiple trips each day between foraging areas and the nest, with distinct peaks of activity in morning and evening hours (Byrd, Day, and Knudtson 1983). Growth rates of chicks have been measured by Sealy (1968), Searing (1977), and Byrd and Knudtson (1978). Searing estimated a 32 day fledging period, and Sealy reported an average of 29.2 days. The maximum chick weight was reached at about 24–25 days in both Searing's and Sealy's samples but by the 18th day in the fairly large sample of Byrd and Knudtson. A loss of anywhere from about 10 to 35 percent of the maximum chick weight then occurs before fledging. Sealy (1981) observed that young hatching earlier in the season fledged at a heavier weight than those hatching later, and the fledging period tended to diminish as the season advanced. He speculated that early breeding might be a selective advantage in this species, especially in years that are phenologically late. The fledging success of chicks was estimated at 56 percent (of 16 young) by Searing and 75 percent (of 12 young) by Byrd and Knudtson. Fledging typically occurs during darkness (in the Aleutians at least), and gulls sometimes take laggard chicks that are still visible on or near shore the following morning (Knudtson and Byrd 1982). After fledging there is a rapid abandonment of the breeding colony as both adult and young birds become progressively pelagic. Nothing is known of the length of parental attachment to chicks during the postbreeding season.

BREEDING SUCCESS AND RECRUITMENT RATES. Overall breeding success rates were estimated as 51 percent (of 28 eggs) by Knudtson and Byrd (1982) and as 34 percent (of 70 eggs) by Searing (1977), the latter probably being biased downward because of increased predation associated with observer influences and the former possibly an overestimate because most egg monitoring began after incubation was well under way. Thus an estimate of the breeding success as about 40 percent may be fairly reasonable. There is no good information on the incidence of nonbreeders in the population, and no independent estimate of recruitment rates based on the percentage of juveniles in the fall migrant population.

Evolutionary History and Relationships

The three species of *Aethia* are probably fairly closely related, and there seems to be no good basis for judging relative relationships among them.

Population Status and Conservation

It is extremely difficult to estimate population numbers and possible trends in this species, largely because of the frequently enormous colony numbers, their hidden nest sites, and the marked temporal fluctuations in colony attendance (Sowls, Hatch, and Lensink 1978). In the Saint Lawrence Island area and perhaps on the Pribilofs the least auklet has increased greatly in recent years, probably because of changes in copepod populations (Springer and Roseneau 1985). However, in some other areas the birds have disappeared as breeders, possibly as a result of predation by introduced arctic foxes, but the birds' use of rock talus rather than soft substrates probably reduces the influence of foxes on breeding success. The least auklet was assigned an oil vulnerability index of 80 by King and Sanger (1979), about average for the family.

Whiskered Auklet

Aethia pygmaea (Gmelin)

OTHER VERNACULAR NAMES: Pygmy auklet; starique pygmee (French); Bartalk (German); shirahige-um-isuzume (Japanese); malaya konyuga (Russian).

Distribution of Species (See Map 23)

BREEDS in the Commander Islands, in the central Kurile Islands, possibly in the Near Islands, and from Buldir Island eastward in the Aleutians to Unimak Pass (Krenitzin Island group: Ugamak, Tigalda, Avatanak, and Rootok islands).

WINTERS in the breeding range south to the Kuriles, casually to Japan.

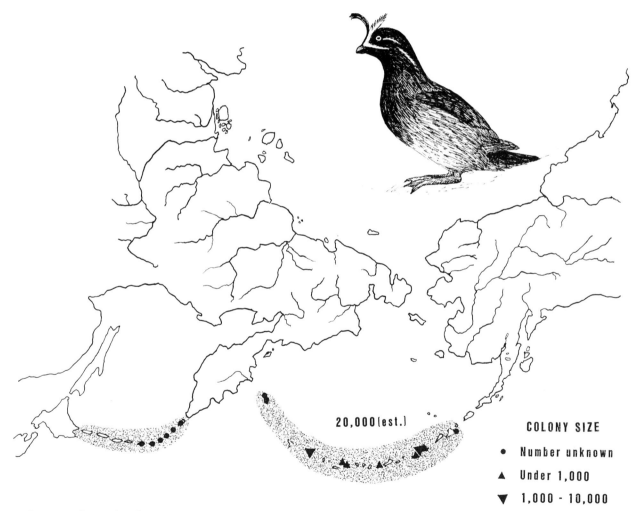

20,000 (est.)

COLONY SIZE

• Number unknown

▲ Under 1,000

▼ 1,000 - 10,000

23. Current inclusive distribution of the whiskered auklet, including colony locations and wintering areas (*shaded*).

Description

ADULTS IN BREEDING PLUMAGE (sexes alike). Upperparts plain grayish dusky (nearest dark neutral gray), darkest on pileum and sides of head, more slaty on scapulars, rump, and upper tail coverts, gradually passing into dusky gray (between chaetura drab and dark quaker drab) on chin, throat, and foreneck, this passing through dull neutral gray on chest into white on vent region and under tail coverts; under wing coverts wholly grayish brown (hair brown); ornamental head plumes white, except the recurved frontal crest, which is dull blackish and grayish dusky; bill bright red (blood red to scarlet), its tip and a narrow space around base of mandible whitish, iris white; legs and feet bluish gray, tinged with violet, the joints darker gray, the webs and soles blackish (Ridgway 1919). In year-old birds the plumes are shorter and the bill is less deep than in adults (Kozlova 1961).

WINTER PLUMAGE. Not materially different from the breeding plumage (all the ornamental plumes being retained), but color of bill duller, the nasal cuirass being dusky instead of red (Ridgway 1919). First-winter birds lack ornamental plumes and have a uniformly brownish bill and a bluish gray iris color (Kozlova 1961).

JUVENILES. Contrary to published descriptions, almost entirely blackish both above and below, with no trace of white on the underparts or facial markings. The iris is bluish gray, and the bill and feet are brownish (F. C. Zeillemaker photos).

DOWNY YOUNG. Densely covered by dark fuliginous down dorsally, becoming lighter and more grayish on the abdomen (Bent 1919).

Measurements and Weights

MEASUREMENTS. Wing: males 104–11 mm (average of 7, 107.2); females 103.5–116.0 mm (average of 4, 109.1). Exposed culmen, males 7–10 mm (average of 7, 8.3); 4 females 9 mm (Ridgway 1919). Eggs: the average of 6 was 46.1 x 31.9 mm (Byrd and Knudtson 1978).

WEIGHTS. The average of 60 breeding adults was 120.7 g (range 102.0–137.5) (Byrd and Knudtson 1978). The calculated egg weight is 26 g (Schönwetter 1967). Newly hatched young average 19.8 g (Byrd and Knudtson 1978).

Identification

IN THE FIELD. This species is likely to be encountered only in the western Aleutian Islands, where the distinctive triple white facial plumes and the forward-pointing crest provide a unique combination of field marks. These plumes are retained even during winter, so only immature birds would pose an identification problem. These are similar to juvenile crested auklets but have three indistinct white facial stripes. Vocalizations are still unstudied in detail but include a harsh, somewhat catlike *mew* note (C. V. Byrd, in litt.).

IN THE HAND. The combination of small size (wings under 120 mm), a red bill that is very short (culmen 7–10 mm), with linear to narrowly oval nostrils and wider than deep at its base, is unique to this species.

Ecology and Habitats

BREEDING AND NONBREEDING HABITATS. This species has a subarctic distribution that during the breeding season is strongly restricted to the coastlines and islands of the northern Pacific where the August surface water temperatures are from about 8°C to 10°C. Coastal cliffs, with talus accumulations, provide the favored breeding habitats, which appear to be almost identical to those of the least and crested auklets. The birds are pelagic in winter but apparently remain as close as possible to their breeding grounds.

SOCIALITY AND DENSITIES. As with the other *Aethia* species, densities on the breeding grounds are probably determined by the distribution and abundance of breeding sites. In the western Aleutian Islands such as Buldir the birds are far less common than least or crested auklets, and on sample census plots their densities ranged from 0 to 1.3 percent of the total auklets present. On a 4.3 hectare area of talus habitat at Buldir Island the maximum estimate of whiskered auklets present varied from 900 to 1,000 birds, depending on the kind of census technique used (Byrd, Day, and Knudtson 1983). In the Fox Islands (between Unimak Pass and Umnak Island) an estimated density (of birds visible on the water from a passing boat) of about 1,000 birds per square kilometer was recorded before and during the breeding season by Byrd and Gibson (1980). In the eastern Aleutian Islands the birds occur as scattered pairs wherever rock crevices in cliffs are available for nesting, and about 200 pairs were found scattered over 33 different islands (Nysewander et al. 1982).

PREDATORS AND COMPETITORS. Although specific information is lacking, it is probable that the same major predators (gulls, eagles, falcons, and possibly voles) affect this species as have been found to influence least and crested auklets in the same areas. Byrd and Knudtson (1978) found a few whiskered auklet remains at the aeries of both peregrines (*Falco peregrinus*) and bald eagles (*Haliaeetus leucocephalus*), in numbers approximately comparable to the species' relative abundance in that area.

Certainly the crested and least auklets are likely to be the major competitors of the whiskered auklet, both for food and for nest sites. Virtually nothing is known in detail of the diet of whiskered auklets, except that it is a plankton forager and thus probably eats much the same foods as the other *Aethia* forms. Similarly, its nest-site requirements (at least as to volume of cavities used) are statistically inseparable from those of the least auklet, while both of these species use cavities significantly smaller than those occupied by the crested auklet (Byrd and Knudtson 1978).

General Biology

FOOD AND FORAGING BEHAVIOR. Almost no published information is yet available on this species, but limited information suggests that it forages primarily on gammarid amphipods, with smaller amounts of other amphipods and a few decapods and gastropods (Stejneger 1885). Cottam and Knappen (1939) examined the stomachs of 5 birds, noting that 3 contained nothing but copepods (*Xanthocalanus*), while 1 consisted mostly of crustaceans (amphipods, isopods, and copepods) as well as a fish (Scorpaenidae) and 1 had a mixture of unidentified crustaceans and possibly some mollusk eggs.

Feeding flocks of birds tend to form at riptides from spring through fall, and the same kind of foraging behavior may also be typical through winter. Up to 5,000 individuals sometimes gather at such locations, where there are upwellings of material from lower levels. Feeding is done during daylight hours, with movements to and from breeding colonies mainly just before dusk and again early in the morning (G. V. Byrd, in litt.).

MOVEMENTS AND MIGRATIONS. There is no evidence for any large-scale migration in this species, at least in North America. Most sightings in the Bering Sea have been from Unimak Pass west to Buldir, and all indications are that this is a very sedentary species (Gould, Forsell, and Lensink 1982). However, there are apparently some movements within the Aleutian chain, for one of the largest flocks seen there (about 9,000 birds) has been in the Andreanof Islands area in spring, where only 800 were later seen during the breeding season, the birds possibly having moved eastward to the Islands of Four Mountains area for nesting (Byrd and Gibson 1980).

Social Behavior

MATING SYSTEM AND TERRITORIALITY. There is no evidence yet on mate retention and nest site fidelity, both of which have been established for the other species of *Aethia.* Very probably territorial behavior is also compa-

rable to that typical of the other species, being limited to defense of the nest site itself. G. V. Byrd (in litt.) noted that pairs of all three of the *Aethia* species called, displayed, billed, and competed for attention in a manner roughly similar to that described by Thoreson (1964) for the Cassin auklet; he also noted that crested auklets were dominant in aggressive encounters with least and whiskered auklets and that whiskered auklets usually were dominant over least auklets.

VOICE AND DISPLAY. G. V. Byrd (in litt.) noted that the voice of the whiskered auklet is a harsh *mew*, a bit like that of a cat, and is uttered frequently. Otherwise, virtually nothing has been noted on the calls of this species. Its displays are evidently very much like those of least and crested auklets, and probably detailed acoustic and postural analyses will be needed to provide any real information on similarities or distinctions among the three.

Reproductive Biology

BREEDING SEASON AND NESTING SUBSTRATE. The breeding season in the Commander Islands evidently begins fairly early, since Stejneger (1885) observed a nestling as early as June 30. Similarly, Byrd and Knudtson (1978) reported that on Buldir Island of the Aleutians nesting occurred at the same time as for least and crested auklets, with most of the eggs of all three species laid over a 10–12 day period. Eggs of all species were found on May 31, and back-dating on the basis of hatching dates indicated that the earliest eggs were laid about May 24. About 80 percent of the whiskered auklet eggs had been laid by June 1, and none were laid after June 4. This resulted in a peak hatching period at about the end of June and early July. Although in some areas of the Aleutians scattered pairs may nest in rock crevices in cliffs, talus slopes provide what is probably the optimum breeding habitat for whiskered auklets. This species is about 40 percent larger than the least auklet and about 50 percent of the size of the crested auklet and thus should be able to occupy intermediate-sized nesting cavities. Knudtson and Byrd (1982) did find a comparable relation between body size and average nest crevice volume. For the whiskered auklet the average nesting cavity volume was 162.5 cubic centimeters, but there was substantial variation and no significant statistical difference from the average they determined for the least auklet. Of 11 nest sites, 9 were on rock substrates, and the eggs were usually placed on depressions in the pebbles.

NEST BUILDING AND EGG LAYING. Except for possibly digging out a depression in the substrate to receive the egg, this species builds no nest. The time between colony occupation and initial egg laying is not known but is likely to be similar to that for the least and crested auklets. Only one egg is laid, and probably little if any renesting occurs after failure of the first egg.

INCUBATION AND BROODING. Evidently incubation begins almost immediately after the egg is laid, and undoubtedly it is performed by both sexes. The only available estimate of the incubation is that of Byrd and Knudtson (1978), who determined (within a 1–3 day spread) a 35–36 day period for this species. During incubation and brooding the birds are largely nocturnal, and it is likely that there are 24-hour shifts by the two parents. In a small (7 egg) sample, 6 hatched, representing a hatching success of 86 percent (Byrd and Knudtson 1978).

GROWTH AND SURVIVAL OF YOUNG. The only information available on growth of the young comes from Byrd and Knudtson (1978). They determined an average weight at hatching of 19.8 grams for 4 birds. From then until the 18th day there was a rapid rise in average weights to a peak of 112.3 grams (for 3 birds). By the 21st day the weight had dropped to 101.5 grams, or about 84 percent of adult weight. It is likely that fledging occurs not long afterward, for in the other species of *Aethia* fledging typically occurs when the young are at about 88 percent of adult weight (Sealy 1968). Byrd and Knudtson (1978) noted that 3 of 6 hatched chicks survived to fledging, a 50 percent fledging success rate. Feeding of young is probably more or less continuous through the daylight hours after the chicks have hatched, although Byrd, Day, and Knudtson (1983) observed a lull in activity of crested and least auklets in mid-afternoon and noted that whiskered auklets seemed to have a similar periodicity of daily activities, except that they tend to leave the colony earlier in the morning and arrive back later in the evening. Probably fledging in this species also takes a form similar to that described for least and crested auklets, though actual observations on it are still lacking.

BREEDING SUCCESS AND RECRUITMENT RATES. Too few eggs and young were followed by Byrd and Knudtson (1978) for them to establish any estimate of breeding success rates in this species, but it is likely to have rates very similar to those of least and crested auklets.

Evolutionary History and Relationships

Very little can be said of the evolutionary affinities of this species. It is interesting that in size the least, whiskered, and crested auklets make up a neat series of size-

types, each about 50 percent larger than the next smaller one, and these morphological relationships are likely to be of ecological significance if not of evolutionary interest. It is also of some interest that these three broadly sympatric species of auklets have more elaborate facial and bill characteristics than does the allopatric Cassin auklet.

Population Status and Conservation

The most recent information suggests that there are something in the neighborhood of 25,000 birds in Alaskan waters, with breeding probably occurring on at least ten islands. The largest known colony of about 3,000 birds is on Buldir Island, but possibly 2,000 occur at Yanaska Island, and fewer than 1,000 per colony are probably present on Koniuji, Chagulak, and Herbert islands (Sowls, Hatch, and Lensink 1978). Actual nesting records are present only from Umnak, Chagulak, Atka, and Buldir islands, but it is very probable that at least 25,000 birds occur in the Aleutian Island area (Byrd and Gibson 1980). The species was assigned an oil vulnerability index of 88, the highest of any alcid or indeed of any bird (King and Sanger 1979). This fact, together with its very small North American population size, makes the whiskered auklet a special concern for conservationists. Control or removal of introduced arctic foxes from the few known nesting sites is one possible means of assisting the species, and reducing human disturbance at these sites is another.

Crested Auklet

Aethia cristatella (Pallas)

OTHER VERNACULAR NAMES: Crested stariki; sea quail; canooskie (Aleut); starique cristatella (French); Schopfalk (German); etorofu umisuzume (Japanese); bolshaya konyuga (Russian); sukispuk (Saint Lawrence Island).

Distribution of Species (See Map 24)

BREEDS on the eastern end of the Chukotski Peninsula, the Diomede Islands, Sakhalin, and the central Kurile Islands, and in North America from the Pribilof and Aleutian islands east to the Shumagin Islands, Alaska.

WINTERS in open waters within the breeding range, south to Hokkaido and sometimes Honshu, and east to the vicinity of Kodiak Island.

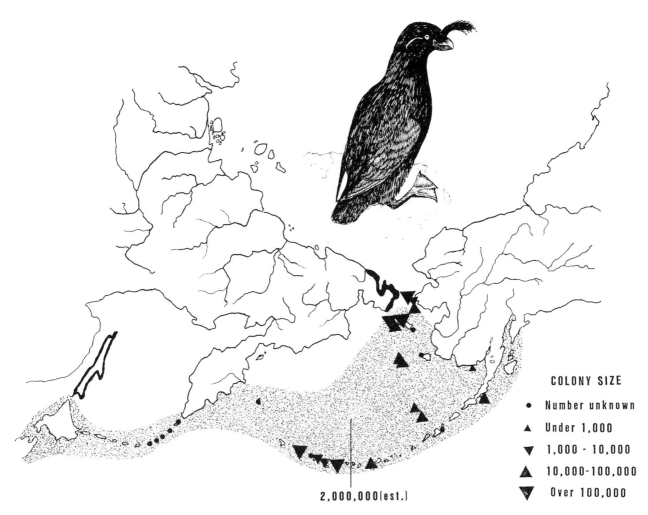

24. Current inclusive distribution of the crested auklet, including colony locations (Alaska) and general breeding range (Asia). The wintering areas are shaded.

COLONY SIZE

- • Number unknown
- ▲ Under 1,000
- ▼ 1,000 - 10,000
- ▲ 10,000-100,000
- ▼ Over 100,000

2,000,000(est.)

Description (Modified from Ridgway 1919)

ADULTS IN BREEDING PLUMAGE (sexes alike). Upperparts, including recurved frontal crest, plain slate blackish; forehead and entire underparts plain brownish gray, paler on posterior underparts; narrowly pointed auricular plumes white; bill, including suprarictal plate, orangy red or reddish orange, the tip more or less whitish or yellowish; iris white, legs and feet pale violet gray, the joints darker, webs blackish, and soles black. First-year birds have smaller and probably duller bills than adults (Kozlova 1961).

WINTER PLUMAGE. Similar in coloration to the breeding plumage, but bill smaller (through shedding of supranasal cuirass, suprarictal plate, and other parts) and grayish, with a pale orange tip. In first-year birds the crest begins growing in November and attains full size by mid-winter (Kozlova 1961).

JUVENILES. Similar to winter adults, but crest only about 5 mm long, iris pearl gray, bill blackish gray, and tarsi gray, becoming darker below. The postocular stripe is gradually developed during the first fall of life.

DOWNY YOUNG. Uniform sooty grayish black, slightly paler below. Iris initially black, becoming blackish gray, bill medium gray, tarsi and feet light gray above and almost black below. This plumage is lost after 30–32 days (Bedard and Sealy 1984).

Measurements and Weights

MEASUREMENTS. Wing: males 125–43 mm (average of 10, 134.8); females 131–37 mm (average of 7, 134). Exposed culmen: males 10–12 mm (average of 10, 11.2); females 10.5–11.5 mm (average of 7, 10.9) (Ridgway 1919). Eggs: average of 30, 54.2 x 37.9 mm (Bent 1919).

WEIGHTS. The average of 192 breeding adults was 286 g (Sealy 1968). The average weight of 10 eggs was 40.5 g (Sealy 1968); estimated fresh egg weight is 41 g (Schönwetter 1967). Newly hatched young average 29.3 g (Sealy 1968).

Identification

IN THE FIELD. This small auklet is the only one that has both a forward-tilted black crest and a single narrow white stripe extending back from the eyes. The smaller whiskered auklet has somewhat similar traits but also has a white "mustache" and flaring "eyebrows." The crested auklet also has a heavier, bright orange- to yellow-tipped bill. In winter adults the crest is smaller and the white facial plumes are somewhat less visible. Immatures lack distinct crests and have reduced facial striping; at that stage they most resemble immature whiskered auklets, but juveniles of that species have three indistinct white facial streaks. Crested auklets are extremely noisy on the breeding ground, uttering loud chirping, grunting, or honking sounds while in their nesting burrows. When in flight or while perched on nesting cliffs they often emit a distinctive doglike yelping note that is easily distinguished from the more croaking calls of parakeet auklets or the high-pitched chirping of least auklets.

IN THE HAND. This auklet has the combination of a moderately large size (wing 125–43 mm) and a red bill that is short (culmen 10–12 mm) and during the breeding season has an enlarged lower mandible and a conspicuous concave horny plate in the rictal area.

Ecology and Habitats

BREEDING AND NONBREEDING HABITATS. This is a high-arctic breeder with an overall range and habitat requirements very similar to those of the least and whiskered auklets. The breeding range includes rocky coastlines of the arctic that have adjoining surface water temperatures in August of about 5°C to 10°C. Precipitous coastal slopes, with associated talus accumulations at their bases, provide optimum breeding habitats; under some conditions of clifftop accumulations of talus the birds may also nest as far as a kilometer from the coast (Bedard 1969b). During the nonbreeding season the birds are pelagic, but little detailed information is available on their actual distribution and habitats at that time. Like the other *Aethia* species, they probably

stay as close to their breeding colonies as winter conditions permit, presumably near the edge of the sea ice boundary.

SOCIALITY AND DENSITIES. Bedard (1969b) estimated breeding densities of crested auklets on Saint Lawrence island to range from 4 to 47.1 birds per 200 square meters of nesting habitat, averaging 23 birds. Searing (1977) made similar estimates for the colonies in Kongkok Bay and found an average of 23.6 birds per 200 square meters in two different habitats, with the lowest numbers on an inland slope of Owalit Mountain. He found that the density of breeding birds was positively correlated with the orientation of the sample plot relative to the nearest water (plots facing the coast having the highest density of birds), with the depth, volume, and diameter of the boulders (higher numbers being present in areas of deeper scree composed of larger boulders), and with the percentage of the plot covered by scree. These results differed somewhat from those of Bedard (1969b), who found a high correlation between boulder size and crested auklet density on his sample plots. Thus crested auklet density was highest where average boulder size ranged from about 0.6 to 0.8 meter and declined rapidly in areas of smaller rocks (which have cavities too small for the birds to use), and the species was nearly absent in areas with boulder diameters above 1.0 meter, where the cavities are typically occupied by larger alcids such as puffins.

PREDATORS AND COMPETITORS. Both the peregrine (*Falco peregrinus*) and the bald eagle (*Haliaeetus leucocephalus*) have been implicated as predators of crested auklets on Buldir Island of the Aleutians, based on the evidence provided by prey remains found at eyries (Byrd and Knudtson 1978). These authors also observed glaucous-winged gulls (*Larus glaucescens*) actively hunting and catching adult auklets, taking birds (species unspecified) from the air, from land, and from the sea, and possibly also taking a few auklet eggs. Fred Zeillemaker (pers. comm.) has also seen these gulls take both adult and fledgling crested auklets. Sealy (1968) found that free-ranging dogs were often seen hunting in auklet colonies, and he noted that auklet remains, predominantly of crested auklets, were often found near their spoor. As with the other auklets, voles (*Microtus, Cleithrionomys,* or both) were found by Sealy to be predators of eggs and chicks, taking from 1.9 to 3.4 percent of the eggs in two different years and killing 5.3 percent of the nestlings in each of the two years.

Competitors include the other similar-sized auklets, especially those of the genus *Aethia.* Bedard (1969a) did a thorough study of foods and foraging behavior in least, crested, and parakeet auklets and found that although all species feed in the same areas and apparently use the same depth range, ecological segregation between least and crested auklets is achieved by variations in bill size that impose obligatory foraging differences. Segregation between the crested and parakeet auklets apparently is based on innate differences in prey preferences, minor differences in the foraging apparatus, and behavioral differences in foraging activities. Apparently all species can dive to similar depths (of about 35–40 meters), but the least auklet seems to feed in shallower waters, and the crested and parakeet auklets have substantial differences in the shapes and areas of their wings that presumably influence their underwater swimming ability. Competition between the crested auklet and the other auklets also exists for nest sites, especially between the most similar-sized forms. Byrd and Knudtson (1978) reported that the average volume of crested auklet nest sites was significantly larger than that of those used by least and whiskered auklets, while Sealy (1968) found only minor differences in entrance perimeters in the nest sites of crested and parakeet auklets and slightly larger differences in the average distances between the nest opening and the perching/landing site. He found no obvious physical differences in the nest sites of these two species but noted that parakeet auklets are very lethargic or cautious in entering their nests after landing on nearby or more distant perches and typically enter from above. Crested auklets usually enter the nest rapidly after landing nearby, usually from the side or below. Thus there may be minor differences in the orientation and relative distance of suitable perching and landing sites preferred by the two species that promote segregation between them.

General Biology

FOOD AND FORAGING BEHAVIOR. The study by Bedard (1969a) provides the best information on this species and is based on the analysis of 107 gullet samples obtained up to the time of hatching and 135 neck-pouch samples obtained during the chick-rearing period. As in the least and parakeet auklets, crustaceans dominated all the samples, and there were marked seasonal variations in the relative abundance of particular types, with a strong shift in the case of the crested auklet to concentrate on euphausiids during the chick-raising period and a secondary use of *Calanus* copepods. During the prehatching period, amphipods (Hyperiidea, Gammaridea) and Mysidacea were important food components, while decapods, larval fish, and other groups remained unimportant throughout the entire period. Searing (1977) made a few additional observations in the same area, but with far smaller sample sizes. He exam-

ined 12 neck pouches of adults during the chick-rearing period and found that virtually all of the contents were copepods, especially *Neocalanus*, and very few euphausiids were present, perhaps because of a "bloom" of *Neocalanus* that year. A sample of 20 throat pouches from the Pribilof Islands in summer indicated that there the birds concentrate on euphausiids during this period, while amphipods were of secondary importance (Hunt et al. 1980).

Searing observed that crested auklets mostly fed in the vicinity of the Northwest Cape of Saint Lawrence Island (about 30 miles from his study area), though some also foraged in an area about 3 miles offshore from the nesting colony. Bedard (1969a) estimated that a crested auklet pair must feed its young an average of 80 grams a day and that the average load carried in its neck pouch was only about 21 grams, with an additional small amount carried in the bird's gullet. Thus at least two trips per day by each member of the pair would probably be needed to keep the chick supplied with food. Like the other auklets, crested auklets tend to have a bimodal activity cycle associated with foraging, with early morning and late evening peaks (Byrd, Day, and Knudtson 1983).

MOVEMENTS AND MIGRATIONS. Very little information is available on migrations of this species, which only rarely strays south of Alaskan waters. It is rare that groups have been seen far from their breeding islands in Bering Sea waters, although there seems to be an eastward winter movement of some birds into the Kodiak archipelago, and wintering rarely occurs as far east as Prince William Sound. There is apparently also a pelagic dispersal during winter southward from Unimak Pass to at least 33° N (Gould, Forsell, and Lensink 1982). In the Pribilof Islands area large numbers overwinter, with the winter numbers apparently several times greater than during the summer, so birds from other nesting colonies probably overwinter in this general region (Preble and McAtee 1923).

Social Behavior

MATING SYSTEM AND TERRITORIALITY. Mate retention and nest site tenacity were proved by Sealy (1968), who noted that one pair of banded birds returned to the same nest site the following year. He also reported (1975d) that 4 birds captured and marked in June, while waiting for the snow covering nesting sites to melt, were later seen in the same locations, either incubating eggs or carrying food to their chicks. He was unable to obtain any direct evidence on territoriality but judged that defense was probably limited to the entrance or interstice of the nest site itself.

VOICE AND DISPLAY. Thoreson (in press) has provided the only description of vocal variations in this species. He noted that single birds on land utter barking or trumpeting (*awka*) or crowing (*coooh-awka-coo*) notes while waiting for their mates to land. Trilled variations of these sounds produce a cackling sequence that is the most frequent vocalization of grouped birds. When cackling, the bird holds its head high, its throat swells, and its breast vibrates (fig. 45A,B). Cackling may be preceded by a faint squeaking note. Duetting is performed by paired birds during billing sequences, and also is stimulated by the arrival of a third bird on a ledge already occupied by a pair.

Kharitonov (1980) has provided the only available account of display posturing in the crested auklet. During threatening posturing the neck is stretched and the head is directed so that the crest is pointed toward the opponent, with the body assuming the shape of the letter **S** (fig. 45C). Fighting may occur between as well as within sexes. When preparing to utter his call the male often shakes his head and preens his neck feathers, alternating this behavior with a stretched neck posture (fig. 45D). The call posture (fig. 45A,B) serves both as a mating signal and as a threat toward other males. Paired birds often sit opposite one another and bill, uttering low trilled sounds (fig. 45E). During the male's mating call the female sometimes assumes a low posture with the neck feathers slimmed (fig. 45G). A somewhat similar posture may at times be assumed by grouped males, in which the birds seem to investigate the substrate; this sometimes occurs during male conflict behavior as an apparent displacement posture.

Copulation evidently occurs both on water and on land. Thoreson (in press) saw what he considered attempted copulations on the water by pairs that had separated themselves from grouped birds. In this behavior one of the birds would vigorously and repeatedly plow through the water toward the other. Sealy (1968) observed two cases of copulation within the nest cavities. These were seen by illuminating the cavity with a flashlight, which caused the birds to immediately "freeze" and cease their sexual behavior. He observed no attempted copulations by birds standing on the snow while waiting for their nest sites to become available.

Reproductive Biology

BREEDING SEASON AND NESTING SUBSTRATE. Records of egg laying from Buldir Island, Aleutian Islands, extend from May 24 to June 7, and hatching dates range from July 1 to 16, with a laying peak about the beginning of June and a hatching peak between July 5 and 12 (Byrd and Knudtson 1978). Sealy's (1968) records of egg

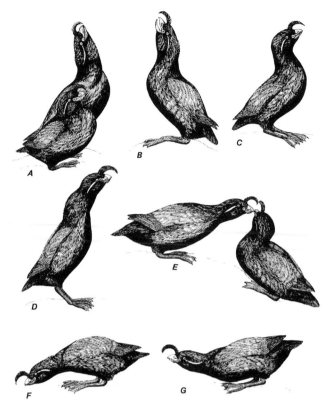

45. Social behavior of the crested auklet (after Kharitonov 1980 and photos by author): A, B, calling posture; C, threat posture; D, preliminary to calling posture; E, billing; F, investigation of substrate posture; G, female courtship posture.

laying for two years at Saint Lawrence Island range from June 14 to July 14, with a substantial year-to-year variation in mean laying date but a small (4–6 day) range in the standard deviation of laying in a single year. Bedard (1967) and Searing (1977) determined that the egg period on Saint Lawrence Island extended from June 14 (back-dated estimate) to mid-August, with probable laying peaks (in two years) of late June and early July and hatching peaks (in three years) ranging from July 29 to August 11. Egg dates for the Pribilofs range from June 16 to July 10 (Preble and McAtee 1923). The nesting substrate invariably consists of the crevices and cavities associated with scree and talus fields. Knudtson and Byrd (1982) reported that the average nesting cavity of this species was 117.6 cubic centimeters, significantly larger than those of least and whiskered auklets. Sealy (1968) found an average nest-entrance perimeter of 40.3 centimeters for 16 nests of this species, only very slightly larger than the average he reported for the slightly smaller parakeet auklet. As noted earlier, only very subtle differences, if any, exist in the nest-site preferences of these two species. About half of the 52 nest sites Knudtson and Byrd examined had rock rather than soil substrates, and in most nest sites the eggs were deposited in depressions among the pebbles or in the soil. In one area they found groups of 10–20 birds nesting in cavelike chambers in the talus, with eggs of adjacent pairs placed as close as 15 centimeters apart where visual isolation occurred and a meter or more apart where there was no such isolation.

NEST BUILDING AND EGG LAYING. These birds do little if any nest building, though a certain amount of scratching to make a substrate depression is evidently typical. On Saint Lawrence Island the birds appear on nesting slopes about 10 days after they are initially sighted on offshore leads and about 5 days after they become visible from shore. Arrival on the nesting grounds may occur from a month to 6 weeks before the first eggs are laid, depending on the rate at which snow and ice disappear from the talus slopes and scree fields where the birds nest (Searing 1977). Much of this time is probably spent in establishing or reestablishing territorial ownership of potential nesting sites and in waiting for these sites to become suitable for egg laying. Only one egg is laid, and Sealy (1968) observed only 2 possible cases of renesting following the abandonment of the initial egg. In one case the egg was addled and in the other it had apparently been chilled by runoff water. Sealy judged the egg weight of fresh eggs as 14.2 percent of average adult weight and that of pipped eggs as 11.5 percent, indicating an average 19 percent weight loss in the course of incubation.

INCUBATION AND BROODING. Incubation is performed by both sexes, probably in about equal amounts. Very little is known of the lengths of incubation shifts, but 24-hour incubation cycles are probably typical of the auklets, with shifts occurring each evening. Thus birds returning to the colony in the evening probably relieve their mates at that time or at least before the next morning, when the latter leave to spend the day foraging (Sealy 1972; Byrd, Day, and Knudtson 1983). Sealy (1968) determined the average incubation period to be 35.6 days (range 34 to 37 days for 6 eggs), including an approximate 2 day pipping period. Byrd and Knudtson (1978) noted that crested auklets incubated for 41 days in their study area, a rather surprising difference from Sealy's estimate. Losses of eggs during the incubation period amounted to 24 percent (of 36 eggs) in Byrd and Knudtson's study area; most were the result of early embryo death or infertility, while cracking accounted for a small percentage of the losses. Searing (1977) reported that only 11 of 48 (23 percent) of the eggs he had under observation hatched successfully, with high losses resulting from infertility or embryo death and progressively lower losses resulting from predation and breakage.

GROWTH AND SURVIVAL OF YOUNG. Information on the rate of chick growth has been provided by Sealy (1968), Searing (1977), and Byrd and Knudtson (1978). Sealy reported that chicks fledged an average of 33 days after hatching, when the primary length was 84.5 percent of adult length and the chicks were about 80 percent of adult weight. Typically the birds reached about 91 percent of adult weight by their 27th day and then lost about 11 percent of their weight. Byrd and Knudtson reported an average maximum chick weight at the 19th day and a drop of about 12 percent by the 23d day, suggesting an earlier fledging period in that area. Searing was able to measure only 2 chicks after the 17th day, and thus his data are of limited value. He was also unable to calculate fledging success because of his small sample. Byrd and Knudtson estimated a fledging success of 67 percent for 21 chicks. Fledging of auklet chicks is apparently not preceded by a period of starvation as in puffins, and Sealy (1968) further noted that it usually occurred during the night or early morning hours. Upon taking flight the chicks would head directly toward the sea, usually landing about half a kilometer from shore. He observed that the adults were not directly involved in this process and believed that the chicks become totally independent upon fledging. Chicks that are blown off course and land inland on the tundra are at least sometimes taken by predators or die from other causes before reaching the sea. However, most chick mortality probably occurs shortly after hatching and before the birds develop effective thermoregulation at 3 or 4 days of age (Knudtson and Byrd 1982). On Buldir Island most crested auklets fledged during the first 10 days of August, whereas on Saint Lawrence Island fledging occurred from mid-August to early September.

BREEDING SUCCESS AND RECRUITMENT RATES. Knudtson and Byrd (1982) estimated a breeding success of 51 percent, based on an initial sample of 36 eggs. Except for this estimate, no specific information is available on reproductive success of the species, nor are there any estimates of adult survival rates.

Evolutionary History and Relationships

As noted earlier, the crested, whiskered, and least auklets all seem to form a close-knit evolutionary group, with no obvious closer or more distant relationships evident among them given the structural and behavioral information now available.

Population Status and Conservation

Breeding in North American waters is confined to Alaska, where an estimated 2 million birds occur in 38 known colonies, but many of the major colonies are still only very incompletely surveyed (Sowls, Hatch, and Lensink 1978). There is no way of knowing whether the population trend is up or down, and differing censusing methods result in quite different estimates of numbers for the same colony (Byrd, Day, and Knudtson 1983). King and Sanger (1979) assigned the species an oil vulnerability index of 76, about average for the family Alcidae.

Rhinoceros Auklet

Cerorhinca monocerata (Pallas)

OTHER VERNACULAR NAMES: Horn-billed auk; unicorn auk; macareus rhinoceros (French); Nashornlund (German); utou (Japanese); dlinnoklyoryi tupik (Russian); alcuela rinoceronte (Spanish).

Distribution of Species (See Map 25)

BREEDS from USSR's Maritime Province, southern Sakhalin, and the southern Kurile Islands south to Korea and northern Honshu; and from at least as far west as Kenai peninsula, southeastern Alaska, to northwestern Washington, southwestern Oregon, and California (at least to Farallon Islands and possibly Point Arguello). Probably breeds locally in the Aleutians (Buldir) and along the Alaska Peninsula.

WINTERS from the southern part of its breeding range southward off the coasts to Korea, Japan, and Baja California.

Description (Modified from Ridgway 1919)

ADULTS IN BREEDING PLUMAGE (sexes alike). Upperparts sooty blackish, the scapulars, interscapulars, and feathers of rump indistinctly tipped with dark sooty grayish; sides of head deep hair brown (or between hair brown and fuscous), passing gradually into lighter hair brown or mouse gray on malar region; chin, throat, chest, sides, and flanks white, more or less clouded with brownish gray, especially on breast, the posterior under tail coverts brownish gray; axillaries and under wing coverts uniform brownish gray, like sides and flanks; a line of straight, elongated, lanceolate white feathers originates at posterior angle of eye and extends backward along sides of occiput to nape, and another broader series starts at the rictus and extends backward beneath suborbital and auricular regions to or beyond posterior end of the latter; bill orangy yellow or dull orange with culmen and both anterior and posterior edges of the hornlike supranasal appendage black; iris brown to pale amber; legs and feet whitish yellow, darker on joints of toes, the inner tarsi and soles blackish. First-year birds have shorter white facial plumes than adults, and their bills are shorter and more slender, with only a small basal knob.

WINTER PLUMAGE. Coloration as in summer, but with reduced head plumes and corneous supranasal horn and gonydeal ridge lacking. First-winter birds lack brown in the throat and breast area and have only a few hairlike brown plumes behind the eye and at the base of the bill (Kozlova 1961).

JUVENILES. Similar to winter adults but lacking white head streaks and with bill smaller and darker. The chest and abdomen feathers are white, with pale brownish tips (Kozlova 1961).

DOWNY YOUNG. Uniform sooty grayish brown, slightly paler on underparts of body. Iris brown, bill and legs blackish.

Measurements and Weights

MEASUREMENTS. Wing: males 175–83 mm (average of 6, 177.8); females 169–81 mm (average of 6, 175.6). Exposed culmen: males 32.5–36.0 mm (average of 6, 34); females 32.5–36.0 mm (average of 6, 34) (Ridgway 1919). Eggs: average of 39, 68.5 x 46.2 mm (Bent 1919).

WEIGHTS. The average of 48 breeding adults was 520 g (Vermeer and Cullen 1982), and 51 adults averaged 521 g (Leschner 1976). The calculated egg weight is 77 g (Schönwetter 1967). Newly hatched chicks average 54 g (Wilson 1977).

Identification

IN THE FIELD. This species is intermediate in size between the puffins and the typical auklets and has a bill that is heavier than in the auklets but more slender than in puffins. The iris is amber, the bill is mostly orange, with a distinct "horn" at its base during the breeding season, and there are white plumes that produce a white "mustache" and a long white crest reaching from above the eye to behind the nape. In winter these white feathers are shorter and relatively faint, but the bill is

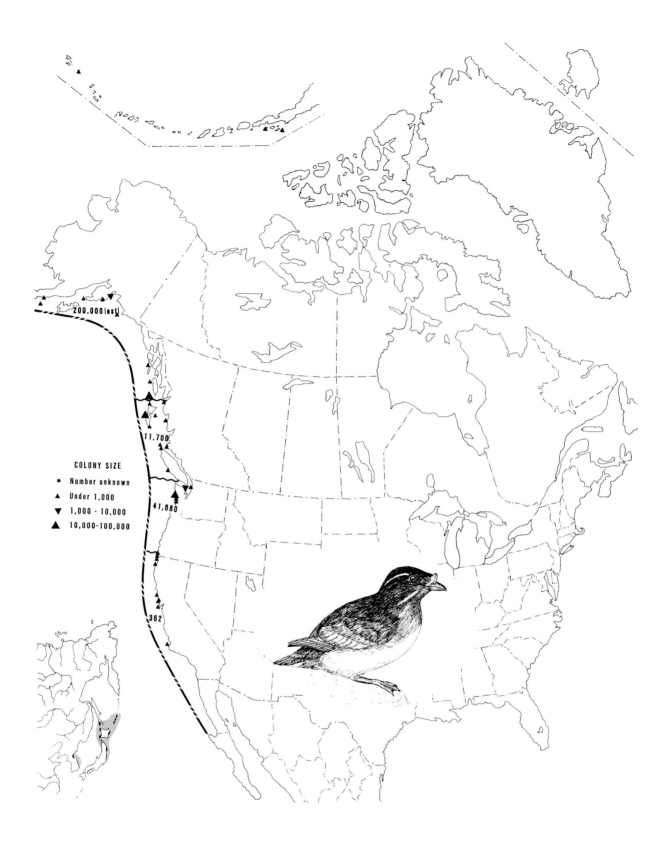

25. Current North American distribution of the rhinoceros auklet, including
colony locations and limits of nonbreeding range (*broken line*).

still distinctly orange tinted, becoming blackish along the culmen ridge. First-winter birds lack white plumes and have smaller bills than adults, but the white markings begin to appear between December and March. Adults utter growling and moaning calls, and chicks make shrill piping sounds.

IN THE HAND. This species can be recognized by the combination of its fairly large size (wing 169–83 mm) and an orangy yellow bill that is fairly long (culmen 32–40 mm), highly compressed with a variably large knob present at the base of the culmen, and edged with blackish. Young birds and nonbreeding birds lack a distinct knob but show a swelling in the appropriate area, at least in winter adults. A blackish culmen seems typical of birds in all age categories.

Ecology and Habitats

BREEDING AND NONBREEDING HABITATS. During the breeding season this species occupies coastal areas having adjacent August surface water temperatures ranging from 12°C to 15°C in North America (and up to about 23°C in Asia), primarily on sloping coasts that are free of digging predators and have a thick soil mantle and variable (usually grassy) vegetational cover, but rarely extending to steep cliffs. The species' southern breeding limits in North America appear to have been rather variable in historical times, perhaps as a reflection of varying environmental conditions, including probable changes in abundance of favored prey species. It is pelagic in winter, usually occurring in nearshore habitats but rarely coming close enough to shore to be visible from land.

SOCIALITY AND DENSITIES. Vermeer (1979) investigated nesting densities of this species on Triangle Island, British Columbia, and found that some 13,380 pairs nested in an area of apparently less than 5 hectares (my estimate), while another 1,500 pairs were within an approximate 1 hectare area. Nesting densities thus must have been in the vicinity of 1,500–3,000 nests per hectare, or 0.15–0.3 nest per square meter. Richardson (1961) estimated burrow densities on one area of Destruction Island to be 132 in 14,040 square feet, or about 0.1 per square meter. Wilson (1977) estimated burrow densities of up to 0.67 per square meter on Protection Island, Washington, and Leschner (1976) indicated densities of 0.20–1.3 burrows per square meter on Destruction Island. Heath (1915) reported a much higher nesting density of up to 400 burrows in 600 square feet (7.2 per square meter!) on Forrester Island, Alaska.

PREDATORS AND COMPETITORS. Certainly this species is vulnerable to various mammalian and avian predators. Wilson (1977) listed the great horned owl (*Bubo virginianus*) as the major cause of adult mortality, accounting for nearly 60 percent of 44 known-cause mortalities on Protection Island. Dogs or cats accounted for a few birds as well. Leschner (1976) judged that gulls killed some nestlings, apparently by pulling them from burrows or catching them outside. The widespread occurrence of introduced arctic foxes (*Alopex lagopus*) is believed to be why this species is absent from nearly the entire length of the Aleutian Islands, except for the fox-free Buldir Island (Sowls, Hatch, and Lensink 1978).

Competition with puffins for food may be a significant factor in the distribution and abundance of this species. Vermeer (1979) noted a substantial overlap in the kinds of foods adults brought to their chicks in rhinoceros auklets and tufted puffins at Triangle Island, although the two species nested in different parts of the island. Generally, puffins tend to nest in relatively open and steeper areas of slopes and cliffs where takeoff is easy (Leschner 1976; Richardson 1961). Rhinoceros auklets typically occur in areas of more gentle slopes and often in fairly wooded sites, where they sometimes even land in trees and then flutter down and walk the remaining distance to their nest burrows (Vermeer 1979). However, the two can occur together, as they apparently do at Goat Island, Oregon (F. Zeillemaker, pers. comm.).

General Biology

FOOD AND FORAGING BEHAVIOR. Most of the available information comes from studies of samples of prey being delivered to chicks. Thus Leschner (1976) analyzed a total of 119 food samples over a two-year period, finding that night smelt (*Spirinchus starksi*), sand launce (*Ammodytes hexapterus*), northern anchovy (*Engraulis mordax*), surf smelt (*Hypomesus pretiosus*), and Pacific herring (*Clupea herengus*) were the major items fed to nestlings. Wilson (1977) analyzed 212 samples, also taken during the chick-feeding period, and found a rather different array of prey species, although sand launce, herring, and anchovies were also present and predominated in the samples. In weight these three food types made up from 63 to 83 percent of the foods in Leschner's samples and from 91 to 97 percent in Wilson's samples. Wilson reported that the average load carried varied from 29.5 to 32.3 grams for the two years, and the average number of fish per load averaged 5.6 in both years. Vermeer (1979) noted a different group of prey fish fed to chicks in Triangle Island, the most important forms being sand launce, bluethroat argentines (*Nansenia candeda*), Pacific saurys (*Cololabris saira*), and rockfish (*Sebastes entomelas*), but almost no use being made of herring. Wilson noted that in nine studies of foods provided chicks, sand launce or capelin (*Malotus*) occurred in all, while in four studies of foods eaten by adults sand launce appeared in only one and euphausiids appeared in two. That summary did not include the analysis of Baltz and Morejohn (1977), who analyzed the foods of 26 wintering adults and found a high proportion of squid (*Loligo*) as well as a rather wide variety of fish, especially anchovies and rockfish (*Sebastes*).

Evidently this species is opportunistic in its foraging behavior, feeding on a geographically and seasonably variable food supply. The birds also seem to eat a higher proportion of crustaceans than do the typical puffins, but nearly all of these species (the tufted puffin being an exception) tend to feed heavily on inshore, subtidal prey (Wehle 1980).

MOVEMENTS AND MIGRATIONS. These birds are at least somewhat migratory, occurring rather commonly during winter along the entire California coast, but especially around the Santa Barbara Islands. However, some wintering probably occurs as far north as the Kodiak Islands area of Alaska (Forsell and Gould 1981). Some birds also winter in southern Puget Sound of Washington and in areas such as Georgia Strait in British Columbia.

Social Behavior

MATING SYSTEM AND TERRITORIALITY. Richardson (1961) stated that banding returns, although few, show that the birds retain their mates year after year. Returns of 6 birds indicated that 3 returned to the same burrows the next year. Two of these 6 returned a year later, and 1 returned two years later, to adjacent burrows, the original ones having been destroyed. Leschner (1976) found that mate retention occurred in 1 of 4 pairs, but she was unable to establish the point for the remaining 3 pairs. She also observed a return to the same nest site for at least 4 of 11 pairs. Territorial defense in the rhinoceros auklet probably includes the nesting burrow and its entrance, the approach path to the burrow, and a specific raised area near the burrow that is used for taking off, landing, and as a resting place. The actual area defended may vary with population density but is typically rather small, and the intensity of defense is strongest in the prelaying and egg-laying stages (Wehle 1980).

VOICE AND DISPLAY. Richardson (1961) attempted to describe the calls of this species, including a commonly heard one consisting of five to seven rather high-pitched, groaning notes, with the accent and longest pause usually on the second or third and the last few notes fainter and dying away. The calls of different indi-viduals varied in pitch, stress, and number of notes, which Richardson thought might help birds locate their mates. Single, low-pitched notes were sometimes heard from within burrows, and likewise a single groanlike call note and a rasping squeak were heard from birds on the water. Wehle (1980) considers the single-note call a threat call. A preflight call was noted by Manuwal and Manuwal (1979).

Displays of this species have been described by Thoreson (1983), who noted that billing between pairs helps maintain pair bonds, and this behavior was observed both on water and on land. As in the typical auklets, the bills pass each other closely as the birds make slow and deliberate movements, but they do not usually touch. Territorial ownership of a burrow was apparently indicated by an "upward huff" stance in which the bird stood erect, with its bill open and pointed upward (similar to the crested auklet posture shown in fig. 45B) and with the wings often partly spread. During this posture the air was blown though the throat in distinctive "huffs." At times birds adopted an immobile posture, as if staring off into space. An aggressive posture was the hunch walk, in which the neck is stretched forward and the body is somewhat hunched as the bird walks slowly and deliberately toward its opponent. A similar "low neck-forward profile" posture was adopted when walking near the burrow (similar to that shown in fig. 45G). A bowed-head display has been seen in this species and also the typical puffins, in which the bird holds its head low and horizontal to the ground and additionally tilts its bill down so that it nearly touches the feet. The function of this display is still not known for rhinoceros auklets, but in puffins it may be a billing invitation display or an aggressive display (Wehle 1980). Behavior patterns leading to copulation as well as copulation itself remain undescribed.

Reproductive Biology

BREEDING SEASON AND NESTING SUBSTRATE. Egg records from southern Alaska range from May 10 to June 22, with a peak between June 9 and 20 (Bent 1919). Leschner (1976) summarized egg-laying dates for a variety of localities, noting that at Forrester Island, Alaska, egg-laying occurs from the last week of May to mid-June. At three sites in British Columbia it ranged from the second week of May to June 7. Three Washington sites ranged from April 30 to June 17. At Destruction Island there is considerable synchrony in egg laying, with a total span of 37 and 46 days in the two years under study by Leschner, but with 80 percent of the eggs laid during the first 18 and 15 days respectively. Wilson (1977) concluded that there are significant differences among the populations of three islands (Smith, Destruction, and Protection) in Washington, with temporal variations between these populations in the same breeding season and also variations within populations between breeding seasons. Wilson considered these the probable result of annual and local variations in the environment that affect the distribution and abundance of food resources.

The nest is invariably in a burrow, typically excavated by the birds themselves. Although some burrowing may occur during the prelaying period, most newly dug burrows are not used in the same season, probably because there is not enough time both to dig a new burrow and to complete the breeding cycle (Wehle 1980).

NEST BUILDING AND EGG LAYING. In Washington, digging is at a peak from late March through April but continues through July. The birds use both their feet and their bills, with the higher areas excavated by the bill and a lower channel dug by the feet. The length of the tunnel probably varies with soil condition, but tunnels as long as 25 feet have been described, and the average is probably 8–10 feet. Some burrows have a blind side passage, and many have wide areas that may allow the mate to pass. The last few feet tend to slope downward, and there may be a slight drop-off from the end of the burrow to the nest chamber itself. A week or two may be sufficient time to dig a 6–8 foot burrow in suitable soil, with the birds working nightly. Evidently both sexes take part in burrow excavation (Richardson 1961).

The egg is laid in a small chamber that may be unlined or have a small accumulation of vegetation. There is fairly good evidence that replacement eggs are laid when the first is destroyed. Leschner (1976) found that two eggs that had been expected to hatch within known incubation periods hatched 12 and 14 days late, and a third that was lost just before hatching was 20 days late. In another case an adult was seen incubating a fresh egg 9 days after its first egg had been deserted.

INCUBATION AND BROODING. Although incubation presumably begins at the laying of the egg, this is not invariable, and delays of up to 8 days before initiation of incubation (at least during the daytime) have been reported (Wilson 1977). Both sexes incubate, with nocturnal shifting of mates but with apparently no very regular lengths of incubation shifts, and it is possible that mate shifting does not occur every night (Richardson 1961). Wilson (1977) stated that during the early phases of incubation the incubating bird might not be relieved for as long as 4 days, and furthermore that eggs may be deserted for as long as 3 days. He reported an average incubation period of 44.9 days, with 28 cases ranging from 39 to 52 days. A similar 45.6 day

period was found by Leschner (1976) for 10 cases, with a range of 42 to 49 days. The time between pipping and hatching is typically less than 24 hours. Wilson (1977) reported that during two years a sample of 162 burrows with eggs had an overall hatching success of 63.5 percent, but a high rate of nest desertion apparently was induced by the study itself.

GROWTH AND SURVIVAL OF YOUNG. Nestling periods in this species are fairly prolonged and rather variable, apparently as a result of local or yearly variations in food supplies. The average period has ranged from as short as 48.3 days to as long as 56 in various studies, but with reported ranges of 35 to 60 days (Wehle 1980). Active brooding is done only for a few days, rarely as many as 9. Until a week or so before fledging the young are fed once or twice a night, and adults typically carry back rather substantial loads of fish in their bills—as many as 13 fish but averaging 6.4 in 37 cases. Collectively the loads may weigh up to about 30 grams (Richardson 1961), but they average about 16.8 percent of the bird's body weight (Sealy 1973b). Vermeer and Cullen (1979) judged that the average meal fed to the chicks was 29.6 grams, or 11.4 percent of adult weight, and that the mean fledging weight was 361 grams, or 69.4 percent of adult weight. They found that chicks hatching later in the season grew more slowly and reached a lower weight peak before fledging than did those hatching earlier. Leschner (1976) similarly found major differences in average number of fish per load and average total load of fish brought to nestlings in two different years, which she believed was responsible for differences in growth rates and asymptotes reached by chicks during the two years. Vermeer (1980) reported that the relative timing and types of prey available for feeding chicks may thus influence the reproductive success of this species. On the other hand, Leschner observed a fairly high fledging success rate of 46 fledged chicks from 56 hatched eggs, or 82.1 percent, without major differences in the two years of her study. Fledging occurs at night and has not yet been described in detail, but it is likely that the chicks flutter or walk to water in the manner of other puffins.

BREEDING SUCCESS AND RECRUITMENT RATES. Wilson (1977) judged that there was a fairly high rate of nest desertion as a result of his study, and he estimated that the hatching success of undisturbed burrows was 30–40 percent higher than in disturbed burrows, or about 86 percent. He estimated a chick mortality of 3.1–7.4 percent before fledging, suggesting an overall breeding success rate of about 82 percent for undisturbed burrows. Leschner (1976) reported a very low hatching success rate of 29.7 to 44.0 percent in the two years of her study, with similar high rates of desertion. Excluding deserted nests, the hatching success would have been 50 percent. She found a prefledging mortality rate of 18 percent as noted, with most chick losses apparently caused by peck wounds from strange adults whose burrows the young might have wandered into. This suggests a minimum breeding success of 41 percent for undisturbed nests. A third estimate of nesting and fledging success is that of Leschner and Burrell (1977), from Chowiet Island, Alaska, where 32 chicks hatched from 45 eggs (71.1 percent) and 23 fledged young were produced (71.9 percent), providing an overall breeding success rate of 51.1 percent. Similarly, Summers and Drent (1979) reported that on Cleland Island, British Columbia, 49 eggs resulted in 44 chicks (90 percent) and 13 of 18 hatched chicks fledged (72 percent), for an overall breeding success of 66 percent. However, in some twinning experiments only 8 of 26 young survived to fledging, and in no case of artificially twinned broods did both chicks survive to fledging. All of these studies suggest that under undisturbed conditions an approximate 50–60 percent breeding success might be typical of rhinoceros auklets. There do not appear to be any estimates of the incidence of nonbreeders in the population or of recruitment rates.

Evolutionary History and Relationships

All recent workers seem to agree that this species is more closely related to the puffins than to the other auklets. Its bill structure is somewhat less modified for digging than that of the typical puffins, but its hind limbs are similarly modified for terrestrial locomotion (Hudson et al. 1969; Kozlova 1961). Storer (1945) considered the rhinoceros auklet the most primitive of the puffins, with the other species having more highly modified bills, claws of the second toe, and coloration of the feet and mouth linings. There are also similarities in the sternum, pelvis, and esophageal structures of the puffins and the rhinoceros auklet (Kuroda 1954).

Population Status and Conservation

In California the rhinoceros auklet was extirpated from the Farallon Islands in the mid-1800s but has recently returned and is probably increasing there. Most California breeding occurs on Castle Rock, where the birds are also possibly now increasing (Sowls et al. 1980). The species may be increasing its range to a limited extent elsewhere in California as well (Scott et al. 1974). The Oregon population appears to be a healthy one (Varoujean 1979), although no trends have been mentioned. In British Columbia the population trend is probably upward, but it is uncertain in both Washington and southeastern Alaska (Manuwal and Campbell 1979). About half of Washington's population nests on Protection Island, and only a third of this island's nesting birds are found within a bird sanctuary. However, removal of sheep and other grazing animals from the area in the early 1960s has allowed for a substantial population increase since that time (Wilson 1977). In the Aleutian Islands and elsewhere in Alaska the rhinoceros auklet has apparently been eliminated from much of its original breeding range because of the introduction of arctic foxes into many of its breeding islands, and Forrester Island has the only remaining large colony in the state (Sowls, Hatch, and Lensink 1978). King and Sanger (1979) have assigned the species an oil vulnerability index of 74, less than the average for the entire family Alcidae.

Tufted Puffin

Fratercula cirrhata (Pallas)

OTHER VERNACULAR NAMES: Sea parrot; old man of the sea; toporkie (Aleut); macareux huppé (French); Schoplund (German); etopirika (Japanese); toporik (Russian); pugharuwuk (Saint Lawrence Island).

Distribution of Species (See Map 26)

BREEDS from the Kolyuchin Islands, East Cape, and the Diomede Islands to Kamchatka, the Commander Islands, Kurile Islands, the Sea of Okhotsk, Sakhalin, Maritime Province, and Hokkaido; and from Cape Lisburne south through the Bering Sea to the Aleutian Islands, Kodiak Island, Kenai Peninsula, southeastern Alaska, British Columbia, Washington, Oregon, and central California (Farallon Islands, formerly to Anacapa Island).

RESIDENT except in the Far North; wanders north to Point Barrow, Alaska, and south to Honshu, Japan, and San Nicolas Island, California.

Description (Modified from Ridgway 1919)

ADULTS IN BREEDING PLUMAGE (sexes alike). Upperparts slightly glossy sooty black, passing into uniform dark sooty brown (dark clove brown) on sides of head and neck, the chin, throat, and foreneck very slightly

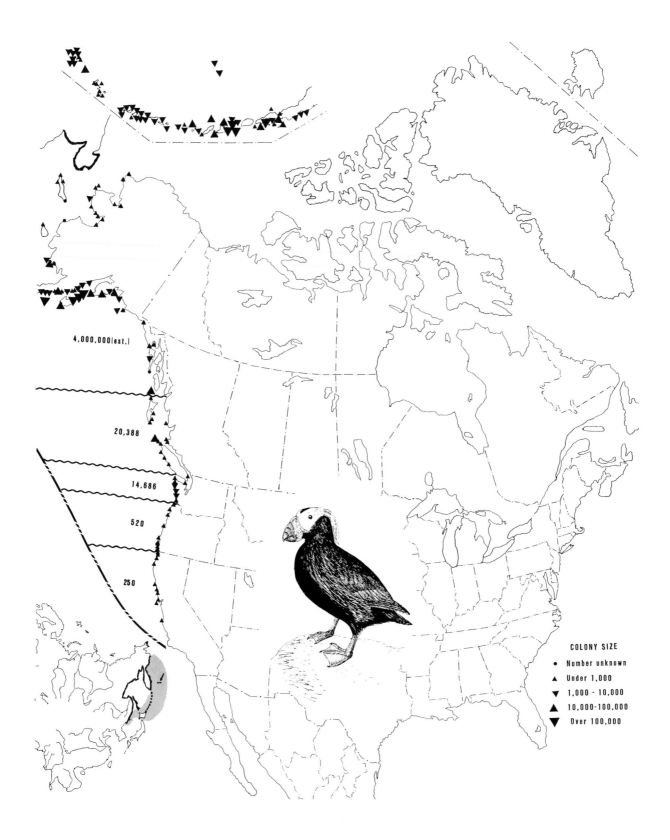

COLONY SIZE

- Number unknown
- Under 1,000
- 1,000 - 10,000
- 10,000-100,000
- Over 100,000

4,000,000(est.)

20,388

14,686

520

250

26. Current North American distribution of the tufted puffin, including colony locations and limits of nonbreeding range (*broken line*). The Asian range is shown on the inset map.

lighter, passing into deep grayish brown (between fuscous and benzo brown or hair brown) on underparts of body; under wing coverts uniform deep brownish gray or hair brown; anterior portion of forehead, whole of loral, orbital, and rictal regions, and anterior portion of malar region and chin immaculate white; elongated postocular tufts naples yellow to creamy buff; distal half (approximately) of bill salmon, basal portion light olive green, the cylindrical ridge more apple green; rictal rosette mostly purplish flesh color, the narrow line of skin between base of bill and feathering of head, rictus, part of the rosette, and naked eyelid vermilion; iris creamy white; legs and feet bright salmon, the soles reddish brown, claws black. Two-year-old birds have white feathers on the sides of the head, short white tufts behind the eyes, and a white bill ring. The bill is shallowly grooved toward the tip and is usually smaller than in adults (Kozlova 1961).

WINTER PLUMAGE. Sides of head wholly dusky, but lighter in region of insertion of nuptial plumes, which are wholly absent; horny nasal cuirass, basal lamina, and other deciduous parts covering basal half of bill absent and replaced by dusky brown membrane; otherwise as in summer, but legs and feet paler red. Basal portion of bill covering brownish, terminal part red. The first-winter plumage is very similar to the juvenal plumage but has more distinct brownish gray edging on the white chest and underpart feathers. By the following spring these edges wear away, resulting in white underparts (Kozlova 1961).

JUVENILES. A vague gray stripe behind eye, chest and abdomen white with brownish gray markings, otherwise upperparts similar to those of adults in winter. The iris is brownish gray, the bill is brownish, and the feet are light gray (Kozlova 1961).

DOWNY YOUNG. Uniform dark sooty grayish brown, paler below, especially on abdomen, where inclining to sooty gray. A small percentage of birds have white underparts (Wehle 1980).

Measurements and Weights

MEASUREMENTS. Wing: males 187–206 mm (average of 11, 194.4); females 179.5–196.5 mm (average of 9, 189.2). Culmen: males 53–65 mm (average of 11, 57.1); females 54.6–60.0 mm (average of 9, 57.1) (Ridgway 1919). Eggs: average of 43, 72 x 49.2 mm (Bent 1919).

WEIGHTS. A sample of 62 adult males during the breeding season averaged 792.3 g (range 732–986 g), and 51 adult females averaged 734.9 g (range 643–862). A sample of 117 eggs averaged 94.3 g (Wehle 1980). Estimated fresh egg weight, 91 g (Schönwetter 1967). Newly hatched chicks average 58 g (Thoreson, in press).

Identification

IN THE FIELD. This Pacific-coast puffin is the only one with blackish underparts and the only puffin with a hanging tuft of feathers behind the eyes. Breeding adults have whitish eyes, and the reddish bill is tinged with greenish basally. In winter the face becomes completely dark brown, the tufts are lost or are rudimentary, and the underparts become more whitish but remain distinctly darker than in the horned puffin. This species is relatively silent but produces growling notes near the nest.

IN THE HAND. This is the only puffinlike species that has a dark gray to blackish underpart coloration, especially in adult breeding birds. Immature birds are paler underneath but are still distinctly dusky, and this feature plus their bills (with inward-curving grooves) separates them from young of the other puffin species.

Ecology and Habitats

BREEDING AND NONBREEDING HABITATS. This species is widely distributed along the western coast of North America and the eastern coast of Asia between the areas having August surface water temperatures of about 5°C and 18°C. However, population densities are greatest toward the northern portions of its North American range, and relatively few birds breed south of Alaskan waters. The species has a similarly broad tolerance for diverse nesting habitats, including grassy slopes, rocky slopes, sodded boulder rubble, cliff faces, and cliff edges, and it is able either to use natural crevices in rocks or to excavate burrows in softer substrates. During the nonbreeding seasons the tufted puffin is probably the most pelagic of all alcids, with birds often occurring as far as 800 kilometers from land in the Gulf of Alaska. There the densest winter concentrations have been found to occur in a band about 100–200 kilometers offshore, while in California the highest numbers have been seen within 50 kilometers of land. Even during summer periods the birds have been seen as far as 300 kilometers from land, though most birds are much closer to land at such times (Sanger 1975).

SOCIALITY AND DENSITIES. In some areas the density of breeding birds is extraordinarily high; the largest Alaskan population may be that of Kaligagan Island (a tiny island south of the western end of Unimak Island), where about 375,000 birds breed. The average number of birds reported at all the 502 known Alaskan breeding sites is nearly 8,000 (Sowls, Hatch and Lensink 1978). On Triangle Island, British Columbia, over 16,000 pairs were estimated to nest in an area of less than 9 hectares, about 2,000 pairs per hectare (Vermeer 1979). Highest average breeding densities of 29.3 burrows per 25 square meter plot were found in areas of clifftop with hairgrass (*Deschampia*) cover, especially where there were steep slopes and narrow strips of sloping cliff perimeter just above cliff faces. Amaral (1977) estimated densities of from 0.19 to 0.69 burrow per square meter in various habitat types on East Amatuli Island, Alaska. On 13 hectare Destruction Island, Washington, where some 350 pairs nest (27 pairs per hectare), most nests are within a few meters of the tops of the steepest and least vegetated cliff faces (Burrell 1980).

PREDATORS AND COMPETITORS. Amaral (1977) found that on East Amatuli Island, Alaska, potential predators included common ravens (*Corvus corax*), gulls, peregrines (*Falco peregrinus*), bald eagles (*Haliaeetus leucocephalus*), and river otters (*Lutra canadensis*). No evidence of predation was found for ravens or gulls, but peregrines evidently occasionally took adult birds, and bald eagles were found to be feeding their chicks in part on tufted puffins. Likewise, river otters were found to prey on adults and especially on nestlings when they were available. Wehle (1980) reported egg losses to glaucous-winged gulls (*Larus glaucescens*) and ravens but considered these losses minimal because of the inaccessibility of nest sites. Fred Zeillemaker (pers. comm.) saw a parasitic jaeger (*Stercorarius parasiticus*) chase a tufted puffin out to sea, where it finally escaped by plunging into the ocean.

Probably the major competitor of the tufted puffin over most of its range is the horned puffin. Wehle (1976) judged that competition between these two species is mitigated by partially asynchronous breeding cycles, temporal differences in the rhythm of diurnal activity, spatial segregation in foraging areas of fledged birds, differences in adult diet, and differences in foods brought to nestlings. Similarly, Amaral (1977) found that the major basis for niche segregation between these species was differences in preferred nesting habitats (tufted were predominantly burrowers, horned were crevice nesters), with less substantial differences in temporal breeding cycles (tufted earlier nesters) and diurnal rhythms. Nestling foods and activity patterns of adults during the nestling periods overlapped considerably, though in general tufted puffins tended to rely more on benthic fish species than did horned puffins. Horned puffins also have a smaller wing loading than tufted puffins and thus can probably use lower nesting sites

and more gentle slopes (Lehnhausen 1980). On Triangle Island, where tufted puffins nest in large numbers together with rhinoceros auklets, these two species differ in nesting sites (tufted nesting at higher elevations and steeper slopes, for easier takeoff), in circadian activity rhythms (tufted are diurnal, rhinoceros auklets are nocturnal), foraging locations (tufted feed mostly offshore, rhinoceros auklets feed inshore and offshore), and in foods brought to their young (in part based on relative diurnality of foraging behavior in the two species).

General Biology

FOOD AND FORAGING BEHAVIOR. Wehle (1980) has analyzed the foods of all the species of puffin based on his own data and that of other investigators. In 280 samples of adult tufted puffins he summarized, the relative frequencies of major food types of adults included fish (52 percent), squid (37.8 percent), crustaceans (7 percent), and polychaetes (2.9 percent). In general, a larger number of pelagic or offshore species of fish are utilized by the tufted puffin than by the other species, although as in the others there is a good deal of seasonal, yearly, and geographic variation in its diet. Hunt, Burgeson, and Sanger (1981) examined 23 adult birds from the vicinity of the Pribilof Islands and found a predominance of fish in their diets, especially walleye pollock (*Theragra chalcogramma*), which made up almost half of the total food. Subtidal, inshore species were lacking, and nereid worms (Polychaeta) were the major invertebrate food. A sample of 19 adults from south of the Aleutians and the eastern Bering Sea contained a mixture of fish, squid, and amphipods (Sanger 1975), while 89 birds from the Kodiak Island area exhibited a high rate of use by osmerids, especially capelin (*Mallotus villosus*), a small incidence of squid (13.5 percent), and no crustaceans. Wehle (1980) reported that during seven colony-years of study either Pacific sand launce (*Ammodytes hexapterus*) or capelin was the most common prey species brought to nestlings, with these species constituting over 90 percent of the food delivered in five of the seven years. Other foods brought to nestlings included squid, octopus, sculpins (*Hemilepidotus jordani*), greenling (*Hexagrammus stelleri*), Atka mackerel (*Pleurogrammus monopoterygius*), and cod (*Eleginus gracilis*). Squid and octopus may be the most important food types for tufted puffin nestlings exclusive of sand launce and capelin, whereas in the horned puffin greenling and cod seem to be the prime subsidiary foods.

Prey are captured by extended dives, often in very deep waters. Wehle (1980) summarized a variety of data on average bill loads of this species, finding a range of mean weights varying from 7.5 to 20.4 grams and a mean number of prey items varying from 3.4 to 10.1. Chicks up to a week old were fed an average of 1.6 times per day, and those from 4 to 6 weeks old were fed an average of 3.8 times per day. Early in the chick-raising period, adult tufted puffins were found by Amaral (1977) to carry an average of 1.3 fish per load, while later the average increased to 3.8. Although older chicks are fed more often than younger ones, there is no apparent tendency for the adults to select larger prey for these older nestlings (Amaral 1977).

MOVEMENTS AND MIGRATIONS. Evidently tufted puffins can be found well out to sea at all times of the year; summer observations of such birds probably represent immature nonbreeders. The fall and winter movements of the birds evidently occur within the overall latitudes of the species' breeding distribution, although a few individuals may penetrate farther south. Generally the birds winter from the northern limits of open water southward, including the Aleutian Islands and occasionally also as far north as the Pribilof Islands (Wehle 1980). Probably most birds wintering in bay habitats are immatures; adults tend to occur in deeper waters and farther offshore. The birds also show a tendency to be solitary when at sea; Sanger (1975) noted that of 170 birds seen during winter in the Gulf of Alaska 65 percent were single birds and 25 percent were pairs. Limited Alaskan observations suggest that the distribution of the birds may be limited by water temperatures of less than 4°C to 6°C (Gould, Forsell, and Lensink 1982).

Social Behavior

MATING SYSTEM AND TERRITORIALITY. All the available evidence indicates that prolonged monogamy is typical of this species. Wehle (1980) noted that the same birds were present in at least 2 of 7 burrows where marked birds had nested the previous year. According to him, tufted puffins defend a territory that includes the burrow entrance, the path to the burrow, and a specific area used for landing and resting within the colony, often an earthen mound or protruding rock. This territory usually had a radius of less than half a meter from the burrow entrance. Territorial defense is strongest during the prelaying stage of reproduction and gradually declines through the breeding season.

VOICE AND DISPLAY. Amaral (1977) reported that this species is rather silent, uttering little more than a low growl when caught or harassed, though the young birds vocalize frequently, especially when being fed. Wehle (1980) recognized four vocalizations, the most common of which was a short, low-pitched *errr*, a threat or warning note uttered upon being disturbed. He also described

a softer, purring call similar to the threat call but lasting much longer and of unknown function. A third vocalization was the "bisyllabic call," which consists of a short *er* followed by a second higher-pitched note. This call was heard most often among birds in the colony but was sometimes uttered on the water as well. Finally, a multinoted call was recorded, consisting of at least three syllables, with the final syllable repeated many times, varying rhythmically in frequency and intensity and producing a sirenlike effect. This call was heard most often during the prelaying and incubation phases and might have some sexual function.

According to Wehle (1980), courtship in tufted puffins occurs on waters near the breeding colony, among flocks of rafting birds. Courting males typically lower the back of the head to the shoulders while holding the bill parallel. The bill is then repeatedly raised nearly to the vertical while being opened, and lowered to the resting position while being closed (fig. 46E). A vocalization possibly is associated with this movement. The performing male usually follows a female at a distance of several meters, and if she is receptive to copulation she will swim rapidly ahead of him and assume a crouched position on the water. As the male approaches, he increases the rate of his head jerking, and the movements become more exaggerated. When he is within a meter or so of her he flaps his wings, rises out

46. Social behavior of tufted puffin (mostly after Wehle 1980): A, wing flapping; B, yawning; C, gaping; D, head bowing; E, head jerking; F, copulation; G, postcopulatory wing flapping.

of the water, and lands on her back (fig. 46F). During copulation the female sinks down so that only her head is above water, while the male continues to flap his wings and sometimes also continues his head jerking. He may also peck the female's nape. Copulation normally ends with the female's diving. Afterward one or both of the participants usually wing flap (fig. 46G). Several other displays or possible displays occur at other times and are unrelated to the sex of the individual. These include bill dipping, wing flapping (fig. 46A), billing, bowed head posture (fig. 46D), and bill gaping (fig. 46C). Bill dipping and wing flapping may simply be comfort movements rather than actual displays, but bill gaping is the most important threat display of the species. It is rather similar to yawning (fig. 46B), but during yawning the neck feathers are not strongly ruffled and the tongue is usually not visible as it typically is during gaping. During the bowed head display the bird stands on the ground with its head held low and the body almost horizontal and tilts its bill strongly downward. The head is then slowly swung from side to side, the body is sometimes convulsed, and a vocalization may possibly be uttered. This display is most often performed near a burrow entrance, usually facing inward. Often the behavior has the effect of drawing the bird's mate to the burrow, at which point mutual billing may occur. As in other puffins, billing is an important social display between mates, and it often occurs after one bird lands next to its mate in the colony. It often also occurs just before both birds enter their nest site, after an aggressive interaction with an intruder, and after a bowed head display by one of the partners. In all these cases it seems to function as a pair-bonding or pair-maintaining display. One last display is a landing display, performed immediately after landing in a colony. Once on the ground, the bird holds its body low while stretching its wings upward and holding its head in line with the body or bending it down to varying degrees. It may then take several exaggerated steps while in this posture before adopting a normal stance (Wehle 1980). As noted earlier, somewhat similar postlanding postures occur in guillemots and murres.

Reproductive Biology

BREEDING SEASON AND NESTING SUBSTRATE. California egg records extend from April 30 to July 8, with a peak between May 27 and June 17 (Bent 1919). Egg records cited by Bent from Washington are from May 30 to July 23, but Burrell (1980) indicated a mean egg-laying date for Destruction Island of May 16, with extremes of May 6 and June 8, and a hatching range from June 21 to July 24, with a mean of July 1. Various

egg records from British Columbia are from June 12 to July 24, and unfledged birds have been observed as late as August 25. On East Amatuli Island, Alaska, egg laying occurred from late May to late June, with 90 percent of the eggs laid between June 1 and 15 (Amaral 1977), and a study on Ugaiushak and Buldir islands indicated that on Ugaiushak Island the peak of the laying occurred 7–10 days earlier, although the onset of laying was at about the same time. However, the period of egg laying, including replacement clutches, generally lasted about a month, and the egg laying period in tufted puffins was 1–3 weeks earlier than in horned puffins. Over the entire geographic range egg laying in tufted puffins tends to peak between the last week of May and mid-June (Wehle 1980). The tufted puffin prefers to nest in burrow sites on grassy slopes, though the birds also often nest in rocky crevices, especially where burrow sites are unavailable. Amaral (1977) reported the highest nesting densities in steep sea-facing slopes with grassy and herbaceous cover, and he noted progressively lower densities on cliff edges adjacent to vertical slopes and with grassy cover, in rock crevices, and on gradual slopes of about 45°. Lehnhausen (1980) judged that slope angle, vegetation cove, and the presence of other birds are major factors influencing nest site choice in tufted puffins. The birds favored steep slopes and low or sparse vegetation and avoided vegetation forming dense mats or having dense, intertwining root systems. Tufted puffins can probably outcompete the smaller horned puffins for nest sites, and, though colonial, they tend to maintain some spacing relative to other tufted puffins. Other factors that might influence nest site choice include proximity to landing and takeoff sites, soil particle size, soil moisture, and soil depth.

NEST BUILDING AND EGG LAYING. Tufted puffins spend considerable time excavating and cleaning nest sites, sometimes beginning with their return to the colony and at other times variably later, evidently depending on the damage sustained over the winter and on when the burrows become ice-free. Both members participate, but most is done by the larger bird, presumably the male. Excavation is done with both the bill and the feet, the bill being used mainly as a chisel for excavating or as pliers for tearing and wrenching. Rocks as much as twice the weight of the bird might be removed during excavation. Burrows are probably not used the same year in which they are initially excavated; thus subadults probably excavate burrows during the year before initial breeding. Egg laying was found by Wehle (1980) to begin 3–4 weeks after initial arrival and probably was dependent upon relative accessibility of nest sites. Although two incubation patches are present, the birds lay only single eggs and will apparently incubate only one even if a second egg is provided them, according to Wehle. Wehle also removed 10 freshly laid eggs from burrows and found that 7 of these burrows subsequently contained replacement eggs. From 10 to 21 days elapsed between egg removal and egg replacement.

INCUBATION AND BROODING. Both sexes incubate, typically exchanging places in the morning, again in midafternoon, and usually again before darkness (Amaral 1977). However, birds may at times incubate for an entire day and at other times may leave the egg unattended for a day or longer (Wehle 1980). The average incubation period was determined by Amaral (1977) to be 45.2 days (range 43 to 53 days for 11 eggs) and by Wehle (1980) to be 46.5 days (range 42 to 53 days for 35 eggs). Continuous incubation may not begin for as long as 4 days after the egg is laid, and there is also considerable variation in the hatching (initial cracking or pipping to emergence) period, which averages 3–4 days but may vary from 1 to 12 days (Wehle 1980).

SURVIVAL AND GROWTH OF YOUNG. Brooding of the chick is more or less continuous for the first several days after hatching, though after 3–5 days the adults are normally not present in the burrow except while feeding their chicks. Wehle (1980) reported a total average nestling period of 43.6 days for 13 chicks, and Amaral (1977) noted a 47 day average for 9 chicks. Variations in fledging periods result from differential feeding abilities of the adults. Amaral noted an average of 1.6 to 4 feedings (the larger numbers typical of older age-classes) per 24 hour period and an average food load of 14.9 grams, with about 95 percent of the prey brought to chicks being capelin. Usually the first feeding is shortly after sunrise, with a second feeding peak at midday and a final peak before sunset. In some cases chicks were fed up to 6 times a day, but normally a trimodal pattern of colony attendance was typical in the colony Amaral studied. He reported a chick weight of 69.7 grams shortly after hatching and a fledging weight of 550 grams, and Burrell (1980) reported a fledging weight of 496.8 grams and a maximum average prefledging weight of 551.8 grams. Immediately before fledging the chicks spend increasing amounts of time at the burrow entrance, sometimes exercising their wings. Fledging occurs at night or during early morning hours, and evidently most birds are still flightless at the time of nest departure. The birds apparently walk or flutter down the nesting slopes toward the sea without parental involvement, and on entering the sea they gradually work their way offshore, occasionally diving (Amaral 1977).

BREEDING SUCCESS AND RECRUITMENT RATES. About half of the active tufted puffin burrows never have eggs laid in them at all; many of these presumably are inhabited by subadult birds. Wehle (1980) has summarized hatching and fledging success rates for a variety of areas and studies. These studies suggest an average hatching success of about 50 percent, but with substantial variations that at least in part are the result of differential human disturbance. Probably most egg losses under natural conditions are caused by infertility or death of the chick at the time of hatching. Wehle estimated that a natural hatching success of 75–90 percent is probably typical of undisturbed birds. He also judged that fledging success is probably 60–70 percent, with major variations the result of weather, food availability, and predation or kleptoparasitism (stealing by gulls of fish being brought to chicks). Evidently most chick losses occur during the first 2 weeks after hatching, with the major cause of death in older chicks being lack of food. There are no available estimates of recruitment rates for this species, but if only half the active burrows contain eggs and there is an average reproductive success of about 0.5 young per adult pair, it is likely to be no more than about 10 percent, assuming that a substantial if unmeasurable part of the total population consists of immature nonbreeders.

Evolutionary History and Relationships

Clearly this species is a close relative of the typical *Fratercula* puffins, though it lacks such features as the epidermal adornments around the eyes and has a unique "roll" along the crest up the basal portion of the upper mandible. It also has longer hind limbs than the typical puffins, and there are some minor differences in bill and pterygoid structure (Kozlova 1961). Hudson et al. (1969) judged that *Lunda* and *Fratercula* are very closely related, and Strauch (1977) considered the puffins (including *Cerorhinca*) a sister group to all other alcids. It seems possible that *Lunda* is an evolutionary link between *Cerorhinca* and *Fratercula* on the basis of its burrowing rather than crevice-nesting tendencies and some similarities in display posturing (such as the head bowing display), but it is questionable whether a separate genus for it is warranted.

Population Status and Conservation

The highest populations of tufted puffins in North America occur in Alaska, where more than 500 colonies have been reported and an estimated population of 4,000,000 birds occurs (Sowls, Hatch, and Lensink

1978). The largest population south of Alaska is at Triangle Island, British Columbia, which contains about 80 percent of the British Columbian population of this species (Vermeer 1979). The British Columbian population is probably stable, but that in Washington may be declining (Manuwal and Campbell 1979). The population in Oregon is relatively small, and its trends have apparently not been characterized. In California the species' population has declined greatly from those known to be present in historical times, and its range has contracted northward from an original limit in the Channel Islands. Oil pollution and a crash in the Pacific sardine population have been cited as possible causes (Sowls et al. 1980). The species has been assigned an oil vulnerability index of 72 (King and Sanger 1979).

Atlantic Puffin

Fratercula arctica (Linnaeus)

OTHER VERNACULAR NAMES: Common puffin; sea parrot; lunde (Danish); macareux moine (French); Papageitaucher (German); qilangag (Greenland); lundi (Icelandic); tupik (Russian); lunnefågel (Swedish).

Distribution of North American Subspecies (See Map 27)

Fratercula arctica arctica (Linnaeus)
BREEDS from western Greenland south along the coasts of western Greenland and Labrador to southeastern Quebec, Newfoundland, southern New Brunswick, and eastern Maine; and on Iceland, Bear Island, and northern Norway.

WINTERS in western Atlantic waters from the ice line south to Massachusetts, casually to southern New Jersey; on the European side to the Faeroes and western Sweden, rarely to Denmark.

Fratercula arctica naumanni Norton
BREEDS in northern Greenland, on Jan Mayen, Spitsbergen, Novaya Zemlya, and the Murmansk coast, intergrading with *F. a. arctica* in Finland and east along the Kola Peninsula.

WINTERS in adjacent seas.

Description (Modified from Ridgway 1919)

ADULTS IN BREEDING PLUMAGE (sexes alike). Pileum uniform deep grayish brown (deep fuscous or chaetura drab) passing into light brownish gray on anterior portion of forehead; rest of upperparts, together with sides of neck and a broad band across foreneck, uniform black, the band across foreneck, however, inclining toward color of pileum, the black of nape sharply defined against the dark grayish brown of pileum; sides of head very pale brownish gray or almost grayish white on loral and suborbital regions, passing into light mouse gray posteriorly (including supra-auricular region), the malar region mostly mouse gray passing into grayish white next to base of mandible (narrowly) but posteriorly separated from the blackish of neck by a space of pale brownish gray; underparts of body, together with lower foreneck, immaculate white; under wing coverts light brownish gray; bill with basal lamina of maxilla and first ridge of both maxilla and mandible dull yellow, the nasal cuirass and basal portion of mandible grayish blue or bluish gray, the remainder vermilion, the tip of mandible and terminal grooves yellowish; rictal rosette gamboge yellow; inside of mouth, together with tongue, yellow; iris dark grayish to grayish brown; eyelids vermilion, the callosities bluish gray or grayish blue; legs and feet vermilion or coral.

WINTER PLUMAGE. Similar to the breeding plumage except for color and form of basal portion of bill, color of anterior portion of sides of head, and other minor details; nasal cuirass and basal lamina of bill absent and replaced by membrane of brownish black; rictal rosette much reduced and dull purplish red instead of yellow; eyelids dull purplish red and destitute of the callous appendages; whole of loral and orbital regions blackish; legs and feet paler red. First-winter birds are darker around the eyes and on the lores than are adults, and second-winter birds may be distinguished from adults by the bill shape, which is more slanted and less bent toward the tip (Kozlova 1961).

JUVENILES. Similar to winter adults, but with bill much smaller, much duller in color, and without grooves or ridges.

DOWNY YOUNG. Plain dark sooty grayish brown, paler below, the breast and upper abdomen dull grayish white or pure white. The legs and feet are black, the iris is brown, and the mouth is pale flesh color. The bill is dark reddish gray.

Measurements and Weights

MEASUREMENTS (of *arctica*). Wing: males 158–68 mm (average of 14, 162.8); females 157–68 mm (average of 14, 162.9). Culmen: males 45.0–53.5 mm (average of 14, 49.8); females 44–51 mm (average of 14, 48). Eggs: average of 41, 63 x 44.2 mm (Bent 1919).

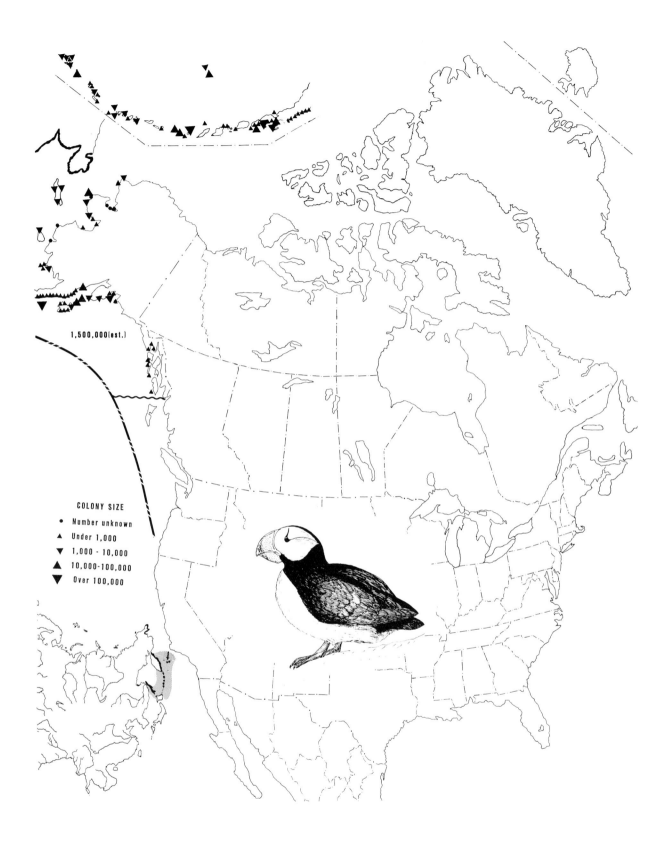

COLONY SIZE
- • Number unknown
- ▲ Under 1,000
- ▼ 1,000 - 10,000
- ▲ 10,000-100,000
- ▼ Over 100,000

1,500,000(est.)

27. Current North American distribution of the Atlantic puffin, including colony locations and limits of nonbreeding range (*broken line*). The European range is shown on the inset map.

WEIGHTS (of *arctica*). Nettleship (1972) reported that 39 males from Newfoundland averaged 479 g (range 429–524), while 57 females averaged 445.5 g (range 386–511). These weights are somewhat less than averages reported for *arctica* in the USSR by Dementiev and Gladkov (1968), especially for more northerly populations. The average of 150 eggs from Newfoundland was 65.3 g (Nettleship 1972). Newly hatched young average 48 g (Glutz and Bauer 1982).

Identification

IN THE FIELD. In breeding plumage this species is unmistakable; it is the only Atlantic-coast puffin, characterized by a rounded white face and a semicircular and colorfully banded bill that is dark basally. The horned puffin is similar, but their ranges do not overlap, and the bill of the horned puffin lacks a black band near the base. In winter the two species are very similar, but in the horned puffin the area around the lores and chin is darker, almost as dark as the crown. In both species the white facial pattern is largely lost during winter. Low purring notes are uttered in flight, and a low, deep *awe* note is also produced.

IN THE HAND. The distinctive puffin bill separates the species from all others except the horned puffin. At least adults of the two species can be separated on the basis of tail length (no more than 53 mm in the common puffin, compared with at least 60 mm in the horned), and by the more oblique grooving of the bill of the Atlantic puffin. Distinction of immature birds is more difficult, but in all ages and plumages the Atlantic puffin has a light gray to whitish chin and throat, while that of the horned puffin tends toward brownish black.

Ecology and Habitats

BREEDING AND NONBREEDING HABITATS. This species occupies rocky coastal areas of the North Atlantic from temperate to arctic seas having August surface temperatures of from about 0°C to 17°C (Voous 1960). Coastlines that have sharp cliff edges where the slope angle is too high for most predators to reach and that provide landing sites near the burrow so it is difficult for gulls to rob adults returning with food for the chicks (Nettleship 1972) are preferred for nesting over those with gradual slopes or with landing sites well separated from the nesting burrows. Outside the breeding season the birds are pelagic, spending much of the winter period along pack ice areas of the North Atlantic well away from coastlines.

SOCIALITY AND DENSITIES. This is a highly social species, and the distribution of burrows in colonies tends to be strongly aggregated with respect to particular habitat features (Grant and Nettleship 1971). Thus densities are highest on steep slopes adjacent to the sea and having associated turfs for easy burrowing. In Great Britain there are local densities of burrows as high as 1.7 per square meter, while in Iceland the average breeding density over a large area (of 22.5 hectares) was 0.66 burrow per square meter, with local maxima as high as 2.72.

PREDATORS AND COMPETITORS. Various predators of nestlings or newly fledged young have been reported as possibly affecting this species, including domestic cats (*Felis cattus*), crows (*Corvus* spp.), and various gulls including herring gulls (*Larus argentatus*). It is possible that in some areas fledglings may at times be killed by common ravens (*Corvus corax*) as the chicks fly out to sea (Myrberget 1962), and it is probable that some adults are taken by the larger falcons or other raptors. Eggs, nestlings, and adults may also be taken by foxes (*Alopex* and *Vulpes*), larger weasels (*Mustela*), and otters (*Lutra*), while rats (*Rattus*) may locally steal and destroy eggs (Kartashev 1960). In many areas gulls, crows, jaegers (*Stercorarius* spp.), or jackdaws (*Corvus monedula*) that steal from adults returning to feed their young may have an important effect on rearing success (Corkhill 1973; Nettleship 1972). Additionally, egg loss resulting from disturbance by nest-prospecting Manx shearwaters (*Puffinus puffinus*) has been found to be locally significant (Ashcroft 1979).

General Biology

FOOD AND FORAGING BEHAVIOR. Relatively little information on the food of adult Atlantic puffins is available for North America, and even the data on European birds are limited by comparison with information on tufted and horned puffins. Wehle (1980) summarized information obtained from stomach analysis of 117 Atlantic puffins and reported a concentration on fish, which based on frequency of occurrence represented 83.1 percent of the collective sample, while polychaetes and crustaceans composed 11.8 and 5.1 percent respectively. Most of the fish taken during the chick-rearing period are schooling species such as sand launce (*Ammodytes*), sprats (*Sprattus sprattus*), capelin (*Mallotus villosus*), and herring (*Clupea harengus*), with supplemental use of various forms of cod (*Pallachius, Gadus, Merlangius, Ciliata, Gaidropsarus*) (Glutz and Bauer 1982). In Great Britain young puffins have been found to be reared primarily on sand launce, sprats, and herring (Harris and Hislop 1978). Most individual fish that are brought to chicks are 10–15 centimeters long and rarely may range up to 33 grams in weight. Typically from 5 to 12 rather small prey items (averaging about a gram) are brought back per load by adults, but as many as 62 prey in a single load have been reported by Harris and Hislop. These authors also observed a usual variation of feeding frequencies of from 3.8 to 15.7 trips per day, with a maximum of 24 trips recorded. Pearson (1968) observed a considerable range in weights of individual prey fish and in the length of sand launce captured, and also a substantial overlap in these respects with prey taken by eight other species of seabirds breeding in the same general area. However, Swennen and Duiven (1977) reported that hand-raised puffins preferred prey fish (of *Clupea* and *Trisopterus*) offered in an experimental situation that averaged 4–6 grams and had an average length of 15 millimeters. This prey size was identical to that preferred by razorbills, which are substantially larger birds.

MOVEMENTS AND MIGRATIONS. The migration routes off eastern North America are still not well documented. After the breeding season the birds are most commonly found in the Labrador Sea and off southeastern Labrador, although their breeding origins are still uncertain. In any case, migration away from the arctic breeding areas is almost complete by October. Movements after that period are undocumented (Brown et al. 1975). Breeding birds and young from Greenland evidently begin to move south directly after the breeding season, with part of the population remaining in the southern parts of the Davis Strait (as far north as Sukkertoppen) but most birds leaving Greenland waters and probably following the Labrador Current south to Labrador and Newfoundland. There is some dispersal eastward to the open Atlantic Ocean, where some subadults may remain through the summer. There is also a westward dispersal of some European birds that may rarely winter as far west as the southernmost post of Greenland's west coast. Birds arrive back on the Greenland breeding grounds during the first half of May (Salomonsen 1967).

Social Behavior

MATING SYSTEM AND TERRITORIALITY. Mate retention and nest site tenacity have been proved typical of this species. Ashcroft (1979) found that over 142 pair-years accumulated for banded birds, there was a 7.8 percent "divorce" rate per year, exclusive of cases where one of the members died. In more than half of these cases one member of the pair was displaced by another bird, and in several cases pairs broke up after both members were evicted from their burrow. Of 7 females whose mates had disappeared, 7 bred in the same burrows with new mates the following year, and in 3 cases where one of

the pair members (the female) was late in returning to the colony the male in each case paired with a new female but re-paired with his original mate when she returned. In general mate fidelity is high, but birds are quick to re-pair when their original mates disappear, and birds evidently only rarely (1 observed case in 13) miss a year's breeding because of losing a previous mate. Apparently obtaining a new mate or changing burrows does not measurably affect breeding success. In 502 bird-years, only 7.8 percent of burrow owners left their burrows each year, either voluntarily or because they were evicted by other birds. Birds that failed as breeders the previous year were found by Ashcroft (1979) to be more than twice as prone to move as were successful pairs. Nettleship (1972) reported an overall 77 percent rate of nest site tenacity for 61 marked birds in two different habitats, with no observations of any of the missing birds in other locations. Territorial defense in Atlantic puffins evidently includes the burrow, its entrance, the path to the burrow, and a raised area used for landing, taking off, and resting (Wehle 1980).

VOICE AND DISPLAY. A full comparative study of puffin vocalizations remains to be done, but Wehle (1980) concluded that Atlantic puffins have four adult vocalizations comparable to those of tufted and horned puffins. These include a short, harsh *urrrr* that functions as a threat call, a deeper, purring *arrr* that corresponds to the purring call of tufted and horned puffins, a bisyllabic call that sounds like *haa-haa* or *co-o-or-aa*, and a long, multinote call that consists of a prolonged *haa-aa . . . aa-aa-aa-aa-aa*. Two call notes of young Atlantic puffins have been described and illustrated by Glutz and Bauer (1982), as have the adult multinote and a short *orr* call (probably the purring call mentioned above) that may serve as a contact note.

Displays of the Atlantic puffin consist of both aquatic and terrestrial courtship and agonistic postures. Although attempts at copulation have been observed on land, all successful copulations evidently occur on the water. As in the other puffins, courtship is social and often involves one or more birds' following a female. At such times the males perform a rapid head jerking display that has a hiccoughlike quality, and in which the bill is quickly raised toward the vertical and returned again to the horizontal, as illustrated in figure 47F for the horned puffin. It is likely that soft vocalizations accompany these head jerking movements. There are at least two types of head raising movements; one appears to be used primarily in various social including aggressive situations, while the other is used as a sexual display. Myrberget (1962) described these two displays as social nodding and sexual nodding. During sexual nodding the bill is sometimes directed vertically upward at

47. Social behavior of Atlantic and horned puffins (after Lockley 1953 and Wehle 1980): A–C, billing sequence, D, aggressive bowing, and E, head nodding by Atlantic puffin; F, head jerking, G, copulation, H, postcopulatory wing flapping by horned puffin.

its maximum intensity, and the display serves as an invitation to mating. This display may be performed on land as well as in the usual courtship situation on water, but when performed on land it is always directed only toward the mate (fig. 47E). Social nodding is performed only on land and is less intense than sexual nodding, with the bill less strongly oriented upward and moved less rapidly back and forth. During gaping the tongue is usually raised and visible, and the neck feathers are strongly ruffled. Aggressive gaping is used frequently and is usually performed with the opened bill oriented toward the opponent, although the bill may also be directed downward in an aggressive bowing display (fig. 47D). The same or a very similar bowing posture may indicate a desire for billing (Myrberget 1962). Billing is frequently performed between members of a pair and often is initiated when one of the birds approaches its mate with a lowered bill (fig. 47A), then raises its bill to meet the lower part of its mate's bill, which is typically pointed somewhat downward. With

the bills in lateral contact, the birds' heads are rapidly moved from side to side for a variable period (fig. 47B,C). The tail may be raised somewhat during the billing ceremony, but the bills remain closed and thus are not clenched together (Lockley 1953). Behavior associated with copulation is apparently almost identical to that described and illustrated for the horned puffin.

Reproductive Biology

BREEDING SEASON AND NESTING SUBSTRATE. Egg records for the Gulf of Saint Lawrence extend from June 6 to July 10, with a peak between June 15 and 26. A small number of records from Newfoundland and eastern Labrador are from June 8 to July 7, and a few from Maine are from June 19 to July 27. Greenland egg records extend from June 1 to July 16, peaking about June 20 (Bent 1919). Egg-laying records Nettleship (1972) obtained for Newfoundland extend from early May to mid-June, with the median dates for two years falling between May 17 and 24. Grant and Nettleship (1971) determined that in Iceland nest substrate choice and nesting density were related to the cliff edge, with burrow density decreasing with distance from the cliff edge above the cliff and positively correlated with the relative abundance (perimeter) of boulders in study plots below the cliff. That is, at the junction of rock and soil at the foot of the cliff, the relative density of boulders influences burrow site selection, perhaps because these boulders provide landmarks for rapid identification of burrow sites for landing puffins. Nettleship (1972) found that at Great Island, Newfoundland, puffins nesting above the cliffs were at higher densities on sloping habitats close to the cliff edge than on adjacent level ground away from the cliff edge. This seemed to be related to the incidence of egg loss to predatory gulls as well as to the ease with which adults can reach their burrows to provide food for their chicks without danger of having their food stolen by gulls and to a resulting difference in breeding success in these two habitats (twice as high on slopes as on level habitats).

NEST BUILDING AND EGG LAYING. Atlantic puffins do not establish continuous occupancy of burrows until several weeks after they arrive on the breeding areas. The beginning of continuous occupancy in this species is correlated with the start of egg laying. Nettleship (1972) reported that on Great Island the highest frequencies of fighting over burrow sites occurred in early May on the favored slope habitats and from mid-May to mid-June on level habitats, and the median onset of egg laying was in late May on both habitats. Burrow digging and repair is probably performed by both sexes, although some have reported that it is done mostly by the male. Most digging of new burrows is done not during the laying stage but rather in later stages of the breeding season, presumably by subadult birds or by those that have lost or abandoned their earlier burrows (Wehle 1980). Egg laying in colonies tends to be spread out over about a month, with young birds breeding for the first time tending to lay later than experienced birds and with a low level (4–14 percent) of replacement laying by pairs that have lost their first egg (Ashcroft 1979). Replacement eggs are apparently laid only when the first is lost early in incubation, and Ashcroft found that such eggs hatched 13–23 days later than the initial egg would have hatched. Uspenski (1958) reported a re-laying interval of 14–17 days. Nettleship (1972) also found that birds nesting in preferred slope habitats tended to lay sooner and more synchronously than those nesting on level sites.

INCUBATION AND BROODING. Incubation is performed by both sexes, with shifts occurring at least once a day. Myrberget (1962) noted that during incubation most of the nest exchanges occurred at night, and shifts averaged 32.5 hours. As with the other puffins, a substantial amount of egg neglect appears typical, sometimes with both birds present at a burrow but neither incubating the egg (Lockley 1953). The average incubation time has been estimated to be as short as 35 days and as long as 45, but most estimates of averages are 39–42 days (Myrberget 1962; Ashcroft 1976; Lockley 1953). The hatching period from initial shell cracking to emergence averaged 4.3 days in Myrberget's study.

GROWTH AND SURVIVAL OF YOUNG. The young begin to be fed by both parents as soon as hatching is completed. Myrberget (1962) noted that 47 percent of 66 food-carrying puffins were females. He found that the average weight of loads was 10.3 grams, with an average of 5.2 fish per load early in the season and 3.6 later on, although the average load weight increased later in the summer because larger prey were selected, especially herring (*Clupea*). Most feeding of the young was done early in the mornings and again in the afternoon, with an average of 2.5 feedings per day in Myrberget's study. The young gained weight steadily until they were 33–34 days old, then held a nearly constant weight until 41–42 days old, and finally decreased in average weight during the last few days as nestlings, when the parents stopped feeding them. On average the young lay fasting in their nests for 8.2 days (range 5–11 days), becoming restless and irritable and sometimes walking out of their nests at night. In Myrberget's study the average nestling period was 47.7 days (range 43–52 days); most young apparently then flew down to the sea at night,

though some probably walked down. Various other estimates of average fledging periods have ranged from 37.3 to 54.5 days, suggesting substantial variations in fledging times, no doubt reflecting temporal or local differences in food availability, incidence of food stealing by gulls, and resultant feeding rates (Wehle 1980). Harris (1983b) believed that young puffins can influence the number of feedings they receive from adults by their begging calls, based on experimental use of recorded begging calls.

BREEDING SUCCESS AND RECRUITMENT RATES. Estimates of breeding success rates of the Atlantic puffin in North America have been provided by Nettleship (1972), while similar data have been summarized for Skomer and Skokholm Islands, United Kingdom, by Ashcroft (1976, 1979) and Dickinson (1958) and for Lovunden, Norway, by Myrberget (1962). These studies indicate hatching success rates of from 46 to 78 percent and fledging success rates of from 21.4 to 98 percent. Overall breeding success rates of from 10 to 90.5 percent have been reported, indicating a very great range in success rates, even within single areas such as the various nesting habitats reported on by Nettleship (1972). As in the other puffins, hatching success under natural, undisturbed conditions is probably in the range of 75–90 percent, while typical fledging success seems to be about 70 percent (Wehle 1980), resulting in an annual productivity of about 0.5 fledged young per breeding pair. Ashcroft (1979) estimated that each year 20–27 percent of the colony adults were without nesting burrows, and 5–16 percent of those pairs with burrows did not lay any eggs. At least 10–16 percent of the fledglings survived to 4 years old, which she established to be the earliest age of initial breeding in that colony. However, it was not determined whether this survival of young balanced the 5 percent estimated annual adult mortality rate. Harris (1981) found one case of breeding by a 3-year-old, but of 127 known-age breeders only 9 were under 5 years old, and 46 were at least 8 years old. He (1983a) judged that on the Isle of May the annual survival rate of adults is 96 percent, that the average production is 0.56 fledged young per nesting pair, and that 39 percent of the fledged young survive to reach initial breeding age at 5 years. It thus appears that the age structure of Atlantic puffin colonies is strongly skewed toward older and more experienced age-classes.

Evolutionary History and Relationships

The close relationships and probable evolutionary histories of the horned and Atlantic puffins are discussed in the former's species account.

Population Status and Conservation

It has been estimated that in the late 1970s about 624,000 birds were associated with 28 colonies of eastern Canada. Of these, the colonies of Labrador and eastern Newfoundland were stable or of uncertain population trend, while those on the Gulf of Saint Lawrence and the Bay of Fundy region were declining (Nettleship 1977). There is some evidence of a recent minor range extension of the species into Hudson Strait at Digges Sound (Gaston and Malone 1980). At least during this century the breeding population of Maine has been extremely small; in 1904 only a single colony of 300 birds existed (on Machias Seal Island), plus two pairs on Matinicus Rock (Bent 1919). As of 1977 an estimated 125 pairs of puffins were censused on Matinicus Rock (Erwin and Korschgen 1979). On Machias Seal Island, which is jointly claimed by the United States and Canada, the population had increased to about 750 pairs in the 1970s but declined sharply in 1977 (Korschgen 1979). Since 1977 efforts have been under way to restore the species on Egg Rock, Muscongus Bay, Maine. As of 1981, 111 of 530 transplanted puffins had returned to the general area where they had previously been raised and had been seen at Egg Rock, Machias Seal Island, or Matinicus Rock. In 1981 four pairs successfully hatched young on Egg Rock (Kress 1982). By 1984 a total of 14 pairs bred on the island, and that year also marked the first return to the island of a young puffin that had been raised there by wild adults.

Horned Puffin

Fratercula corniculata (Naumann)

OTHER VERNACULAR NAMES: Sea parrot; macareux cornu (French); Hornlun (German); tsunomedori (Japanese); ipatka (Russian); kupruwuk (Saint Lawrence Island).

Distribution (See Map 28)

BREEDS in northeastern Siberia from Kolyuchin Bay to the southeastern Chukotski Peninsula, the Diomede Islands to the Gulf of Kresta, the east coast of Kamchatka, the Commander Islands, the Gulf of Shelekhova, the Shantarskie Islands, Sakhalin, and the northern Kurile Islands; and from the Alaskan coast at Cape Lisburne south through the islands of the Bering Sea to the Aleu-tian Islands, along the Alaska Peninsula east and south to Glacier Bay and Forrester Island, and to the Queen Charlotte Islands of British Columbia, possibly to Triangle Island.

WINTERS in open waters throughout the breeding range, south to British Columbia, Washington, and Oregon; casually to California and central Japan.

Description (Modified from Ridgway 1919)

ADULTS IN BREEDING PLUMAGE (sexes alike). Pileum uniform grayish brown or drab; entire side of head, including superciliary and supra-auricular regions, white; all of the neck and entire upperparts uniform black, the throat more sooty, this passing into brownish gray on chin; underparts, including lower foreneck, immaculate white; under wing coverts brownish gray; tip of bill, to between second and third grooves, salmon along culmen and gonys, elsewhere brownish red; basal portion of bill (including first ridge and basal maxillary lamina) clear light chrome yellow; rictal rosette, tongue, and interior of the mouth bright orange; iris brownish gray; eyelids vermilion, the soft appendages brownish black; legs and feet deep vermilion.

WINTER PLUMAGE. Bill differently shaped, being broader through middle than at base, the deciduous nasal cuirass, basal lamina, and so on having been shed, all this basal portion dusky instead of yellow; the rictal rosette greatly reduced and pale yellow instead of red; superciliary hornlike appendage absent, and eyelids brownish gray instead of red; sides of head gray, becoming sooty blackish on orbital and loral regions, and legs and feet much paler red. Birds in second fall are like winter adults, but the bill has no terminal grooves, and the crest of the upper mandible gradually curves downward (Kozlova 1961).

JUVENILES. Similar in coloration of plumage to winter adults, but bill very different, being much less deep, the culmen much less arched, the terminal portion of both maxilla and mandible destitute of grooves or ridges and horn color or brownish, without reddish tinge.

DOWNY YOUNG. Uniform dark sooty grayish brown, the breast and upper abdomen rather abruptly white.

Measurements and Weights

MEASUREMENTS. Wing: males 170.0–187.5 mm (average of 7, 181.4); females 168–85 mm (average of 12, 177.9). Culmen: males 46–55 mm (average of 7, 50.6); females 47.0–52.5 mm (average of 12, 49.7) (Ridgway 1919). Eggs: average of 5, 66.7 x 45.6 mm (Amaral 1977).

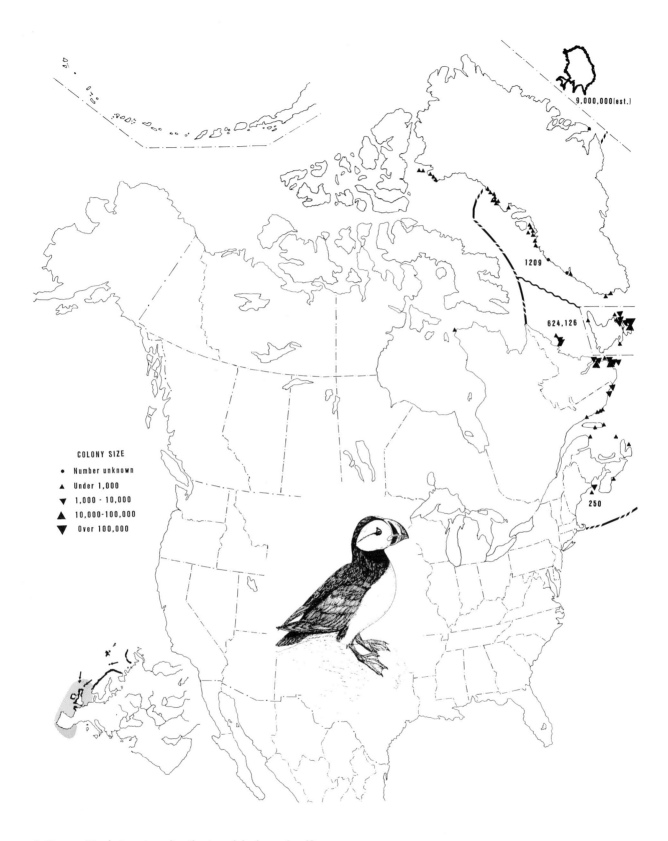

COLONY SIZE
- Number unknown
▲ Under 1,000
▼ 1,000 - 10,000
▲ 10,000-100,000
▼ Over 100,000

9,000,000(est.)

1209

624,126

250

28. Current North American distribution of the horned puffin,
including colony locations and limits of nonbreeding range
(*broken line*). The Asian range is shown on the inset map.

WEIGHTS. A sample of 29 adult males during the breeding season averaged 518.7 g (range 415–602), and one of 57 adult females averaged 490.8 g (range 415–559). A sample of 71 eggs averaged 75.3 g (Wehle 1980). Estimated fresh egg weight, 80 g (Schönwetter 1967). Newly hatched chicks average 57 g (Sealy 1973a).

Identification

IN THE FIELD. In breeding plumage this species greatly resembles the closely related Atlantic puffin, but the bill is uniformly yellowish behind the reddish tip, and the "horn" above the eye is long and pointed. Otherwise the rounded white face and white underparts will serve to distinguish it from the tufted puffin of the same area. In juveniles and during winter these two species are more similar, but the horned puffin always exhibits white flanks and underparts, while the tufted puffin is grayish to blackish in these areas.

IN THE HAND. The distinctly puffinlike bill of even young birds allows for separation from all species except the very similar Atlantic puffin. Adult horned puffins have a bill that shows more nearly vertical grooving than does that of the Atlantic puffin, and in nonbreeding or immature stages the longer tail of the horned puffin (at least 60 mm in adults, compared with no more than 53 mm in Atlantic adults) should provide for separation.

Ecology and Habitats

BREEDING AND NONBREEDING HABITATS. The North American and Asian breeding range of this species extends across a region of high-arctic and low-arctic coastlines with adjoining surface water temperatures ranging from about 5°C to 12°C, primarily where rocky coastlines occur. They tend to be more common at the higher latitudes of their breeding range, where strongly glaciated and rocky coastlines prevail, and are sympatric over much of their range with the tufted puffin, with which they probably compete locally for nest sites. Outside the breeding season the birds tend to occur well away from shore, but they are not as pelagic as tufted puffins and are usually found within the limits of the continental shelf.

SOCIALITY AND DENSITIES. Of the 435 known Alaskan colonies, 15 sites have estimated populations of more than 10,000 birds, and on the Semidi Islands the estimated populations include a colony of more than 100,000 birds on Chowiet Island, while an estimated 140,000 breed at Amagat Island (east of Unimak Island)

in an area of less than a square mile. More specific estimates of nesting densities do not appear to have been made for this species, probably because of the great difficulty of censusing actual nest sites, which are almost always in inaccessible rock crevices.

PREDATORS AND COMPETITORS. The usual array of avian predators affect the horned puffin, including common ravens (*Corvus corax*), glaucous-winged gulls (*Larus glaucescens*), peregrines (*Falco peregrinus*), bald eagles (*Haliaeetus leucocephalus*), and others. Peregrines have been observed taking horned puffins near colonies, and puffin remains have been found near eagle roosts (Amaral 1977). In Wehle's (1980) study areas snowy owls (*Nyctea scandiaca*), peregrines, and bald eagles were found to be the major predators of adult puffins, while elsewhere both arctic foxes (*Alopex lagopus*) and red foxes (*Vulpes fulva*) have been implicated as predators. However, the major actions of these predators may be simply to restrict the birds to more protected nest sites (Wehle 1980). Where foxes are lacking the birds may nest among rocks at the bases of cliffs, but under predation pressure they tend to use the steep cliff faces (Lehnhausen 1980).

Probably the major competitor of the horned puffin over much of its range is the tufted puffin, and it is believed that owing to its smaller size the horned puffin is generally unable to compete equally with tufted puffins for available nest sites. However, it is able to use smaller rock crevices and prefers these to earthen burrows, which it uses only where tufted puffins are rare or absent. In some areas the birds may dig through a layer of soil to reach subsurface rock cavities. Probably they are easily able to displace any of the small auklets that might be using the same or similar nesting habitats. Foods taken by the tufted and horned puffin are very similar, although there is some evidence of ecological segregation between these species. When foods are abundant both may forage in inshore waters, but during food shortages the tufted puffin apparently feeds farther offshore, thus reducing food competition (Wehle 1976, 1980).

General Biology

FOOD AND FORAGING BEHAVIOR. Wehle's (1980) studies on this species included materials from 64 adults collected at Buldir and Ugalushak islands, and he summarized additional information on the species, including that of Hunt, Burgeson, and Sanger (1981) and Swartz (1966). Most of these studies suggest that fish is the predominant food taken by adults, with cephalopods and crustaceans usually secondary. However, the precise types of fish taken tend to vary between colonies as well as exhibiting seasonal variations within colonies. In general, Wehle (1980) concluded that horned puffins tend to supplement their diets with squid more often than do tufted puffins and also take a greater variety of fish, especially inshore, subtidal forms. On the basis of 133 samples of adult stomach contents from various areas and studies, Wehle found the relative frequency of major food types to be fish (47.9 percent), squid (31.4 percent), polychaetes (12.4 percent), and crustaceans (8.3 percent). In a similar summary of information on foods provided to chicks, Wehle concluded that sand launce (*Ammodytes*) and capelin (*Mallotus*) are the primary foods and that greenling (*Hexagrammos*) and cod (*Gadus*) are important subsidiary prey, with squid and sandfish (*Trichodon*) still less important.

Wehle (1980, 1983) reported seeing horned puffins feed during the entire breeding season in inshore waters, usually within 1 or 2 kilometers of shore. The actual depths at which the birds feed is still unknown. During the nestling period the adults bring food to their chicks 2–6 times a day, with an average of 5.2 prey per load (98 samples) and an average load weight of 11.9 grams (74 samples) (Wehle 1983). Other studies have found average load weights of 7.9–17.0 grams, and 1.5–12.6 prey per load (Wehle 1980). The number of prey per load and the average weight of the loads did not appear to change during the nestling period in Wehle's study, although in unpublished observations of G. Burrell cited by Wehle the average load weight did increase during the nestling period.

MOVEMENTS AND MIGRATIONS. Little is known of horned puffin migrations. During fall the birds apparently disperse into oceanic habitats, with only a few remaining in bays and near shore areas; furthermore, they tend to migrate singly or in small groups, making their movements difficult to trace. Estimates of wintering birds in the Gulf of Alaska total fewer than 200,000 in winter, or only a fraction of the known breeding population of the region (Gould, Forsell, and Lensink 1982). There are surprisingly few British Columbian records, but probably at least along the Queen Charlotte Islands the birds are regular winter and spring visitors (Sealy and Nelson 1973). There is also fairly good evidence that the species may be colonizing British Columbia as a breeder (Campbell, Carter, and Sealy 1979). There are also some winter records for Washington, Oregon, and California (Hoffman, Elliott, and Scott 1975), but it is likely that many birds winter along the Aleutian chain and probably also at sea, from the limits of open water southward (Sealy 1973a).

Social Behavior

MATING SYSTEM AND TERRITORIALITY. It is assumed that prolonged monogamy occurs in this species, as is typical of the Atlantic puffin and other alcids, though direct evidence is lacking. Wehle (1980) was unable to obtain evidence on nest site tenacity, but since he found nests in successive years in exactly the same locations in talus slopes and under beach boulders that were isolated by several hundred yards from their nearest neighbors, it is highly likely that such tenacity does exist. He suspected that, unlike the tufted puffin, the horned puffin defends only its cavity rather than the cavity entrance and immediately surrounding areas.

VOICE AND DISPLAY. Wehle (1980) noted four vocalizations of adult horned puffins, all of which were similar to those he observed in tufted puffins. The first was a short, low-pitched *errr* note that seemed to be a threat or warning. The second was a similar but more prolonged purring call that was catlike and in this species trailed off gradually and was not repeated. It was of undetermined function but usually was produced by birds resting in the colony. The third vocalization was a bisyllabic call that consisted of a short, intense *er* note followed by a second and higher-pitched note with an undulating pitch. It was uttered both on land and on water. The final adult call was a multinoted call of six or seven syllables, with the first two notes identical with the bisyllabic call, the third note the highest in pitch, and the remaining syllable matching the first one in acoustic characteristics. It was judged to possibly have sexual significance.

Courtship behavior and displays have been discussed by Amaral (1977) and Wehle (1980). According to Wehle, the ceremonies of the tufted and horned puffins are very similar, with the horned puffin somewhat more social than the tufted puffin. In such social groups males begin display activity by following the female while raising and partially lowering the bill (fig. 47F). This movement is faster in the horned puffin than in the tufted, and the bill is only partially lowered on the downward portion of the display. The bill is apparently opened somewhat on the upward phase and closed on the downward phase, but it is not known whether vocalizations accompany the display. If the female is receptive she allows the male to approach to within about a meter, at which point he flies to her and alights on her back (fig. 47G). Copulation lasts a rather long time; Amaral (1977) reported an average of 35 seconds for horned puffins compared with 48 seconds for tufted puffins, and the male continues to beat his wings during treading. Treading ends with the female's submerging and reappearing about a meter away. Afterward one and usually both birds wing flap (fig. 47H). Besides precopulatory head jerking, horned puffins also perform head jerking in other contexts, sometimes even when alone. Rarely, both members of a courting pair will perform head jerking, and in some cases head jerking between two individuals leads to billing. Billing is probably the most important pair-bonding behavior in puffins and occurs throughout the breeding season. In horned puffins it often follows a bird's landing near its mate before the two enter their nest-site, after aggressive actions toward a third bird, or after head jerking. As in the Atlantic puffin, billing involves pressing or slapping the lateral surfaces of the bills together as the birds face one another, often as one stands erect and the other crouches somewhat (fig. 47A–C). Such bouts of billing sometimes last several minutes. Bill gaping is the major threat display and takes the same form as in the Atlantic puffin, with the bill pointed toward the opponent, wide open, and the tongue often visible. Immediately after landing in the colony, birds usually assume a posture with the wings held up above the back, the body held low, and the head in line with the body or tilted downward. This "landing display" is held longest when the bird performing it lands close to several other birds in the colony (Wehle 1980).

Reproductive Biology

BREEDING SEASON AND NESTING SUBSTRATE. Egg records from Alaska north to the Alaska Peninsula are from June 24 to September 1, with a peak between June 27 and July 9. Records from south of the peninsula are from June 6 to July 11, with a peak between June 17 and July 5 (Bent 1919). On the Pribilof Islands egg laying typically occurs between mid-June and mid-July (Hunt, Burgeson, and Sanger 1981). Generally, peak egg laying over the Alaskan range of this species occurs between mid-June and the first week of July, 1–3 weeks later than the tufted puffin's egg-laying peak. In general there is about a month-long period of egg laying, including replacement clutches, with about two-thirds of the sample populations laying within a week (Wehle 1980). The usual nest substrate of horned puffins consists of natural rock crevices, with secondary use of talus slopes and boulder rubble at the base of cliffs and even more limited use of earthen burrows (Lehnhausen 1980). In some areas the birds may dig through thin layers of soil and vegetation to reach subterranean rock cavities, but there is still no detailed information on optimum cavity or crevice size for this species. Probable effects of competition with tufted puffins for suitable sites where both species breed together has been mentioned earlier.

NEST BUILDING AND EGG LAYING. Although in some areas actual nest excavation does occur, in most cases horned puffins utilize preexisting cavities and crevices for nesting, and so no nest building as such is normally required. However, some nest preparation is typical, including the removal of fallen vegetation, eggshells, and other debris that might have accumulated in the nest since the previous season. Both sexes participate in this, with the operation usually requiring several bouts of a few minutes each (Wehle 1980). Egg laying begins about 2–3 weeks after continuous occupancy of a burrow has been established, which occurs almost immediately after initial arrival on the breeding grounds. This is in contrast to the situation in the tufted puffin, in which continuous occupancy usually begins a week or two after initial arrival, after which egg laying begins within about a week. In a test of egg replacement, Wehle (1980) removed 10 newly laid eggs from horned puffin nests. Eggs were later found in 3 of these nests. Two nests were rebuilt after egg loss, and one of these subsequently contained a replacement egg. All three nests containing replacement eggs were abandoned shortly after the second egg was laid, though in one of these nests a third egg was eventually laid, possibly, but not certainly, by the original pair.

INCUBATION AND BROODING. Both sexes participate in incubation, probably normally exchanging places in the early evening (Amaral 1977). However, additional exchanges might occur during daylight, since the adults make frequent visits to nest crevices at such times. Like tufted puffins, horned puffins sometimes leave their eggs unattended for a day or more and may at times incubate longer than a day without relief. Sealy (1973a) reported an average incubation period of 41.1 days, with observed extremes of 40–43 days for 5 eggs. Amaral (1977) reported an average period of 40.2 days, with a range of 39–42 days for 5 eggs. Wehle (1980) noted considerable variation in the hatching period, averaging 3 days from initial cracking to emergence and varying in 2 cases from 2–4 days.

GROWTH AND SURVIVAL OF YOUNG. Chick weight at hatching was reported by Sealy (1973a) to average 58.6 grams (2 chicks) and by Amaral (1977) to average 54.3 grams (3 chicks). Sealy determined a fledging period of 38 days for one chick, while Amaral (1977) observed an average period of 40 days (range 38–42) and Wehle (1980) an average (for 2 chicks) of 38.5 days. Amaral (1977) noted an average weight gain by chicks of 13.4 grams per day during the period of most rapid growth and reported that maximum average nestling weight was attained at 37.5 days (71 percent of adult weight). The major prey species fed to chicks is sand launce, as with the other puffins, with no general trends apparent in load weight or number of prey per load during the nestling period. As with other puffin nestlings, there is a slight weight loss during the last few days of nestling life, probably as a result of voluntarily reduced feeding by the chick, increased levels of activity, and loss of water during tissue maturation. Although information is limited, chick losses appear to be greatest shortly after hatching, when the young are vulnerable to drowning when nesting crevices flood following heavy precipitation. Food availability probably also influences chick survival, and predation near the time of fledging may also have an effect.

BREEDING SUCCESS AND RECRUITMENT RATES. Wehle (1980) reported that hatching success during various years and in different study areas ranged from 56 to 100 percent and averaged about 80 percent. Amaral (1977) found a similar (79 percent) hatching success rate, and other observers cited by Wehle have provided similar estimates of hatching success in this species, which seems less vulnerable to flooding and perhaps also to fox predation than the tufted puffin. Fledging success rates are rather poorly documented but also appear to be moderately high, judging from estimates of 53–77 percent reported by Wehle for his own studies and other sources. All told, it is thus likely that an overall breeding success of about 0.5 young fledged per nesting pair is fairly typical of horned puffins. The incidence of nonbreeding by subadult birds attending the breeding colonies is still unknown, as is the percentage of immatures that do not visit breeding colonies at all. There are also no estimates of recruitment rates or adult mortality rates for the species.

Evolutionary History and Relationships

The horned and Atlantic puffins obviously constitute a closely related superspecies, and in the opinion of Kozlova (1961) speciation of these two forms resulted from climatic changes during the Quaternary period, with isolated populations evolving in the Bering Sea and North Atlantic respectively. She believed that an earlier separation from the ancestral *Lunda* line had occurred during Pliocene times in the Pacific region. It seems likely that separation of pre*Cerorhinca* stock probably occurred at about the same time as that of "Lunda."

Population Status and Conservation

Almost all the breeding in North America is confined to Alaska, where there are an estimated 1.5 million birds in 435 known nesting grounds (Sowls, Hatch, and Lensink 1978). However, no population estimates are available for 54 of these areas, and thus this overall statewide estimate may prove too conservative. Observations during the 1970s suggest that horned puffins are now extending their breeding range into British Columbia, although apparently in small numbers (Campbell, Carter, and Sealy 1979). No estimates are available on the possible population trends of this species, which has been assigned an oil vulnerability index of 72 (King and Sanger 1979).

Appendix 1

Sources of Scientific and Vernacular Names

Aechmophorus—from the Greek *aichme*, spear, and *phoros*, bearing.

    clarkii—after John Clark, nineteenth-century American ornithologist.

    occidentalis—from Latin, western.

Aethia—from the Greek *aithuia*, an Aristotelian and Homeric name for a water bird.

    cristatella—diminutive of the Latin *cristatus*, crested.

    pusilla—from Latin, small or petty.

    pygmaea—from Latin, pygmy.

Alca—Latinized form of Scandinavian vernacular (*alk, alka, alke*) names for these birds.

    torda—from a Swedish vernacular name (*tordmule*) for the razorbill.

Alle—possibly from the Latin *allex*, referring to the lack of a hallux.

Auk—from the Danish and Norwegian names (*alke*) for the razorbill.

Brachyramphus—from the Greek *brachys*, short, and *ramphos*, bill.

    brevirostre—from the Latin *brevis*, short, and *rostrum*, bill. The vernacular name Kittlitz is for F. H. Kittlitz (1779–1874), a German naturalist on a Russian expedition to Kamchatka, who collected the first specimens.

    marmoratum—from Latin, marbled.

Cepphus—from the Greek *kepphus*, seabird.

    columba—from Latin, dove or pigeon.

    grylle—probably from Greek, meaning "I grunt."

Cerorhinca—from the Greek *keras*, horn, and *rynchos*, beak.

    monocerata—from the Greek *monos*, one, and *ceras*, horn.

Cyclorrhynchus—from the Greek *kycklos*, circle, and *rynchos*, beak.

    psittacula—from the Latin *psittacus*, small parrot, in reference to the beak.

Dovekie—a diminutive of dove.

Endomychura (see also *Synthliboramphus*)—from the Greek *endomychos*, secret or hidden, and *oura*, tail, referring to the short tail.

Fratercula—from the Latin *fraterculus*, meaning "little friar" or "little brother," in reference to the general appearance.

    arctica—from Latin, of the arctic.

    cirrhata—from the Latin *cirratus*, curled hair, in reference to the nuptial tufts.

    corniculata—from Latin, horned.

Gavia—from Latin, sea mew, as used by Pliny.

    adamsii—after Edward Adams (1824–56), an English naval surgeon and naturalist on an arctic voyage during which he collected the first specimens.

    arctica—from Latin, of the arctic.

    immer—probably a variant of the English *ember* and the Swedish *immer* or *emmer*, gray or ashlike. Possibly also from the Latin *immersus*, to immerse.

    stellata—from Latin, starred, in reference to the speckled back.

Grebe—from French, of uncertain meaning, but perhaps from *griabe*, a Savoyard word for a sea mew (or gull), or from the Breton *krib* meaning crest.

Guillemot—A pet form of Guillaume, which is derived from the Old French name Willem, and echoic of the juvenile's "will" call.

Loon—corruption of Shetland *loom* and related to the Icelandic *lōmr* and Swedish *lom*, lame or clumsy, referring to its helplessness on land.

Lunda—from the Scandinavian *lunde* (Swedish *lunne*), a vernacular name for puffins.

Murre—apparently imitative of the murmuring sound produced by the birds. Possibly also related to *marrot, morrot,* dialect English terms for guillemots. The vernacular name murrelet, coined by E. Coues, is a diminutive of murre.

Pinguinis—New Latin for penguin; a combination of the Welsh words *pen,* head, and *gwyn,* white.

impennis—from the Latin *in,* negative, and *pinna,* feather, meaning flightless. The vernacular "garefowl" is from the Icelandic *geirfugl.*

Plautus—from Latin, flat-footed. See also *Alle.*

Podiceps—from the Latin *podicus,* rump, and *pes,* foot, or "rump-footed."

auritus—from Latin, eared.

dominicus—see under *Tachybaptus.*

grisegena—from the Latin *griseus,* gray, and *gena,* cheek. The earlier vernacular name is after Carl Peter Holboell (1795–1856), Danish governor of South Greenland in the 1820s.

nigricollis—from the Latin *niger,* dark or black, and *collum,* neck.

Podilymbus—an abbreviation of the Latin *podiceps,* rump-footed, combined with the Greek *kolymbus,* diver.

podiceps—see *Podiceps* above.

Ptychoramphus—from the Greek *ptychos,* fold, and *ramphos,* beak.

aleutica—Latin, of the Aleutian islands. The vernacular name is after John Cassin (1813–69), American ornithologist, for whom the Cassin sparrow, Cassin finch, and Cassin kingbird were also named.

Puffin—from the Middle English *poffin* or *pophyn,* apparently in reference to the fat or "puffy" appearance of the adults and young.

Synthliboramphus—from the Greek *synthlibo,* to compress, and *rhamphos,* beak.

antiquum—from Latin, ancient, or gray-headed.

craveri—after Fredrico Craveri (1815–90), Italian meteorologist thus honored by Tomasso Salvadori, who described the species.

hypoleucus—from the Greek *hypo,* under or less than, and *leukos,* white, referring to its absence of a white scapular stripe. The vernacular name Xantus is after John Xantus de Vesey (1825–94), the Hungarian naturalist who first collected the species.

Tachybaptus—from the Greek *tachys,* swift, and *bates,* treading or climbing.

dominicus—from Santo Domingo, in the West Indies.

Tystie—a common vernacular name for the black guillemot, apparently based on the species' twittering notes.

Uria—from the Greek *ouria,* a kind of water bird.

aalge—a Danish word for the murre.

lomvia—Faroese for a kind of diving bird. The vernacular name Brünnich's murre refers to M. T. Brünnich (1737–1827), a Danish zoologist.

Appendix 2

Keys to Identification of Loons, Grebes, and Auks

A. Hind toe large.
  B. Anterior toes webbed, rectrices (tail feathers) extend beyond tail coverts . . . Gaviidae (loons).
  BB. Anterior toes separately lobed; rectrices small and hidden by tail coverts . . . Podicipedidae (grebes).
AA. Hind toe absent, front toes webbed . . . Alcidae (auks).

KEY TO SPECIES OF NORTH AMERICAN LOONS (GAVIIDAE)

A. Culmen at least 75 mm long (over 52 mm from nostril to tip) in adults, wing over 318 mm.
  B. Bill blackish and slightly decurved along culmen, which is 75–90 mm long and has chin feathers terminating well posterior to nostrils . . . Common loon
  BB. Bill never blackish beyond middle of culmen, rest yellowish to straw colored, the culmen (83–96 mm) nearly straight and the chin feathers reaching a point directly below the nostrils . . . Yellow-billed loon.
AA. Culmen no more than 70 mm long (under 52 mm from nostril to tip) in adults, wing under 315 mm.
  B. Tarsus longer than middle toe including claw; culmen slightly decurved and lower mandible not distinctly angulated . . . Arctic loon.
  BB. Tarsus about as long as middle toe excluding its claw; culmen nearly straight and lower mandible somewhat angulated . . . Red-throated loon.

KEY TO SPECIES OF NORTH AMERICAN GREBES (PODICIPEDIDAE)

A. Length of bill about twice its depth; secondaries all brownish on outer webs . . . Pied-billed grebe.
AA. Length of bill at least three times its depth, white on both webs of secondaries.

B. Length of bill about three times its depth; neck not as long as body.
  C. Wing under 100 mm; iris yellowish . . . Least grebe.
  CC. Wing over 120 mm; iris reddish.
    D. Wing no more than 150 mm; bill under 8 mm deep at base.
      E. Bill deeper than wide, culmen slightly decurved, and lower mandible not distinctly angulated; tarsus at least 44.5 mm . . . Horned grebe.
      EE. Bill wider than deep, culmen nearly straight, and lower mandible distinctly angulated; tarsus up to 44.5 mm . . . Eared grebe.
    DD. Wing over 150 mm; bill over 12 mm deep at base . . . Red-necked grebe.
BB. Bill about five times as long as deep; neck about as long as body . . . Western grebe.
  C. Black of crown not reaching lores or eyes; bill of adults yellow to orange . . . dark phase (*occidentalis*).
  CC. Black of crown reaching below eyes and lores; bill of adults dull greenish yellow . . . light phase (*clarkii*).

KEY TO SPECIES OF NORTH AMERICAN AUKS (ALCIDAE)

A. Forehead feathering extends forward to nostrils, often hiding them; bill never with accessory pieces in adults.
  B. Wing under 140 mm; culmen under 25 mm; nostrils oval to circular (murrelets and dovekie).
    C. Tarsus with entirely reticulated (networklike) scale pattern.
      D. Tarsus much shorter than middle toe without claw; scapulars never black.

E. Total culmen length at least 25 mm; lateral rectrices brownish or only narrowly edged with white . . . Marbled murrelet.

EE. Total culmen length under 15 mm; lateral rectrices white or mostly white . . . Kittlitz murrelet.

DD. Tarsus at least as long as middle toe without claw; scapulars black.

E Under wing coverts white, and inner webs of distal primaries, sides, and flank feathers tipped with white . . . Xantus murrelet.

EE. Under wing coverts and primaries variably brownish gray; sides and flank feathers lacking white tips . . . Craveri murrelet.

CC. Scales on lower front of tarsus scutellate (aligned vertically).

D. Bill deeper than wide; 14 rectrices . . . Ancient murrelet.

DD. Bill as wide as deep; 12 rectrices . . . Dovekie.

BB. Wing over 140 mm; exposed culmen at least 25 mm; nostrils linear (auks, murres, and guillemots).

C. White upper wing coverts; feathers not extending beyond anterior edge of nostrils, thus exposing them.

D. Under wing coverts white, normally 12 rectrices . . . Black guillemot.

DD. Under wing coverts brownish gray; 14 rectrices . . . Pigeon guillemot.

CC. Black upper wing coverts; nostrils entirely hidden by feathering on upper mandible.

D. Exposed culmen 30–40 mm; wing over 175 mm.

E. Depth of bill at base nearly equal to the exposed culmen; bill relatively short (to 35 mm) and blunt tipped . . . Razorbill.

EE. Depth of bill at base less than a third of the exposed culmen; bill longer (over 38 mm) and sharply pointed.

F. Depth of bill at nostrils less than a third of exposed culmen; crown and hindneck brownish; no white stripe on mandible . . . Common murre.

FF. Depth of bill at nostrils more than a third of exposed culmen; crown and hindneck blackish; white stripe present on edge of upper mandible . . . Thick-billed murre.

DD. Exposed culmen at least 75 mm; wing under 175 mm . . . Great auk.

AA. Forehead feathering well separated from nostrils; bill usually with seasonal accessory pieces in adults (puffins and auklets).

B. Wing under 160 mm; tarsus with reticulate (networklike) scale pattern; iris whitish in adults (typical auklets).

C. Width of bill at base greater than its basal depth.

D. Bill grayish and depressed basally; adults never with white "mustache" or crest . . . Cassin auklet.

DD. Bill reddish and not depressed; white "mustache" and crest present in adults . . . Whiskered auklet.

CC. Width of bill at base less than its depth.

D. Tip of lower mandible bluntly truncated; underparts grayish; adults crested . . . Crested auklet.

DD. Tip of lower mandible pointed; underparts white; adults never crested.

E. Bill pointed and gradually tapering; cheeks and throat variably white . . . Least auklet.

EE. Bill strongly rounded, with lower mandible upturned; cheeks and throat pale gray to dusky . . . Parakeet auklet.

BB. Wing at least 160 mm; iris usually brownish or yellowish; scales on lower front of tarsus scutellate (aligned vertically).

C. Depth of bill no more than twice its length; breast brown to grayish black.

D. Culmen under 40 mm; inner toe claw normally shaped . . . Rhinoceros auklet.

DD. Culmen over 50 mm; inner toe claw curved inward . . . Tufted puffin.

CC. Depth of bill at least three times its width; breast and abdomen entirely white.

D. Tail no more than 53 mm; grooves on sides of bill very oblique . . . Atlantic puffin.

DD. Tail at least 60 mm; grooves on sides of bill nearly vertical . . . Horned puffin.

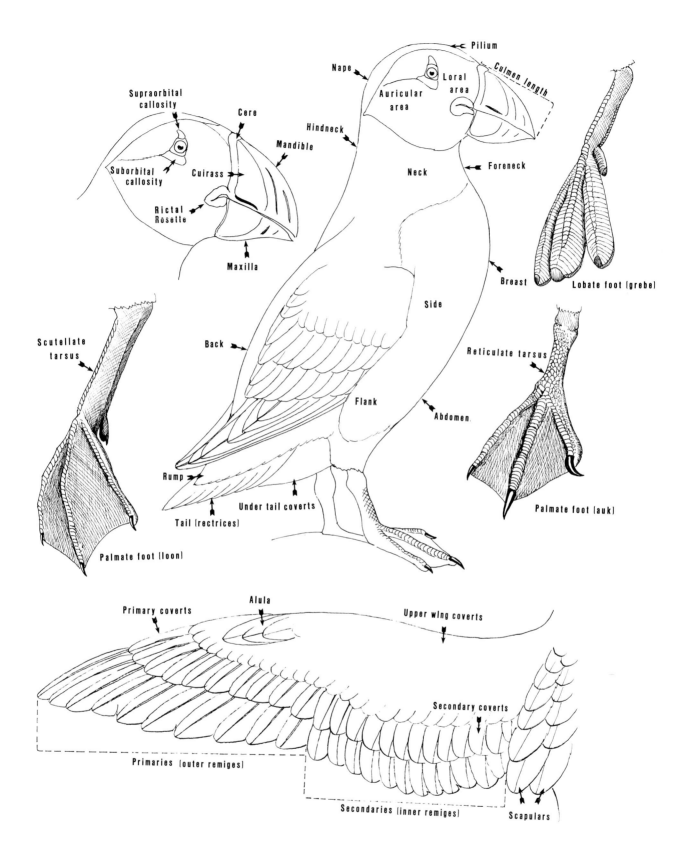

48. External features of loons, grebes, and auks.

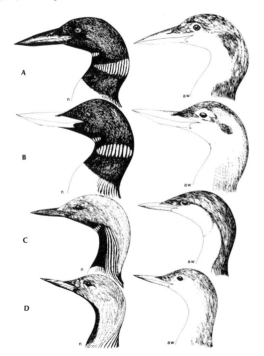

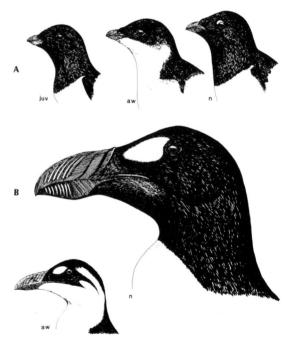

49. Representative downy young: A, common loon; B, dovekie; C, common murre; D, razorbill; E, pigeon guillemot; F, Cassin auklet; G, Kittlitz murrelet; H, ancient murrelet; I, Xantus murrelet; J, least auklet; K, parakeet auklet; L, rhinoceros auklet; M, tufted puffin.

51. Head profiles of North American grebes: A, least, B, pied-billed, C, eared, D, horned, E, red-necked, and F, western, showing juvenal (juv), adult winter (aw), nuptial (n), and adult (ad) male and female plumages.

50. Head profiles of adult loons: A, common, B, yellow-billed, C, arctic, and D, red-throated, showing nuptial (n) and adult winter (aw) plumages.

52. Head profiles of A, dovekie and B, great auk, showing juvenal (juv), adult winter (aw), and nuptial (n) plumages. The adult winter plumage illustrated for the great auk is somewhat tentative and shown in reduced size.

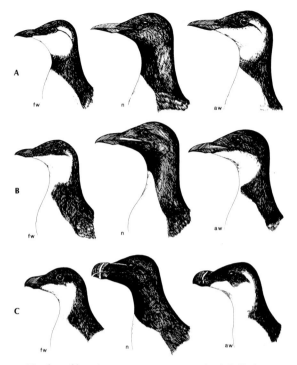

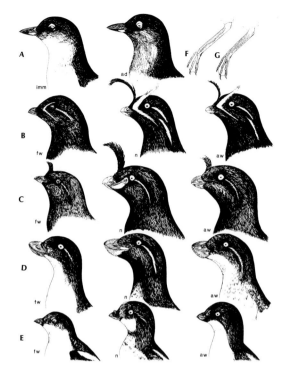

53. Head profiles: A, common murre, B, thick-billed murre, and C, razorbill, showing first-winter (fw), nuptial (n), and adult winter (aw) plumages.

55. Head profiles: A, Cassin auklet, B, whiskered auklet, C, crested auklet, D, parakeet auklet, and E, least auklet, showing first-winter (fw), immature (imm), nuptial (n), adult (ad), and adult winter (aw) plumages. Also shown are leg scaling patterns: F, typical murrelets and G, auklets.

54. Head profiles: A, black guillemot, B, pigeon guillemot, C, marbled murrelet, D, Kittlitz murrelet, E, ancient murrelet, F, Xantus murrelet, and G, Craveri murrelet, showing first-winter (fw), nuptial (n), adult winter (aw), and adult (ad) plumages.

56. Head profiles: A, rhinoceros auklet, B, tufted puffin, C, horned puffin, and D, Atlantic puffin, showing first-winter (fw), nuptial (n), and adult winter (aw) plumages.

Appendix 3

Major North American Auk Colonies (Shown in Figure 4)

Index No.	Specific Location	State or Region	Major Species	Total Species	Estimated Total Auk Population
1	Channel Islands	California	Cassin auklet, Xantus murrelet	3	23,000
2	Farallon Islands	California	Cassin auklet, common murre	5	127,000
3	Castle Rock	California	Common murre, Cassin auklet	5	44,000
4	Three Arch Rocks	Oregon	Common murre	3	80,000
5	Copalis Beach	Washington	Rhinoceros auklet	5	22,000
6	Destruction Island	Washington	Rhinoceros auklet	5	15,000
7	Protection Island	Washington	Rhinoceros auklet	3	34,000
8	Triangle Island	British Columbia	Cassin auklet, tufted puffin	5	250,000
9	Queen Charlotte Island	British Columbia	Ancient murrelet, Cassin and rhinoceros auklets	5	80,000
10	Forrester Island	Alaska	Tufted puffin, rhinoceros auklet, ancient murrelet	7	100,000
11	Chiswell Islands	Alaska	Common murre, tufted puffin	5	60,000
12	Middleton Island	Alaska	Tufted puffin, common murre	6	20,000
13	Barren Islands	Alaska	Tufted puffin, common murre, horned puffin	8	310,000
14	Kodiak and Trinity islands	Alaska	Common murre, tufted puffin	6	280,000
15	Cape Newenham, Hagemeister islands	Alaska	Common murre, tufted puffin	6	1,350,000
16	Nunivak Island	Alaska	Common murre, tufted puffin	7	300,000
17	Saint Lawrence Island	Alaska	Least and crested auklets, murres	10	2,000,000
18	King Island	Alaska	Murres, least, crested, and parakeet auklets	9	240,000

(continued)

Index No.	Specific Location	State or Region	Major Species	Total Species	Estimated Total Auk Population
19	Little Diomede Island	Alaska	Least and crested auklets, murre, horned puffin	10	1,200,000
20	Cape Thompson	Alaska	Common and thick-billed murres	3	10,000
21	Cape Lisburne	Alaska	Common and thick-billed murres	6	100,000
22	Attu and Semichi islands	Alaska	Murres, tufted puffin	6	45,000
23	Agattu Island	Alaska	Murres, tufted puffin	6	30,000
24	Buldir Island	Alaska	Crested and least auklets, murres, puffins	12	430,000
25	Kiska Island	Alaska	Crested and least auklets	4	25,000
26	Segula and Semisopochnoi islands	Alaska	Least auklet, murres, puffins	9	170,000
27	Delarof Island	Alaska	Least, parakeet, and crested auklets, puffins	10	660,000
28	Adak and Kanaga islands	Alaska	Puffins, pigeon guillemot	8	33,000
29	Atka Island	Alaska	Tufted puffin, crested auklet	9	120,000
30	Chagulak Island	Alaska	Murres, parakeet auklet	5	120,000
31	Kagamil Island	Alaska	Murres	3	286,000
32	Umnak Island	Alaska	Murres, puffins	6	100,000
33	Unalaska and Baby islands	Alaska	Tufted puffin	8	100,000
34	Krenitzen Islands	Alaska	Tufted puffin	5	500,000
35	Saint Paul (Pribilofs)	Alaska	Murres, parakeet and least auklets	7	200,000
36	Saint George (Pribilofs)	Alaska	Murres, parakeet and least auklets	7	2,000,000
37	Sandman Reefs	Alaska	Puffins, Cassin auklet	8	150,000
38	Stepovak Bay area	Alaska	Puffins, murres, auklets	10	700,000
39	Shumagan Islands	Alaska	Puffins, murres	9	400,000
40	Semidi Islands	Alaska	Puffins, murres	9	1,000,000
41	Prince Leopold Island	Franklin district	Thick-billed murre	2	150,000
42	Cape Hay	Bylot Island	Thick-billed murre	1	800,000
43	Cape Graham Moore	Bylot Island	Thick-billed murre	1	40,000
44	Cambridge Point	Coberg Island	Thick-billed murre	1	400,000
45	Saunders Island	Greenland	Thick-billed murre	1	400,000
46	Agpat	Greenland	Thick-billed murre	3	200,000
47	Agparssuit (Kap Shakleton)	Greenland	Thick-billed murre	3	970,000

(continued)

Index No.	Specific Location	State or Region	Major Species	Total Species	Estimated Total Auk Population
48	Sagdleg Island	Greenland	Thick-billed murre	1	300,000
49	Arveprinsens Island	Greenland	Thick-billed murre	1	100,000
50	Reid Bay	Baffin Island	Thick-billed murre	1	400,000
51	Hantzch Island	Franklin district	Thick-billed murre	1	100,000
52	Cape Wolstenholme and Digges Island	Quebec	Thick-billed murre	1	2,000,000
53	Akpatok Island	Quebec	Thick-billed murre	1	1,200,000
54	Gannet Islands	Labrador	Atlantic puffin, common murre, razorbill	4	160,000
55	Bonaventure Island	New Brunswick	Common murre	3	35,000
56	Funk Island	Newfoundland	Common murre	4	1,000,000
57	Green and Great islands	Newfoundland	Atlantic puffin, common murre	4	500,000

Major sources: CALIFORNIA: Varoujean 1979; Sowls et al. 1980. OREGON: Varoujean 1979; Varoujean and Pitman 1979. WASHINGTON: Varoujean 1979; Manuwal and Campbell 1979. BRITISH COLUMBIA: Manuwal and Campbell 1979; Drent and Guiguet 1961. ALASKA: Manuwal and Campbell 1979; Sowls, Hatch, and Lensink 1978. EASTERN CANADA: Brown et al. 1975; Nettleship 1980. ATLANTIC COAST: Erwin and Korschgen 1979. Additional data on Atlantic Coast colonies may be found in Nettleship and Birkhead (1985), which appeared after this manuscript went into production.

Appendix 4

Summer Abundance and Breeding Status of Grebes at Selected National Wildlife Refuges

National Wildlife Refuge	State	Grebe Species					
		Least	Pied-billed	Horned	Red-necked	Eared	Western
Agassiz	Minnesota		C	C	C	C	C
Arrowwood	North Dakota		C	U	R	C	C
Audubon	North Dakota		C	o	o	C	C
Bear Lake	Idaho		C			C	A
Bear River	Utah		C			O	A
Benton Lake	Montana		U			A	o
Bowdoin	Montana		C			C	C
C. M. Russell	Montana		C			C	C
Crescent Lake	Nebraska		C	r		A	C
Klamath Basin[a]	Oregon/California		C	r	U	A	A
Lacreek	South Dakota		C			C	C
Laguna Atascosa	Texas	C	C	o			
Malheur	Oregon		A	O		A	A
Medicine Lake	Montana		C	C	r	A	A
Ninepipe/Pablo	Montana		U	U	C	U	C
Ruby Lake	Nevada		C			C	R
Sand Lake	South Dakota		A			A	A
Souris Loop[b]	North Dakota		C	U	U	C	C
Turnbull	Washington		C	R	r	A	r
Valentine	Nebraska		C			C	C
Waubay	South Dakota		C	O	c	C	C

NOTE: Letters indicate relative abundance: a, abundant; c, common; u, uncommon; o, occasional; r, rare. Upper-case letters indicate known breeding at the specified location.

[a]Includes Upper and Lower Klamath, Bear Valley, Tule Lake, and Clear Lake refuges.

[b]Includes Lostwood, J. Clark Salyer, Des Lacs, and Upper Souris refuges.

Literature Cited

Ainley, D. G. 1976. The occurrence of seabirds in the coastal region of California. *Western Birds* 7:33–68.

Ainley, D. G., and Sanger, G. A. 1979. Trophic relations of seabirds in the northeastern Pacific Ocean and Bering Sea. In Bartonek and Nettleship 1979, pp. 95–122.

Alexander, L. L. 1985. Trouble with loons. *Living Bird Quarterly* 4(2):10–13.

Amaral, M. J. 1977. A comparative breeding biology of the tufted and horned puffin in the Barren Islands, Alaska. M.S. thesis, University of Washington, Seattle.

American Ornithologists' Union (AOU). 1983. *Checklist of North American birds.* 6th ed. Lawrence, Kans.: Allen Press.

Andersson, A.; Lindberg, P.; Nilsson, S.; and Pettersson, A. 1980. [Breeding success of the black-throated diver *Gavia arctica* in the Swedish lakes.] *Var Fagelvarld* 39:85–94. In Swedish, English summary.

Asbirk, S. 1979. The adaptive significance of the reproductive pattern of the black guillemot, *Cepphus grille. Videnskabeleige Meddelelser fra Dansk Naturhistorisk Forening i Kobenhaven* 141:29–80.

Ashcroft, R. E. 1976. Breeding biology and survival of puffins. Ph.D. dissertation, Oxford University.

———. 1979. Survival rates and breeding biology of puffins on Skomer Island, Wales. *Ornis Scandinavica* 10:100–110.

Bailey, A. 1922. Notes on the yellow-billed loon. *Condor* 24:204–5.

———. 1948. *Birds of arctic Alaska.* Popular Series, no. 8. Denver: Colorado Museum of Natural History.

Bailey, E. P. 1973. Discovery of a Kittlitz's murrelet nest. *Condor* 75:457.

Baltz, D. M., and Morejohn, G. V. 1977. Food habits and niche overlap of sea birds wintering on Monterey Bay, California. *Auk* 94:526–43.

Bancroft, G. 1927. Notes on the breeding coastal and insular birds of central Lower California. *Condor* 29:188–97.

Bannerman, D. A. 1963. *Birds of the British Isles.* Vol. 12. Edinburgh: Oliver and Boyd.

Barklow, W. E. 1979. Graded frequency variations of the tremolo call of the common loon (*Gavia immer*). *Condor* 81:53–64.

Barr, J. F. 1973. Feeding biology of the common loon (*Gavia immer*) on oligotrophic lakes of the Canadian shield. Ph.D. dissertation, University of Guelph, Ontario.

Bartonek, J. C., and Nettleship, D. N., eds. 1979. Conservation of marine birds of northern North America. Research Report 11. Washington, D.C.: U.S. Fish and Wildlife Service.

Bateson, P. P. G. 1961. Studies of less familiar birds. 112. Little auk. *British Birds* 54:272–76.

Bauer, K. M., and Glutz, U. N. von Blotzheim, eds. 1966. *Handbuch der Vogel Mitteleuropas.* Vol. 1. Frankfurt: Akademische Verlagsgesellschaft.

Bedard, J. 1967. Ecological segregation among plankton-feeding Alcidae (*Aethia* and *Cyclorrhynchus*). Ph.D. dissertation, University of British Columbia, Vancouver.

———. 1969a. Feeding of the least, crested and parakeet auklets around St. Lawrence Island, Alaska. *Canadian Journal of Zoology* 47:1025–50.

———. 1969b. The nesting of the least, crested and parakeet auklets around St. Lawrence Island, Alaska. *Condor* 71:386–98.

———. 1969c. *Histoire naturelle du gode, Alca torda L., dans de Golfe Saint-Laurent, Province de Quebec, Canada.* Report Series, no. 7. Ottawa: Canadian Wildlife Service.

———. 1969d. Adaptive radiation in Alcidae. *Ibis* 111:189–98.

———. 1976. Coexistence, coevolution and convergent evolution in seabird communities: A comment. *Ecology* 57:177–84.

Bedard, J., and Sealy, S. G. 1984. Moults and feather generations in the least, crested and parakeet auklets. *Journal of Zoology* 202:461–88.

Belopolski, L. O. 1957. *Ecology of sea colony birds of the Barents Sea.* Washington, D.C.: U.S. Department of Commerce. Translated from Russian, Israel Program for Scientific Translations.

Bengtson, S.-A. 1984. Breeding ecology and extinction of the great auk (*Pinguinus impennis*): Anecdotal evidence and conjectures. *Auk* 101:1–12.

Bent, A. C. 1919. *Life histories of North American diving birds, Order Pygopodes.* Bulletin of the U.S. National Museum, no. 107.

Bergman, R. D., and Derksen, D. V. 1977. Observations on arctic and red-throated loons at Storkersen Point, Alaska. *Arctic* 30:41–51.

Bianki, V. V. 1977. *Gulls, shorebirds and alcids of Kandalaksha Bay.* Washington, D.C.: U.S. Department of Commerce. Translated from Russian, Israel Program for Scientific Translations.

Binford, L. C.; Elliott, B. G.; and Singer, S. W. 1975. Discovery of a nest and the downy young of the marbled murrelet. *Wilson Bulletin* 87:303–19.

Binford, L. C., and Remsen, J. V., Jr. 1974. Identification of the yellow-billed loon (*Gamvia adamsii*). *Western Birds* 5:111–26.

Birkhead, T. R. 1974. Movements and mortality rates of British guillemots. *Bird Study* 21:241–54.

———. 1976. Breeding biology and survival of guillemots. Ph.D. dissertation, Oxford University.

———. 1977a. Adaptive significance of the nestling period of guillemots, *Uria aalge. Ibis* 119:544–49.

———. 1977b. The effect of habitat and density on breeding success in the common guillemot (*Uria aalge*). *Journal of Animal Ecology* 46:751–64.

———. 1978. Behavioural adaptations to high density nesting in the common guillemot *Uria aalge. Animal Behaviour* 26:321–31.

———. 1980. Timing of breeding of common guillemots *Uria aalge* at Skomer Island, Wales. *Ornis Scandinavica* 11:142–45.

Birkhead, T. R., and Hudson, P. J. 1977. Population parameters for the common guillemot *Uria aalge. Ornis Scandinavica* 8:145–54.

Bleich, V. C. 1975. Diving times and distances in the pied-billed grebe. *Wilson Bulletin* 87:278–80.

Boertmann, D. 1980. Lommernes (Gaviidae) fylogeni. Doctoral dissertation, Copenhagen University. Not seen; cited by Glutz and Bauer 1982.

Bolam, G. 1921. Rate of progress of great crested grebe under water. *British Birds* 14:189.

Bowes, A. L. 1965. An ecological investigation of the giant pied-billed grebe, *Podilymbus gigas* Griscom. *Bulletin of the British Ornithologists' Club* 85:14–19.

———. 1969. The life history, ecology and management of the giant pied-billed grebe (*Podilymbus gigas*), Lake Atitlan, Guatemala. Ph.D. dissertation, Cornell University, Ithaca.

Bradstreet, M. S. W. 1979. Thick-billed murres and black guillemots in the Barrow Strait area, N.W.T., during spring: Distribution and habitat use. *Canadian Journal of Zoology* 57:1789–1802.

———. 1980. Thick-billed murres and black guillemots in the Barrow Strait area, N.W.T., during spring: Diets and food availability along ice edges. *Canadian Journal of Zoology* 58:2120–40.

———. 1982a. Pelagic feeding ecology of dovekies, *Alle alle*, in Lancaster Sound and western Baffin Bay. *Arctic* 35:126–40.

———. 1982b. Occurrence, habitat use and behavior of seabirds, marine mammals and arctic cod at the Pond Inlet ice edge. *Arctic* 35:28–40.

Brandt, H. 1943. *Alaska bird trails.* Cleveland: Bird Research Foundation.

Brodkorb, P. 1953. A review of the Pliocene loons. *Condor* 55:211–14.

———. 1963. Catalog of fossil birds. Part 1. *Florida State Museum Biological Sciences Bulletin* 7(4):179–293.

———. 1967. Catalog of fossil birds. Part 3. *Florida State Museum Biological Sciences Bulletin* 11(3):99–122.

Broekhuysen, G. J., and Frost, P. G. H. 1968. Nesting behaviour of the black-necked grebe *Podiceps nigricollis* in southern Africa. 2. Laying, clutch size, egg size, incubation and nesting success. *Ostrich* 39:242–52.

Brown, R. G. B. 1976. The foraging range of breeding dovekies. *Alle alle. Canadian Field-Naturalist* 90:166–68.

Brown, R. G. B.; Nettleship, D. N.; Germain, P.; Tull, C. E.; and Davis, T. 1975. *Atlas of eastern Canadian seabirds.* Ottawa: Canadian Wildlife Service.

Brun, E. 1958. Notes on the fledging-period of the razorbill. In *Skokholm Bird Observatory Report for 1958,* pp. 23–26. Bristol: West Wales Field Society.

———. 1966. The breeding population of puffins in Norway. *Sterna* 7:100–110.

Bundy, G. 1976. Breeding biology of the red-throated diver. *Bird Study* 23:249–56.

———. 1978. Breeding red-throated divers in Shetland. *British Birds* 71:199–208.

Burn, D. M., and Mather, J. R. 1974. The white-billed diver in Britain. *British Birds* 67:257–82.

278

Burrell, G. 1980. Some observations on nesting tufted puffins, Destruction Island, Washington. *Murrelet* 61:92–94.

Bylin, K. 1971. [Courtship and calls in the red-throated diver]. *Var Fagelvarld* 30:79–83. In Swedish, English summary.

Byrd, G. V.; Day, R. H.; and Knudtson, E. P. 1983. Patterns of colony attendance and censusing of auklets at Buldir Island, Alaska. *Condor* 85:274–80.

Byrd, G. V., and Gibson, D. D. 1980. Distribution and population status of whiskered auklet in the Aleutian Islands, Alaska. *Western Birds* 11:135–40.

Byrd, G. V.; Gibson, D. D.; and Johnson, D. L. 1974. The birds of Adak Island, Alaska. *Condor* 76:288–300.

Byrd, G. V., and Knudtson, E. P. 1978. Biology of nesting crested, least and whiskered auklets at Buldir Island, Alaska. Typescript report, in files of U.S. Fish and Wildlife Service, Anchorage, Alaska.

Cairns, D. K. 1978. Some aspects of the biology of the black guillemot (*Cepphus grille*) in the estuary and Gulf of St. Lawrence. M.S. thesis, Laval University, Quebec.

———. 1980. Nesting density, habitat structure and human disturbance as factors in black guillemot reproduction. *Wilson Bulletin* 92:352–61.

———. 1981. Breeding, feeding and chick growth of the black guillemot (*Cepphus grille*) in southern Quebec. *Canadian Field-Naturalist* 95:312–18.

Cairns, D. K., and deYoung, B. 1981. Back-crossing of a common murre (*Uria aalge*) and a common murre-thick-billed murre hybrid (*U. aalge* x *U. lomvia*). *Auk* 98:847.

Campbell, R. W.; Carter, H. R.; and Sealy, S. G. 1979. Nesting of horned puffins in British Columbia. *Canadian Field-Naturalist* 93:84–86.

Carter, H. R., and Sealy, S. G. 1984. Marbled murrelet mortality due to gill-net fishing in Barkley Sound, British Columbia. In Nettleship, Sanger, and Springer 1984, pp. 212–20.

Chabreck, R. H. 1963. Breeding habits of the pied-billed grebe in an impounded coastal marsh in Louisiana. *Auk* 80:447–52.

Chamberlin, M. L. 1977. Observations on the red-necked grebe nesting in Michigan. *Wilson Bulletin* 89:32–46.

Clapp, R. B.; Banks, R. C.; Morgan-Jacobs, D.; and Hoffman, W. A. 1982. *Marine birds of the southeastern United States and Gulf of Mexico. Part 1. Gaviiformes through Pelecaniformes.* National Coastal Ecosystems Project Report. Washington, D.C.: U.S. Fish and Wildlife Service.

Cody, M. L. 1973. Coexistence, coevolution and convergent evolution in seabird communities. *Ecology* 54:31–44.

Corkhill, P. 1973. Food and feeding ecology of puffins. *Bird Study* 20:207–20.

Cottam, C., and Knappen, P. 1939. Foods of some uncommon North American birds. *Auk* 56:138–69.

Coues, E. 1882. *Key to North American birds.* 5th ed. Boston: Page.

Cracraft, J. 1981. Toward a phylogenetic classification of the Recent birds of the world. *Auk* 98:681–714.

———. 1982. Phylogenetic relationships and monophyly of loons, grebes and hesperornithiform birds, with comments on the early history of birds. *Systematic Zoology* 31:35–56.

Cramp, S., and Simmons, K. E. L., eds. 1977. *The birds of the western Palearctic.* Vol. 1. Oxford: Oxford University Press.

———. 1985. *The birds of the western Palearctic.* Vol. 4. Oxford: Oxford University Press.

Cross, P. A. 1979. Status of the common loon in Maine during 1977 and 1978. In Sutcliffe 1979a, pp. 73–80.

Cyrus, D. P. 1975. Breeding success of red-throated divers on Fetlar. *British Birds* 68:75–76.

Daan, S., and Tinbergen, J. 1979. Young guillemots (*Uria lomvia*) leaving their arctic breeding cliffs: A daily rhythm in numbers and risk. *Ardea* 67:196–200.

Davis, D. G. 1961. Western grebe colonies in northern Colorado. *Condor* 63:164–65.

Davis, R. A. 1971. Flight speed of arctic and red-throated loons. *Auk* 88:169.

———. 1972. A comparative study of the use of habitat by arctic loons and red-throated loons. Ph.D. dissertation, University of Western Ontario, London.

Day, R. H.; Oakley, K. L.; and Barnard, D. R. 1983. Nest sites and eggs of Kittlitz's and marbled murrelets. *Condor* 85:265–73.

Dean, W. R. J. 1977. Breeding of the great crested grebe at Barberspan. *Ostrich*, suppl. 12, 43–48.

Dementiev, G. P., and Gladkov, N. A., eds. 1968. *Birds of the Soviet Union.* Vol. 2. Jerusalem: Israel Program for Scientific Translations.

Derksen, D. V.; Rothe, T. C.; and Eldridge, W. D. 1981. *Use of wetland habitat by birds in the National Petroleum Reserves—Alaska.* Resource Publication 141. Washington, D.C.: U.S. Fish and Wildlife Service.

Devillers, P. 1972. The juvenal plumage of Kittlitz's murrelet. *California Birds* 3:33–38.

de Vos, A., and Allin, A. E. 1964. Winter mortality among red-necked grebes (*Colymbus grisegena*) in Ontario. *Canadian Field-Naturalist* 78:67–69.

Dewar, J. M. 1924. *The bird as a diver.* London: Witherby.

DeWeese, L. R., and Anderson, D. W. 1976. Distribution and breeding biology of Craveri's murrelet. *Transac-*

tions of the San Diego Society of Natural History
18:155–68.

Dick, M. H., and Donaldson, W. 1978. Fishing vessel endangered by crested auklet landings. *Condor* 80:235–36.

Dickerman, R. W. 1963. The grebe *Aechmophorus occidentalis clarkii* as a nesting bird of the Mexican plateau. *Condor* 65:66–67.

———. 1973. Further notes on the western grebe in Mexico. *Condor* 75:131–32.

Dickinson, H. 1958. Puffins and burrows. In *Skokholm Bird Observatory Report for 1958*, pp. 27–34. Bristol: West Wales Field Society.

Divoky, G. J.; Watson, G. E.; and Bartonek, J. C. 1974. Breeding of the black guillemot in northern Alaska. *Condor* 76:339–43.

Drent, R. H. 1965. Breeding biology of the pigeon guillemot, *Cepphus columba. Ardea* 53:99–160.

Drent, R. H., and Guiguet, C. J. 1961. *A catalogue of British Columbia sea-bird colonies.* Occasional Papers 12. Victoria: British Columbia Provincial Museum.

Dunker, H. 1974. Habitat selection and territory size of the black-throated diver, *Gavia arctica* (L.) in south Norway. *Norwegian Journal of Zoology* 22:15–29.

———. 1975. Sexual and aggressive display of the black-throated diver, *Gavia arctica* (L.). *Norwegian Journal of Zoology* 23:149–63.

Dunker, H., and Elgmark, K. 1973. Nesting of the black-throated diver, *Gavia arctica* (L.), in small bodies of water. *Norwegian Journal of Zoology* 21:33–37.

Eaton, E. H. 1910. Birds of New York. Part 1. *Memoirs of the New York State Museum* (Albany), 12:1–501.

Eldridge, W. D., and Kuletz, K. J. 1979. Chick feeding and adult foraging patterns of pigeon guillemots. *Pacific Seabird Group Bulletin* 6:26 (abstract).

Enquist, M. 1978. [Behavior of the black-throated diver *Gavia arctica* in Braviken during spring]. *Var Fagelvarld* 37:325–32. In Swedish, English summary.

Erwin, R. M., and Korschgen, C. E. 1979. *Coastal waterbird colonies: Maine to Virginia, 1977.* Biological Services Program, report FWS/OBS-79/08. Washington, D.C.: U.S. Fish and Wildlife Service.

Evans, P. G. H. 1981. Ecology and behaviour of the little auk *Alle alle* in West Greenland. *Ibis* 123:1–18.

Faaborg, J. 1976. Habitat selection and territorial behavior of the small grebes of North Dakota. *Wilson Bulletin* 88:390–99.

FAO Fisheries Department. 1981. *Atlas of the living resources of the sea.* 4th ed. FAO Fisheries Series. Rome: Food and Agriculture Organization of the United Nations.

Feerer, J. H. 1977. Niche partitioning by western grebe polymorphs. M.S. thesis, Humboldt State College, Arcata, Calif.

Feerer, J. H., and Garrett, R. L. 1977. Potential western grebe extinction on California lakes. *California-Nevada Wildlife Society Transactions* 1977:80–89.

Ferdinand, L. 1969. Some observations on the behaviour of the little auk (*Plautus alle*) on the breeding-ground, with special reference to voice production. *Dansk Ornithologisk Forenings Tidsskrift* 63:19–44.

Ferguson, R. S., and Sealy, S. G. 1983. Breeding ecology of the horned grebe, *Podiceps auritus*, in south-western Manitoba. *Canadian Field-Naturalist* 97:401–8.

Fjeldså, J. 1973a. Feeding and habitat selection of the horned grebe *Podiceps auritus* (Aves), in the breeding season. *Videnskabelige Meddalelser fra Dansk Naturhistorisk Forening i Kobenhaven* 136:57–95.

———. 1973b. Territory and the regulation of population density and recruitment in the horned grebe *Podiceps auritus arcticus* Boie 1822. *Videnskabelige Maddelelser fra Dansk Naturhistorisk Forening i Kobenhaven* 136:177–89.

———. 1973c. Antagonistic and heterosexual behavior of the horned grebe, *Podiceps auritus. Sterna* 12:161–217.

———. 1973d. Distribution and geographic variation of the horned grebe *Podiceps auritus* (Linnaeus 1758). *Ornis Scandinavica* 4:55–86.

———. 1975. *Grebes.* Hvidovre, Denmark: Villadsen & Christensen.

———. 1977 *Guide to the young of European precocial birds.* Tisvildelje, Denmark. Scarv Nature Publications.

———. 1982a. Some behaviour patterns of four closely related grebes, *Podiceps nigricollis, P. gallardoi, P. occipitalis* and *P. taczanowskii,* with reflections on phylogeny and adaptive aspects of the evolution of displays. *Dansk Ornithologisk Forenings Tidsskrift* 76:37–68.

———. 1982b. The adaptive significance of local variations in the bill and jaw anatomy of North European red-necked grebes *Podiceps grisegena. Ornis Fennica* 59:84–98.

———. 1983a. Ecological character displacement and character release in grebes Podicipedidae. *Ibis* 125:463–81.

———. 1983b. Social behaviour and displays of the hoary-headed grebe *Poliocephalus poliocephalus. Emu* 83:129–40.

———. 1985. Displays of the two primitive grebes *Rollandia rolland* and *R. microptera* and the origin of the complex courtship behaviour of the *Podiceps* species (Aves, Podicipedidae). *Steenstrupia* 11:133–55.

Flint, V., and Kishchinski, A. A. 1982. Taxonomische Wechelbeziehungen innerhalb der Gruppe der Prachttaucher (Gaviidae, Aves). *Beiträge zur Angewandten Vogelkunde* 28:193–206.

Forsell, D. J., and Gould, P. J. 1981. *Distribution and abundance of marine birds and mammals wintering in the Kodiak area of Alaska.* Biological Services Program, report FWS/OBS 81/13. Washington, D.C.: U.S. Fish and Wildlife Service.

Fox, G. A.; Yonge, K. S.; and Sealy, S. G. 1980. Breeding performance, pollutant burden and eggshell thinning in common loons *Gavia immer* nesting on a boreal forest lake. *Ornis Scandinavica* 11:43–48.

Franke, H. 1969. Die Paarungsbalz des Schwarzhalstauchers. *Journal für Ornithologie* 110:286–90.

Friley, C. E., Jr., and Hendrickson, G. O. 1937. Eared grebes nesting in northwest Iowa. *Iowa Bird Life* 7:2–3.

Frith, H. J. 1976. *Reader's Digest complete book of Australian birds.* Sydney: Reader's Digest Services.

Gaston, A. J. 1980. *Populations, movements and wintering areas of thick-billed murres (Uria lomvia) in eastern Canada.* Progress Notes 110. Ottawa: Canadian Wildlife Service.

———. 1984. How to distinguish first-year murres, *Uria* spp., from older birds in winter. *Canadian Field-Naturalist* 98:52–55.

Gaston, A. J., and Malone, M. 1980. Range extension of Atlantic puffin and razorbill in Hudson Strait. *Canadian Field-Naturalist* 94:328–29.

Gaston, A. J., and Nettleship, D. N. 1981. *The thick-billed murre on Prince Leopold Island—a study of the breeding ecology of a colonial high arctic seabird.* Monograph Series 6. Ottawa: Canadian Wildlife Service.

Gaukler, A., and Kraus, M. 1968. Zum Vorkommen und zur Brutbiologie des Schwarzhalstauchers (*Podiceps nigricollis*) in Nordbayern. *Anzeiger der Ornithologischen Gesellschaft in Bayern* 8:349–64.

Glover, F. A. 1953. Nesting ecology of the pied-billed grebe in northeastern Iowa. *Wilson Bulletin* 65:32–39.

Glutz, U. N. von Blotzheim, and Bauer, K. M., eds. 1982. *Handbuch der Vogel Mitteleuropas.* Vol. 8, part 2. Wiesbaden: Akademische Verlagsgesellschaft.

Godfrey, W. A. 1966. *The birds of Canada.* Bulletin 203. Ottawa: National Museums of Canada.

Golowkin, A. M.; Selikman, E. A.; and Georgiev, A. A. 1972. [Biology and food habitats of the dovekie in the pelagic community in the north of Novaya Zemlya]. In [*Peculiarities of biological productivity of waters near birds' bazaars in the north of Novaya Zemlya*], pp. 74–84. Leningrad: Nauka. In Russian. Not seen; cited by Glutz and Bauer 1982.

Gotzman, J. 1965. Environment preference in the grebes (Podicipedidae) during breeding season. *Ekologia Polska*, ser. A, 13:289–302.

Gould, P. J.; Forsell, D. J.; and Lensink, C. J. 1982. *Pelagic distribution and abundance of seabirds in the Gulf of Alaska and eastern Bering Sea.* Biological Services Program report FWS/OBS 82/48. Washington, D.C.: U.S. Fish and Wildlife Service.

Grant, P. R., and Nettleship, D. N. 1971. Nesting habitat selection by puffins *Fratercula arctica* L. in Iceland. *Ornish Scandinavica* 2:81–87.

Greenwood, J. 1964. The fledging of the guillemot *Uria aalge* with notes on the razorbill *Alca torda. Ibis* 106:469–81.

Grieve, S. 1885. *The great auk, or garefowl (*Alca impennis *Linn.): Its history, archeology, and remains.* Edinburgh: Grange Publishing Works.

Gross, A. O. 1949. The Antillean grebe at Central Soledad, Cuba. *Auk* 66:42–53.

Guiguet, C. J. 1950. The marbled murrelet. *Victoria Naturalist* 7:37–39.

———. 1971. A list of seabird nesting sites in Barkley Sound, British Columbia. *Syesis* 4:253–59.

Hammond, D. W., and Wood, R. L. 1977. *New Hampshire and the disappearing loon.* Meredith, N.H.: Loon Preservation Committee. Not seen.

Harle, D. F. 1952. Red-throated diver taking off from the ground. *British Birds* 45:331–32.

Harris, M. P. 1970. Differences in the diet of British auks. *Ibis* 112:540–41.

———. 1981. Age determination and first breeding of British puffins. *British Birds* 74:246–56.

———. 1983a. Biology and survival of the immature puffin *Fratercula arctica. Ibis* 125:56–73.

———. 1983b. Parent-young communication in the puffin *Fratercula arctica. Ibis* 125:109–14.

———. 1984. *The puffin.* Berkhamstead: Poyser.

Harris, M. P., and Hislop, J. R. G. 1978. The food of young puffins *Fratercula arctica. Journal of Zoology* 185:213–36.

Harris, M. P., and Murray, S. 1977. Puffins on St. Kilda. *British Birds* 70:50–65.

Harris, R. D. 1971. Further evidence of tree nesting in the marbled murrelet. *Canadian Field-Naturalist* 85:67–68.

Harrison, C. 1978. *A field guide to the nests, eggs and nestlings of North American birds.* London: Collins.

Hartman, F. A. 1961. Locomotor mechanisms of birds. *Smithsonian Miscellaneous Collections* 143:1–91.

Hatch, S. A. 1983. The fledging of common and thick-billed murres on Middleton Island, Alaska. *Journal of Field Ornithology* 54:266–74.

Hatch, S. A., and Hatch, M. A. 1983. Populations and habitat use of marine birds in the Semidi Islands, Alaska. *Murrelet* 64:39–46.

Hatler, D. F.; Campbell, R. W.; and Dorst, A. 1978. *Birds of Pacific Rim National Park.* Occasional Papers 20. Victoria: British Columbia Provincial Museum.

Heath, H. 1915. Birds observed on Forrester Island, Alaska, during the summer of 1913. *Condor* 17:20–41.

Hedgren, S. 1976. [On the food of the guillemot *Uria aalge* at the island of Stora Karlso, Baltic Sea]. *Var Fagelvarld* 35:287–90. In Swedish, English summary.

———. 1980. Ecological aspects of the breeding biology of the guillemot *Uria aalge* in the Baltic Sea. Dissertation summary, Department of Zoology, University of Stockholm, Sweden.

Heilmann, G. 1927. *The origin of birds.* New York: Appleton.

Heinroth, O., and Heinroth, M. 1931. *Die Vogel Mitteleuropas.* Vol. 4. Berlin: Hugo Bermuhler.

Hemming, J. E. 1968. Copulatory behavior of the red-necked grebe on open water. *Wilson Bulletin* 80:326–27.

Herman, S. G.; Garrett, R. L.; and Rudd, R. L. 1969. Pesticides and the western grebe. In *Chemical fallout (current research on persistent pesticides),* ed. M. W. Miller and G. G. Berg, pp. 24–53. Springfield, Ill.: Charles C. Thomas.

Hirsch, K. V.; Woodby, D. A.; and Astheimer, L. B. 1981. Growth of a nestling marbled murrelet. *Condor* 82:264–65.

Hoeman, J. V. 1965. Marbled murrelet breeding record from Kodiak. *Bulletin of the Alaska Ornithological Society* 5:9. Mimeographed.

Hoffman, W.; Elliott, W. P.; and Scott, J. M. 1975. The occurrence and status of the horned puffin in the western United States. *Western Birds* 6:87–94.

Höhn, O. 1982. *Die Seetaucher.* Neue Brehm-Bücherei 546. Wittenberg Lutherstadt: A. Ziemsen.

Howard, H. 1950. Fossil evidence of avian evolution. *Ibis* 92:1–21.

———. 1970. A review of the extinct avian genus *Mancella. Contributions in Science* (Los Angeles County Museum), 203:1–12.

Hudson, G. E.; Hoff, K. M.; Vanden Berge, J.; and Trivette, E. C. 1969. A numerical study of the wing and leg muscles of Lari and Alcae. *Ibis* 111:459–524.

Hudson, P. J. 1979. Survival rates and behaviour of British auks. Ph.D. dissertation, Oxford University.

———. 1982. Nest site characteristics and breeding success in the razor-bill *Alca torda. Ibis* 124:355–59.

Hunt, G. L., Jr. 1977. Reproductive ecology, foods, and foraging areas of seabirds nesting on the Pribilof Islands. In *Environmental assessment of the Alaskan continental shelf, annual reports of principal investigators, biological studies,* 2:196–392. Boulder, Colo.: NOAA–BLM.

Hunt, G. L., Jr.; Burgeson, B.; and Sanger, G. A. 1981. Feeding ecology of seabirds of the eastern Bering Sea. In *The eastern Bering Sea shelf: Oceanography and resources,* ed. D. W. Hood and J. A. Calder, 2:629–47. Office of Marine Pollution Assessment, NOAA. Seattle: University of Washington Press.

Hunt, G. L., Jr.; Eppley, Z.; Burgeson, B.; and Squibb, R. 1980. Reproductive ecology, foods and foraging areas of seabirds nesting on the Pribilof Islands, 1975–1979. In *Environmental assessment of the Alaskan continental shelf, final report of principal investigators, biological studies,* 12:1–256. Boulder, Colo.: NOAA–BLM.

Hunt, G. L., Jr.; Pitman, R. L.; Naughton, M.; Winnett, K.; Newman, A.; Kelly, P. R.; and Briggs, K. T. 1979. Distribution, status, reproductive ecology and foraging habits of breeding seabirds. In *Draft final report to the Bureau of Land Management: Summary of marine mammal and seabird surveys of the southern California Bight area, 1975–1978.* Irvine, Calif.: Regents of the University of California. Not seen.

Hussell, J. T., and Holroyd, G. L. 1974. Birds of Truelove lowland and adjacent areas of northeastern Devon Island, N.W.T. *Canadian Field-Naturalist* 88:197–212.

Huxley, J. S. 1914. The courtship habits of the great crested grebe (*Podiceps cristatus*), with an addition to the theory of sexual selection. *Proceedings of the Zoological Society of London* 25:492–562.

———. 1923. Courtship activities in the red-throated diver, together with a discussion on the evolution of courtship in birds. *Journal of the Linnean Society of London* 35:253–92.

Ingold, P. 1973. Zur lautilichen Beziehung des Elters zu seinem Küken bei Tordalken. *Behaviour* 45:154–90.

———. 1980. [Adaptations of egg characteristics and breeding behavior of guillemots (*Uria aalge aalge* Pont.) to breeding on cliff-ledges]. *Zeitschrift für Tierpsychologie* 53:341–88. In German, English summary.

Islieb, M. E., and Kessel, B. 1973. Birds of the north Gulf Coast–Prince William Sound region, Alaska. *University of Alaska, Biological Papers,* no. 14.

Jehl, J. 1983. Mortality of eared grebes in winter of 1982–83. *American Birds* 37:832–35.

Jehl, J., and Bond, S. I. 1975. Morphological variation and species limits in murrelets of the genus *Endo-*

mychura. Transactions of the San Diego Society of Natural History 18:9–24.

Johnson, R. A. 1938. Predation of gulls in murre colonies. *Wilson Bulletin* 50:161–70.

———. 1941. Nesting behavior of the Atlantic murre. *Auk* 58:153–63.

Johnson, R. A., and Johnson, H. S. 1935. A study of the nesting and family life of the red-throated loon. *Wilson Bulletin* 47:97–103.

Johnston, S. T., and Carter, H. C. 1985. Cavity-nesting marbled murrelets. *Wilson Bulletin* 97:1–3.

Jonkel, G. M. 1979. Banding as a tool in bird studies. In Sutcliffe 1979a, pp. 11–19.

Kartashev, N. N. 1960. *Die Alkenvogel des Nordatlantiks.* Neue Brehm-Bücherei 257. Wittenberg Lutherstadt: A. Ziemsen.

Kerlinger, P. 1982. The migration of common loons through New York. *Condor* 84:97–100.

Kessel, B., and Gibson, D. D. 1976. Status and distribution of Alaska birds. *Studies in Avian Biology* 1:1–100.

Kharitonov, S. P. 1980. [Materials on birds of Iona Island]. *Ornitologiya* 15:10–15. In Russian.

Kiff, L. F. 1981. Eggs of the marbled murrelet. *Wilson Bulletin* 93:400–403.

Kilham, L. 1954. Courtship behavior of the pied-billed grebe. *Wilson Bulletin* 66:65.

King, J. G., and Sanger, G. A. 1979. Oil vulnerability index for marine oriented birds. In Bartonek and Nettleship 1979, pp. 227–39.

King, R. J. 1979. Loon abundance and distribution in the National Petroleum Reserve—Alaska (NPR–A). *Pacific Seabird Group Bulletin* 6:41 (abstract).

Kirby, R. E. 1976. Breeding chronology and interspecific relations of pied-billed grebes in northern Minnesota. *Wilson Bulletin* 88:493–95.

Kishchinski, A. A. 1968. [On the biology of the Kittlitz and marbled murrelets]. *Ornitologiya* 9:208–13. In Russian.

Knudtson, E. P., and Byrd, G. V. 1982. Breeding biology of crested, least and whiskered auklets on Buldir Island, Alaska. *Condor* 84:197–202.

Kop, P. P. A. M. 1971. Some notes on the moult and age determination in the great crested grebe. *Ardea* 59:56–60.

Korschgen, C. E. 1979. *Coastal waterbird colonies: Maine.* Biological Services Program, FWS/OBS 79/09. Washington, D.C.: U.S. Fish and Wildlife Service.

Korzun, L. P. 1981. [On the phylogenetic relations between Gaviiformes and Podicipediformes]. *Zoologischeskii Zhurnal* 60:1523–32. In Russian, English summary.

Kozlova, E. V. 1961. *Charadriiformes, suborder Alcae.* In *Fauna of U.S.S.R.: Birds*, vol. 2, no. 2. Washington, D.C.: U.S. Department. of Commerce. Translated from Russian by the Israel Program for Scientific Translations.

Kress, S. W. 1982. The return of the Atlantic puffin to eastern Egg Rock, Maine. *Living Bird Quarterly* 1:11–14.

Kretchmar, A. W., and Leonovitch, W. W. 1975. Verbreitung und Brut des gelbsnabligen Eistaucher. *Falke* 12:268–72.

Kuroda, N. 1954. On some osteological and anatomical characters of Japanese Alcidae (Aves). *Japanese Journal of Zoology* 11:311–27.

———. 1967. Morpho-anatomical analysis of parallel evolution between diving petrel and ancient auk, with comparative osteological data of other species. *Miscellaneous Reports, Yamashina Institute for Ornithology* 5:111–37.

Kuzyakin, A. P. 1963. [On the biology of *Brachyramphus marmoratus*]. *Ornitologiya* 6:315–20. In Russian.

Lack, D. 1968. *Ecological adaptations for breeding in birds.* London: Methuen.

Ladhams, D. E. 1969. Song of the pied-billed grebe. *Bristol Ornithology* 2:73–74.

Ladhams, D. E.; Prytherch, R. J.; and Simmons, K. E. L. 1967. Pied-billed grebe in Somerset. *British Birds* 60:295–99.

Lawrence, G. E. 1950. The diving and feeding activity of the western grebe on the breeding grounds. *Condor* 52:3–16.

Lehnhausen, W. A. 1980. Nesting habitat relationships of four species of alcids at Fish Island, Alaska. M.S. thesis, University of Alaska, Fairbanks.

Lehtonen, L. 1970. [Biology of the black-throated diver, *Gavia a. arctica* (L.)]. *Annales Zoologica Fennica* 7:25–60. In Finnish, English summary.

Leschner, L. L. 1976. The breeding biology of the rhinoceros auklet on Destruction Island. M.S. thesis, University of Washington, Seattle.

Leschner, L. L., and Burrell, G. 1977. Populations and ecology of marine birds on the Semidi Islands. In *Environmental assessment of the Alaskan continental shelf, annual reports of principal investigators,* 4:13–109. Boulder, Colo.: NOAA–BLM.

Lindberg, P. 1968. [On the ecology of the black-throated diver and red-throated diver]. *Zoologisk Revy* 30:83–88. In Swedish, English summary.

Lindvall, M. L. 1976. Breeding biology of pesticide-PCB contamination of western grebe at Bear River Migratory Bird Refuge. M.S. thesis, Utah State University, Logan.

Lindvall, M. L., and Low, J. B. 1982. Nesting ecology and production of western grebes at Bear River Migratory Bird Refuge, Utah. *Condor* 84:66–70.

Lloyd, C. S. 1974. Movements and survival of British razorbills. *Bird Study* 21:102–16.

———. 1976a. The breeding biology and survival of the razorbill *Alca torda.* Ph.D. dissertation, Oxford University.

———. 1976b. An estimate of the world breeding population of the razorbill. *British Birds* 69:298–304.

———. 1977. The ability of the razorbill to raise an additional chick to fledging. *Ornis Scandinavica* 8:155–59.

———. 1979. Factors affecting breeding of razorbills on Skokholm. *Ibis* 121:165–76.

Lloyd, C. S., and Perrins, C. M. 1977. Survival and age of first breeding in the razorbill (*Alca torda*). *Bird-Banding* 48:239–52.

Lockley, R. M. 1953. *Puffins.* New York: Devin-Adair.

Lovenskiold, H. L. 1964. Avifauna Svalbardensis, with a discussion on the geographic distribution of the birds in Spitsbergen and adjacent islands. *Norsk Polarinstituut Skrifter* 129:1–460.

Luther, D. 1972. *Die ausgestorbenen Vogel der Welt.* Neue Brehm-Bücherei, 424. Wittenberg Lutherstadt: A. Ziemsen.

McAllister, N. M. 1958. Courtship, hostile behavior, nest-establishment and egg-laying in the eared grebe (*Podiceps caspicus*). *Auk* 75:290–311.

McAllister, N. M., and Storer, R. W. 1963. Copulation in the pied-billed grebe. *Wilson Bulletin* 75:166–73.

McIntyre, J. M. W. 1975. Biology and behavior of the common loon (*Gavia immer*) with reference to adaptability in a man-altered environment. Ph.D. dissertation, University of Minnesota, Minneapolis.

———. 1978. The common loon. Part 3. Population in Itasca State Park, Minnesota, 1957–1976. *Loon* 50:38–44.

———. 1979. Minnesota common loon survey report, 1978. In Sutcliffe 1979a, 123–28.

Macpherson, A. H., and McLaren, I. A. 1959. Notes on the birds of southern Foxe Peninsula, Baffin Island. *Canadian Field-Naturalist* 73:63–81.

Madsen, F. J. 1957. On the food habits of some fish-eating birds in Denmark. *Danish Review of Game Biology* 3:19–83.

Mahoney, S. P., and Threlfall, W. 1982. Notes on the agonistic behavior of common murres. *Wilson Bulletin* 94:595–98.

Manuwal, D. A. 1972. The population ecology of Cassin's auklet on southeast Farallon Island, California. Ph.D. dissertation, University of California, Los Angeles.

———. 1974a. The natural history of Cassin's auklet (*Ptychoramphus aleuticus*). *Condor* 74:421–31.

———. 1974b. Effects of territoriality on breeding in a population of Cassin's auklet. *Ecology* 53:1399–1406.

———. 1979. Reproductive commitment and success in Cassin's auklet. *Condor* 81:111–21.

Manuwal, D. A., and Campbell, R. W. 1979. Status and distribution of breeding seabirds of southeastern Alaska, British Columbia, and Washington. In Bartonek and Nettleship 1979, pp. 73–91.

Manuwal, D. A., and Manuwal, N. J. 1979. Habitat specific behavior of the parakeet auklet in the Barren Islands. *Western Birds* 10:189–200.

Marchant, S. 1960. The breeding of some S.W. Ecuadorian birds. *Ibis* 102:584–99.

Martin, P. W., and Myres, M. T. 1969. Observations on the distribution and migration of some seabirds off the outer coasts of British Columbia and Washington State, 1946–49. *Syesis* 2:241–56.

Mather, J. 1967. Pied-billed grebe in Yorkshire. *British Birds* 60:290–95.

Meinertzhagen, R. 1955. The speed and altitude of bird flight (with notes on other animals). *Ibis* 97:81–117.

Merrie, T. D. H. 1978. Relationship between spatial distribution of breeding divers and the availability of fishing waters. *Bird Study* 25:119–22.

Metcalf, L. 1979. The breeding status of the common loon in Vermont. In Sutcliffe 1979a, pp. 101–10.

Miller, L. 1946. The Lucas auk appears again. *Condor* 48:32–36.

Miller, R. F. 1942. The pied-billed grebe, a breeding bird of the Philadelphia region. *Cassinia* 32:23–34.

Mink, L., and Gibson, T. 1976. The red-necked grebe. *North Dakota Outdoors* 39(1):18–19.

Munro, J. A. 1941. *The grebes.* Occasional Papers 3. Victoria: British Columbia Provincial Museum.

———. 1945. Observations on the loon in the Caribou Parklands, British Columbia. *Auk* 62:38–49.

Munyer, E. A. 1965. Inland wanderings of the ancient murrelet. *Wilson Bulletin* 77:235–42.

Murie, O. J. 1959. *Fauna of the Aleutian Islands and Alaska Peninsula.* North American Fauna 61. Washington, D.C.: U.S. Fish and Wildlife Service.

Murphy, E. C.; Roseneau, D. G.; and Bente, P. M. 1984. An inland nest record for the Kittlitz's murrelet. *Condor* 86:218.

Murray, K. G. 1980. Predation by deer mice on Xantus' murrelet eggs on Santa Barbara Island, California. M.S. thesis, California State University, Northridge.

Murray, K. G.; Winnett-Murray, K.; Eppley, Z. A.; Hunt, G. L., Jr.; and Schwartz, D. B. 1983. Breeding biology of the Xantus' murrelet. *Condor* 85:12–21.

Murray, K. G.; Winnett-Murray, K.; and Hunt, G. L., Jr. 1979. Egg-neglect in Xantus' murrelet. *Proceedings of the Colonial Waterbird Group* 3:186–95.

Myrberget, S. 1962. [Contribution to the breeding biology of the puffin, *Fratercula arctica* (L.): Eggs, incubation and young]. *Papers of the Norwegian State Game Research Institute*, 2d ser., no. 11:1–49. In Norwegian, English summary.

Nelson, D. A. 1981. Sexual differences in measurements of Cassin's auklet. *Journal of Field Ornithology* 52:233–34.

Nelson, R. W., and Myres, M. T. 1976. Declines in populations of peregrine falcons and their seabird prey at Langara Island, British Columbia. *Condor* 78:281–93.

Nero, R. W.; Lahrman, F. W.; and Bard, F. G. 1958. Dryland nest-site of a western grebe colony. *Auk* 75:347–49.

Nettleship, D. N. 1972. Breeding success of the common puffin (*Fratercula arctica* L.) on different habitats at Great Island, Newfoundland. *Ecological Monographs* 42:239–68.

———. 1977. Seabird resources of eastern Canada: Status, problems and prospects. In *Proceedings of the symposium, "Canada's endangered species and habitats* (May 20–24, 1976), ed. T. Mosquin and C. Suchal, pp. 96–108. Special Publication 6. Ottawa: Canadian Nature Federation.

———. 1980. *A guide to the major seabird colonies of eastern Canada.* Manuscript report 97. Dartmouth, N.S.: Canadian Wildlife Service. Not seen.

Nettleship, D. N., and Birkhead, T. R., eds. 1985. *The Atlantic Alcidae.* New York: Academic Press.

Nettleship, D. N.; Sanger, G. A.; and Springer, P. F., eds. 1984. *Marine birds: Their feeding ecology and commercial fisheries relationships.* Proceedings of the Pacific Seabird Group Symposium, Seattle, Washington, January 6–8, 1982. Ottawa: Canadian Wildlife Service.

Nilsson, S. G. 1977. Adult survival rate of the black-throated diver *Gavia arctica. Ornis Scandinavica* 8:193–95.

Norberg, A., and Norberg, U. 1971. Take-off, landing and flight speed of *Gavia stellata* Pont. *Ornis Scandinavica* 2:55–66.

———. 1976. Size of fish carried by flying red-throated divers *Gavia stellata* (Pont.) to nearly fledged young in nesting tarn. *Ornis Fennica* 53:92–95.

Norderhaug, M. 1967. Migration, homing instinct and pair formation of little auk (*Plautus alle*) in Svalbard. *Fauna* 20:236–44.

———. 1968. Trekkforhold, stedstrohet og pardennelse hos alkekonge pa Svalbard. *Meddelelser Norsk Polarinstituut* 96:236–44.

———. 1970. The role of the little auk *Plautus alle* (L.) in arctic ecosystems. In *Antarctic ecology*, ed. M. W. Holdgate, 1:558–60. New York: Academic Press.

———. 1980. Breeding biology of the little auk (*Plautus alle*) in Svalbard. *Norsk Polarinstituut Skrifter* 173:1–45.

Nordstrom, G. 1962. Finnische Wiederfunde im Ausland beringter Vogel. *Ornis Fennica* 39:131–51.

———. 1963. Einige Ergebnisse der Vogelberingung in Finnland im den Jahren 1913–62. *Ornis Fennica* 40:81–124.

Norrevang, A. 1957. On the breeding biology of the guillemot (*Uria aalge* [Pont.]). *Dansk Ornithologisk Forenings Tidsskrift* 52:48–74.

Nuechterlein, G. L. 1975. Nesting ecology of western grebes on the Delta Marsh, Manitoba. M.S. thesis, Colorado State University, Fort Collins.

———. 1981a. Courtship behavior and reproductive isolation between western grebe morphs. *Auk* 98:335–49.

———. 1981b. Variations and multiple functions of the advertising display of western grebes. *Behaviour* 76:289–317.

———. 1982. The birds that walk on water. *National Geographic* 161(5):624–37.

Nuechterlein, G. L., and Storer, R. W. 1982. The pair-formation displays of the western grebe. *Condor* 94:350–69.

Nysewander, D. R.; Forsell, D. J.; Baird, P. A.; Shields, D. J.; Weiler, G. J.; and Kogan, J. H. 1982. Marine bird and mammal survey of the eastern Aleutian Islands, summers of 1980–1981. Typescript report, files of U.S. Fish and Wildlife Service, Anchorage.

Oades, R. D. 1974. Predation of Xantus' murrelet by western gull. *Condor* 76:229.

Olson, S. L. 1977. A great auk, *Pinguinus*, from the Pliocene of North Carolina (Aves: Alcidae). *Proceedings of the Biological Society of Washington* 90:690–97.

Olson, S. T., and Marshall, W. H. 1952. *The common loon in Minnesota.* Occasional Papers 5. Minneapolis: Minnesota Museum of Natural History.

Onno, S. 1960. Zur Ökologie der Lappentaucher (*Podiceps cristatus, grisegena* und *auritus*) in Estland. In *Proceedings of the Twelfth International Ornithological Congress* (Helsinki 1958), pp. 577–82.

Otto, J. E. 1983. Breeding ecology of the pied-billed grebe (*Podilymbus podiceps* [Linnaeus]) on Rush Lake, Winnebago County, Wisconsin. M.S. thesis, University of Wisconsin, Oshkosh.

Palmer, R. S., ed. 1962. *Handbook of North American*

birds. Vol. 1. *Loons through flamingos.* New Haven: Yale University Press.

Paludin, K. 1960. Alkefugle. *Nordens Fugle i Farver* 3:207–51.

Parmelee, D. F.; Stephens, H. A.; and Schmidt, R. H. 1967. The birds of southeastern Victoria Island and adjacent small islands. *Bulletin of the National Museums of Canada* 222:1–229.

Payne, R. B. 1965. The molt of breeding Cassin auklets. *Condor* 67:220–28.

Pearson, T. H. 1968. The feeding biology of sea-bird species breeding on the Farne Islands. *Journal of Animal Ecology* 37:521–52.

Peck, G. K., and James, R. D. 1983. *Breeding birds of Ontario: Nidology and distribution.* Vol. 1. *Non-passerines,* Miscellaneous Publications. Toronto: Royal Ontario Museum.

Pennycuick, C. J. 1956. Observations on a colony of Brunnich's guillemot *Uria lomvia* in Spitsbergen. *Ibis* 98:80–99.

Perry, R. 1975. *Watching sea birds.* New York: Taplinger.

Petersen, A. 1981. Breeding biology and feeding ecology of black guillemots. Ph.D. dissertation, Oxford University.

Peterson, M. R. 1976. Breeding biology of arctic and red-throated loons. M.S. thesis, University of California, Davis.

———. 1979. Nesting ecology of arctic loons. *Wilson Bulletin* 91:608–17.

Piatt, J. F., and Nettleship, D. N. 1985. Diving depths of four auks. *Auk* 102:293–97.

Piatt, J. F.; Nettleship, D. N.; and Threlfall, W. 1984. Net-mortality of common murres and Atlantic puffins in Newfoundland, 1951–81. In Nettleship, Sanger, and Springer 1984, pp. 196–206.

Plumb, W. J. 1965. Observations on the breeding biology of the razorbill. *British Birds* 58:449–56.

Pool, E. L. 1938. Weights and wing areas of North American birds. *Auk* 55:511–17.

Portenko, L. A. 1981. *Birds of the Chukchi Peninsula and Wrangel Island.* Vol. 1. New Delhi: Amerind, for the U.S. Department of Commerce.

Preble, E. A., and McAtee, W. L. 1923. *A biological survey of the Pribilof Islands, Alaska.* North American Fauna 46. Washington, D.C.: U.S. Fish and Wildlife Service.

Preston, W. C. 1968. Breeding ecology and social behavior of the black guillemot, *Cepphus grille.* Ph.D. dissertation, University of Michigan, Ann Arbor.

Prinzinger, R. 1974. [Observations on the behavior of the black-necked grebe, *Podiceps n. nigricollis* Brehm (1831)]. *Anzeiger Ornithologische Gesellschaft Bayern* 13:1–34. In German, English summary.

———. 1979. *Der Schwarzhalstaucher.* Neue Brehm-Bücherei 521. Wittenberg Lutherstadt: A. Ziemsen.

Pruess, N. O. 1969. [The distribution and numbers of breeding grebes (*Podiceps*) in Denmark]. *Dansk Ornithologisk Forenings Tidsskrift* 63:174–85. In Danish.

Ratti, J. T. 1977. Reproductive separation and isolating mechanisms between sympatric dark- and light-phase western grebes. Ph.D. dissertation, Utah State University, Logan.

———. 1979. Reproductive separation and isolating mechanisms between sympatric dark- and light-phase western grebes. *Auk* 96:573–86.

———. 1981. Identification and distribution of Clark's grebe. *Western Birds* 12:41–46.

Ratti, J. T.; McCabe, T. R.; and Smith, L. M. 1983. Morphological divergence between western grebe color morphs. *Journal of Field Ornithology* 54:424–26.

Reimchen, T. E., and Douglas, S. 1980. Observations on loons (*Gavia immer* and *G. stellata*) at a bog lake on the Queen Charlotte Islands. *Canadian Field-Naturalist* 94:398–404.

———. 1984. Feeding schedule and daily food consumption in red-throated loons (*Gavia stellata*) over the prefledging period. *Auk* 101:593–99.

Renaud, W. E., and Bradstreet, M. S. W. 1980. Late winter distribution of black guillemots in northern Baffin Bay and the Canadian high arctic. *Canadian Field-Naturalist* 94:421–25.

Renaud, W. E.; McLaren, P. L.; and Johnson, S. R. 1982. The dovekie, *Alle alle,* as a spring migrant in eastern Lancaster Sound and western Baffin Bay. *Arctic* 35:118–25.

Richardson, F. 1961. Breeding biology of the rhinoceros auklet. *Condor* 63:456–73.

Ridgway, R. 1919. The birds of North and Middle America. *U.S. National Museum Bulletin* 50, part 8:1–851.

Roberts, T. S. 1949. *Manual for the identification of the birds of Minnesota and neighboring states.* Minneapolis: University of Minnesota Press.

Roby, D. D.; Brink, K. L.; and Nettleship, D. N. 1981. Measurements, chick meals and breeding distribution of dovekies (*Alle alle*) in northwest Greenland. *Arctic* 34:241–48.

Rudd, R. L., and Herman, S. G. 1972. Ecosystem transferral of pesticide residues in an aquatic environment. In *Toxic effect of pesticide residues on wildlife: Environmental toxicology of pesticides,* ed. F. Matsumara, G. Bonsh, and T. Misato, pp. 471–85. New York: Academic Press.

Rüppell, G. 1969. Beiträge zur Verhalten der Krabben-tauchers (*Plautus alle alle*). *Journal für Ornithologie* 110:161–69.

Sage, B. L. 1971. A study of white-billed divers in arctic Alaska. *British Birds* 66:24–30.

———. 1973. Studies of less familiar birds: Red-necked grebe. *British Birds* 66:24–30.

Salomonsen, F. 1944. The Atlantic Alcidae. *Meddelande fran Goteborgs Musei Zoologiska Avdelning* 108:1–138.

———. 1967. *Fuglene pa Gronland.* Copenhagen: Rhodos. Translated as *Studies on northern seabirds.* Report no. 100. Ottawa: Canadian Wildlife Service, 1981.

Sanger, G. A. 1975. Observations on the pelagic biology of the tufted puffin. *Pacific Seabird Group Bulletin* 2:30–31 (abstract).

Sanger, G. A.; Hironaka, V. F.; and Fukuyama, A. K. 1978. The feeding ecology and trophic relationships of key species of marine birds in the Kodiak Island area, May–September, 1977. In *Environmental assessment of the Alaskan continental shelf, annual reports of principal investigators,* 3:773–834. Boulder, Colo.: NOAA–BLM.

Savile, D. B. O. 1972. Evidence of tree nesting by the marbled murrelet in the Queen Charlotte Islands. *Canadian Field-Naturalist* 86:389–90.

Sawyer, L. E. 1979. Maine Audubon Society loon survey, 1978. In Sutcliffe 1979a, pp. 81–100.

Schneider, D., and Hunt, G. L., Jr. 1984. A comparison of seabird diets and foraging distribution around the Pribilof Islands, Alaska. In Nettleship, Sanger, and Springer 1984, pp. 86–95.

Schönwetter, M. 1967. *Handbuch der Oologie.* Lieferung 8. Berlin: Akademie.

Schorger, A. W. 1947. The deep diving of the loon and oldsquaw in Michigan. *Wilson Bulletin* 50:151–9.

Sclater, P. L. 1880. Remarks on the present status of the Systema Avium. *Ibis* 12:176–80.

Scott, J. M. 1973. Resource allocation in four syntopic species of marine diving birds. Ph.D. dissertation, Oregon State University, Corvallis.

Scott, J. M.; Hoffman, W.; Ainley, D.; and Zeillemaker, C. F. 1974. Range expansion and activity patterns in rhinoceros auklets. *Western Birds* 5:13–20.

Sealy, S. G. 1968. A comparative study of breeding ecology and timing in plankton-eating alcids (*Cyclorhynchus* and *Aethia* spp.) on St. Lawrence Island, Alaska. M.S. thesis, University of British Columbia, Vancouver.

———. 1972. Adaptive differences in breeding biology in the marine bird family Alcidae. Ph.D. dissertation, University of Michigan, Ann Arbor.

———. 1973a. Breeding biology of the horned puffin on St. Lawrence Island, Bering Sea, with zoogeographical notes on the north Pacific puffins. *Pacific Science* 27:99–119.

———. 1973b. Adaptive significance of post-hatching developmental patterns and growth rates in the Alcidae. *Ornis Scandinavica* 4:113–22.

———. 1974. Breeding phenology and clutch size in the marbled murrelet. *Auk* 91:10–23.

———. 1975a. Aspects of the breeding biology of the marbled murrelet. *Bird-Banding* 46:141–54.

———. 1975b. Egg size of murrelets. *Condor* 77:500–501.

———. 1975c. Feeding ecology of the ancient and marbled murrelets near Langara Island, British Columbia. *Canadian Journal of Zoology* 53:418–33.

———. 1975d. Influence of snow on egg-laying in auklets. *Auk* 92:528–38.

———. 1976. Biology of nesting ancient murrelets. *Condor* 78:294–306.

———. 1977. Wing molt of the Kittlitz's murrelet. *Wilson Bulletin* 89:467–69.

———. 1978. Clutch size and nest placement of the pied-billed grebe in Manitoba. *Wilson Bulletin* 90:301–2.

———. 1981. Variation in fledging weight of least auklets *Aethia pusilla. Ibis* 123:230–33.

———. 1982. Voles as a source of egg and nestling loss among nesting auklets. *Murrelet* 63:9–14.

Sealy, S. G., and Bedard, J. 1973. Breeding biology of the parakeet auklet (*Cyclorrhynchus psittacula*) on St. Lawrence Island, Alaska. *Astarte* 6:59–68.

Sealy, S. G., and Campbell, R. W. 1979. Post-hatching movements of young ancient murrelets. *Western Birds* 10:25–30.

Sealy, S. G.; Carter, H. R.; and Alison, D. 1982. Occurrences of the Asiatic marbled murrelet (*Brachyramphus marmoratus perdix* [Pallas]) in North America. *Auk* 92:778–81.

Sealy, S. G., and Nelson, R. W. 1973. The occurrences and status of the horned puffin in British Columbia. *Syesis* 6:51–56.

Searing, G. F. 1977. Some aspects of the ecology of cliff-nesting seabirds at Kongkok Bay, St. Lawrence Island, Alaska, during 1976. In *Environmental assessment of the Alaskan continental shelf, annual reports of principal investigators,* 5:263–412. Boulder, Colo.: NOAA–BLM.

Sergeant, D. 1951. Ecological relationships of the guillemots *Uria aalge* and *Uria lomvia.* In *Proceedings of the Tenth International Ornithological Congress* (Uppsala, 1950), pp. 578–87.

Shufeldt, R. W. 1904. On the osteology and systematic

position of the Pygopodes. *American Naturalist* 38:13–49.

Sibley, C. G., and Ahlquist, J. E. 1972. A comparative study of the egg white proteins of non-passerine birds. *Peabody Museum of Natural History Bulletin* 39:1–276.

Simmons, K. E. L. 1955. Studies on great crested grebes. *Avicultural Magazine* 61:3–13, 93–102, 131–46, 235–53, 294–316.

———. 1970. Duration of dives in the red-necked grebe. *British Birds* 63:300–302.

———. 1974. Adaptations in the reproductive biology of the great crested grebe. *British Birds* 67:413–37.

———. 1975. Further studies on great crested grebes. 1. Courtship. *Bristol Ornithologist* 8:89–107.

Simons, T. R. 1980. Discovery of a ground-nesting marbled murrelet. *Condor* 82:1–9.

Sjolander, S. 1968. [Observations on the ethology of the black-throated diver (*Gavia arctica*)]. *Zoologisk Revy* 30:89–93. In Swedish, English summary.

———. 1978. Reproductive behavior of the black-throated diver *Gavia arctica*. *Ornis Scandinavica* 9:51–65.

———. In press. On the behaviour of the red-throated diver *Gavia stellata*. Not seen.

Sjolander, S., and Agren, G. 1972. Reproductive behavior of the common loon. *Wilson Bulletin* 84:296–308.

———. 1976. Reproductive behavior of the yellow-billed loon. *Condor* 78:454–63.

Slater, P. J. B. 1974. Orientation of fish in the tystie's beak. *Bird Study* 21:238–40.

Slater, P. J. B., and Slater, E. P. 1972. Behaviour of the tystie during feeding of the young. *Bird Study* 19:105–13.

Smail, J.; Ainley, D. G.; and Strong, H. 1972. Notes on the birds killed in the 1971 San Francisco oil spill. *California Birds* 3:25–32.

Snyder, L. L. 1957. *Arctic birds of Canada.* Toronto: University of Toronto Press.

Southern, H. N.; Carrick, R.; and Potter, W. G. 1965. The natural history of a population of guillemots *Uria aalge* (Pont.). *Journal of Animal Ecology* 34:649–65.

Sowls, A. L.; DeGange, A. R.; Nelson, J. W.; and Lester, G. S. 1980. *Catalog of California seabird colonies.* Biological Services Program report FWS/OBS 80/37. Washington, D.C.: U.S. Fish and Wildlife Service.

Sowls, A. L.; Hatch, S. A.; and Lensink, C. J. 1978. *Catalog of Alaskan seabird colonies.* Biological Services Program report FWS/OBS/78. Washington, D.C.: U.S. Fish and Wildlife Service.

Speich, S., and Manuwal, D. A. 1974. Gular pouch development and population structure of Cassin's auklet. *Auk* 91:291–306.

Spring, L. 1971. A comparison of functional and morphological adaptations in the common murre (*Uria aalge*) and thick-billed murre (*Uria lomvia*). *Condor* 73:1–27.

Springer, A. M., and Roseneau, D. G. 1985. Copepod-based food webs: Auklets and oceanography in the Bering Sea. *Marine Ecology—Progress Series* 21:229–37.

Stejneger, L. 1885. Results of ornithological explorations in the Commander Islands and in Kamchatka. *Bulletin of the U.S. National Museum* 29:1–382.

Stempniewicz, L. 1981a. Breeding biology of the little auk in the Hornsund region, Spitsbergen. *Acta Ornithologica* 18:141–65.

———. 1981b. [Factors influencing the growth of the little auk, *Plautus alle* L. nestlings on Spitsbergen]. *Ekologia Polska* 28:557–81. In Polish.

Stettenheim, P. 1959. Adaptations for underwater swimming in the common murre (*Uria aalge*). Ph.D. dissertation, University of Michigan, Ann Arbor.

Steventon, D. J. 1979. Razorbill survival and population estimates. *Ringing and Migration* 2:105–12.

Stewart, R. E. 1975. *Breeding birds of North Dakota.* Fargo: North Dakota Institute for Regional Studies.

Stolpe, M. 1935. *Colymbus, Hesperornis, Podiceps:* Ein Vergleich ihrer hinderer Extremität. *Journal für Ornithologie* 83:115–28.

Storer, R. W. 1945. Structural modifications in the hind limb of the Alcidae. *Ibis* 87:433–56.

———. 1952. A comparison of variation, behavior and evolution in the sea bird genera *Uria* and *Cepphus*. *University of California Publications in Zoology* 52:121–222.

———. 1956. The fossil loon, *Colymboides minutus*. *Condor* 58:413–26.

———. 1960. Evolution in the diving birds. In *Proceedings of the Twelfth International Ornithological Congress* (Helsinki, 1958), pp. 694–707.

———. 1962. Courtship and mating behavior and the phylogeny of the grebes. In *Proceedings of the Thirteenth International Ornithological Congress* (Ithaca, 1962), pp. 562–69.

———. 1963. Observations on the great grebe. *Condor* 65:279–88.

———. 1965. Color phases of the western grebe. *Living Bird* 4:59–61.

———. 1967. Observations on Rolland's grebe. *Hornero* 10:339–50.

———. 1969. Behavior of the horned grebe in spring. *Condor* 71:180–205.

———. 1971. The behavior of the New Zealand dabchick. *Notornis* 18:175–86.

———. 1976a. The behavior and relationships of the least grebe. *Transactions of the San Diego Society of Natural History* 18:113–26.

———. 1976b. The Pleistocene pied-billed grebe (Aves: Podicipedidae). *Smithsonian Contributions to Palaeobiology* 27:147–53.

———. 1978. Systematic notes on the loons (Gaviidae: Aves). *Brevoria* 448:1–8.

———. 1979. Orders Gaviiformes and Podicipediformes. In *Check-list of birds of the world*, 2d ed., ed. E. Mayr and G. W. Cottrell, 1:135–55. Cambridge, Mass.: Museum of Comparative Zoology.

———. 1982. The hooded grebe on Laguna de los Escharchados: Ecology and behavior. *Living Bird* 19:50–67.

Storer, R. W., and Jehl, J. R., Jr. 1985. Moult pattern and moult migration in the black-necked grebe. *Ornis Scandinavica* 16:253–60.

Storer, R. W., and Nuechterlein, G. L. 1985. An analysis of plumage and morphological characters of the two color forms of the western grebe (*Aechmophorus*). *Auk* 102:109–19.

Storer, R. W.; Siegfried, W. R.; and Kinahan, J. 1975. Sunbathing in grebes. *Living Bird* 14:45–56.

Strauch, J. G., Jr. 1977. The phylogenetic relationships of the Alcae. *American Zoologist* 17:972 (abstract).

———. 1985. The phylogeny of the Alcidae. *Auk* 102:520–39.

Sugden, L. G. 1977. Horned grebe breeding habitat in Saskatchewan parklands. *Canadian Field-Naturalist* 91:372–76.

Summers, K. R., and Drent, R. H. 1979. Breeding biology and twinning experiments of rhinoceros auklets on Cleland Island, British Columbia. *Murrelet* 60:16–22.

Sutcliffe, S. A., ed. 1979a. *The common loon.* Proceedings of the Second North American Conference on Common Loon Research and Management. New York: National Audubon Society.

———. 1979b. Common loon status in New Hampshire, 1976–1978. In Sutcliffe 1979a, pp. 111–16.

Swartz, L. G. 1966. Sea-cliff birds. In *Environment of the Cape Thompson region, Alaska*, ed. N. J. Wilimowsky and J. N. Wolfe, pp. 611–78. Washington, D.C.: U.S. Atomic Energy Commission, Division of Technical Information.

Swennen, C., and Duiven, P. 1977. Size of food objects of three fish-eating seabird species: *Uria aalge, Alca torda,* and *Fratercula arctica* (Aves, Alcidae). *Netherlands Journal of Sea Research* 11:92–98.

Thayer, J. E. 1914. Nesting of the Kittlitz murrelet. *Condor* 16:117–18.

Thompson, M. C.; Hines, J. Q.; and Williamson, F. S. L. 1966. Discovery of the downy young of the Kittlitz's murrelet. *Auk* 83:349–51.

Thoreson, A. C. 1964. The breeding behavior of the Cassin's auklet. *Condor* 66:456–76.

———. 1983. Diurnal activity and social displays of rhinoceros auklets on Teuri Island, Japan. *Condor* 85:373–75.

———. In press. *The auks of the world.* Vancouver: Hancock House.

Thoreson, A. C., and Booth, E. S. 1958. Breeding activities of the pigeon guillemot *Cepphus columba columba* (Pallas). *Department of Biological Science and Biological Station Publication* (Walla Walla College), 23:1–37.

Threlfall, W., and Mahoney, S. P. 1980. The use of measurements in sexing common murres from Newfoundland. *Wilson Bulletin* 92:266–68.

Titus, J. R. 1979. Response of the common loon (*Gavia immer*) to recreational pressure in the Boundary Waters Canoe Area, northeastern Minnesota. Ph.D. dissertation, University of Minnesota, Minneapolis.

Titus, J. R., and Van Druff, L. W. 1981. Response of the common loon to recreational pressure in the Boundary Waters Canoe Area, northeastern Minnesota. *Wildlife Monographs* 79:1–60.

Trivelpiece, W.; Brown, S.; Hicks, A.; Fekete, R.; and Volkman, N. J. 1979. An analysis of the distribution and reproductive success of the common loon in the Adirondack Park, New York. In Sutcliffe 1979a, pp. 45–56.

Tschanz, B. 1959. [The breeding biology of the common murre, (*Uria aalge aalge* Pont.)] *Behaviour* 14:1–100. In German, English summary.

———. 1968. Trottellummen. *Zeitschrift für Tierpsychologie*, suppl. 4:1–103.

———. 1972. Beobachtungen on Dickschabel- und Trottellummen auf Vedoy (Lofoten, Norwegen). *Ornithologischer Beobachter* 69:169–77.

Tschanz, B., and Hirsbrunner-Scharf, M. 1975. Adaptations to colony life on ledges: A comparative study of guillemot and razorbill chicks. In *Function and evolution in behaviour*, ed. C. G. Beer and A. Manning, pp. 359–80. Oxford: Clarendon Press.

Tschanz, B., and Wehrlin, J. 1968. Kreuzung zwischen Trottellumme und Dickschnabellumme. *Fauna* 21:53–55.

Tuck, L. M. 1960. *The murres.* Wildlife Series 1. Ottawa: Canadian Wildlife Service.

Udvardy, M. D. F. 1963. Zoogeographical study of the Pacific Alcidae. In *Pacific basin biogeography: A symposium*, ed. J. L. Gressit, pp. 85–111. Honolulu: Bishop Museum Press.

———. 1979. Zoogeography and taxonomic relationships of seabirds in northern North America. In Bartonek and Nettleship 1979, pp. 167–70.

Uspenski, A. M. 1958. *The bird bazaars of Novaya Zemlya.* Translations of Russian Game Reports, vol. 4. Ottawa: Canadian Wildlife Service.

van Oordt, G. J., and Huxley, J. S. 1922. Some observations on the habits of the red-throated diver in Spitsbergen. *British Birds* 16:34–46.

Varoujean, D. H. 1979. *Seabird colony catalog, Washington, Oregon and California.* Report, Region 1. Portland, Oreg.: U.S. Fish and Wildlife Service.

Varoujean, D. H., and Pitman, R. L. 1979. *Oregon seabird colony survey, 1979.* Report, Region 1. Portland, Oreg.: U.S. Fish and Wildlife Service.

Vermeer, K. 1973. Some aspects of the breeding and mortality of common loons in east-central Alberta. *Canadian Field-Naturalist* 87:403–8.

———. 1979. Nesting requirements, food and breeding distribution of rhinoceros auklets (*Cerorhincha monocerata*) and tufted puffins (*Lunda cirrhata*). *Ardea* 67:101–10.

———. 1980. The importance of timing and type of prey to reproductive success of rhinoceros auklets *Cerorhincha monocerata. Ibis* 122:343–50.

Vermeer, K., and Cullen, L. 1979. Growth of rhinoceros auklets and tufted puffins, Triangle Island, British Columbia. *Ardea* 67:22–27.

———. 1982. Growth comparison of a plankton- and a fish-eating alcid. *Murrelet* 63:34–39.

Vermeer, K.; Summers, K. R.; and Bingham, D. S. 1976. Birds observed at Triangle Island, British Columbia, 1974 and 1975. *Murrelet* 57:35–41.

Vermeer, K.; Vermeer, R. A.; Summers, K. R.; and Billings, R. R. 1979. Number and habitat selection of Cassin's auklets breeding on Triangle Island, British Columbia. *Auk* 96:143–51.

Voous, K. H. 1960. *Atlas of European birds.* London: T. Nelson.

———. 1973. List of Recent Holarctic bird species: Non-passerines. *Ibis* 115:612–38.

Voous, K. H., and Payne, H. A. W. 1965. The grebes of Madagascar. *Ardea* 53:9–31.

Wahl, T. R. 1975. Seabirds in Washington's offshore zone. *Western Birds* 6:117–34.

Wehle, D. H. S. 1976. Summer food and feeding ecology of tufted and horned puffins on Buldir Island, Alaska, 1975. M.S. thesis, University of Alaska, College.

———. 1980. The breeding biology of the puffins: Tufted puffin (*Lunda cirrhata*), horned puffin (*Fratercula corniculata*), common puffin (*Fratercula arctica*) and rhinoceros auklet (*Cerorhinca monocerata*). Ph.D. dissertation, University of Alaska, College.

———. 1983. The food, feeding and development of young tufted and horned puffins in Alaska. *Condor* 85:427–42.

Wetmore, A. 1924. *Food and economic relations of North American grebes.* Bulletin 1196. Washington, D.C.: U.S. Department of Agriculture.

Williams, A. J. 1974. Site preferences and interspecific competition among guillemots *Uria aalge* (L.) and *Uria lomvia* (L.) on Bear Island. *Ornis Scandinavica* 5:113–21.

———. 1975. Guillemot fledging and predation on Bear Island. *Ornis Scandinavica* 6:117–24.

Williams, S. O., III. 1982. Notes on the breeding and occurrence of western grebes on the Mexican plateau. *Condor* 84:127–30.

Wilson, U. W. 1977. Reproductive biology and activity of the rhinoceros auklet on Protection Island, Washington. M.S. thesis, University of Washington, Seattle.

Winn, H. E. 1950. The black guillemots of Kent Island, Bay of Fundy. *Auk* 67:477–75.

Winnett, K. A.; Murray, K. G.; and Wingfield, J. C. 1979. Southern race of Xantus' murrelet breeding on Santa Barbara Island, California. *Western Birds* 10:81–82.

Witherby, H. F.; Jourdain, F. C. R.; Ticehurst, N. F.; and Tucker, B. W. 1941. *The handbook of British birds.* Vol. 4. London: Witherby.

Wobus, U. 1964. Der Rothalstaucher (*Podiceps grisegena* Boddaert). Neue Brehm-Bücherei 330. Wittenberg Lutherstadt: A. Ziemsen.

Wolf, K. 1955. Some effects of fluctuating and falling water levels on waterfowl production. *Journal of Wildlife Management* 19:13–23.

Yocom, C. J.; Harris, S. W.; and Hansen, H. A. 1958. Status of grebes in eastern Washington. *Auk* 75:36–47.

Zang, H. 1977. [On the frequency of second broods in the great crested grebe]. *Journal für Ornithologie* 118:261–67. In German, English summary.

Zimmer, G. E. 1982. The status and distribution of the common loon in Wisconsin. *Passenger Pigeon* 44:60–66.

Zimmerman, D. A. 1957. Display in the least grebe. *Auk* 74:390.

Index

This index is limited to the English vernacular and Latin names of species and subspecies of loons, grebes, and auks discussed individually in this book. Complete indexing is confined to entries for the English vernacular names of species as used in this book. The principal account of each species is indicated by italic page numbers.